Trends in applications of mathematics to mechanics

Trends in applications of mathematics to mechanics

Edited by

W Schneider, H Troger and F Ziegler
Technical University Vienna

Copublished in the United States with
John Wiley & Sons, Inc., New York

Longman Scientific & Technical
Longman Group UK Ltd,
Longman House, Burnt Mill, Harlow,
Essex CM20 2JE, England
and Associated Companies throughout the world.

Copublished in the United States with
John Wiley & Sons Inc., 605 Third Avenue, New York, NY 10158

First published 1991

ISSN 0962-4589

British Library Cataloguing in Publication Data
Symposium on Trends in Applications of Mathematics to
Mechanics (8th ; Hollabrunn, Austria ; 1989)
8th Symposium on "Trends in Applications of Mathematics to
Mechanics" - Stamm–8, Hollabrunn, Austria 13–18, 1989.
1. Physical sciences. Applications of mathematics
I. Title II. Schneider, Wilhelm III. Troger, Hans
IV. Ziegler, Franz V. Series
500.20151

ISBN 0-582-05822-8

Library of Congress Cataloging-in-Publication Data
Symposium on Trends in Applications of Mathematics to Mechanics (8th :
1989 : Hollabrunn, Hollabrunn, Austria)
8th Symposium on "Trends in Applications of Mathematics to
Mechanics" : STAMM-8, Hollabrunn, August 13–18, 1989 / edited by
Wilhelm Schneider, Hans Troger, Franz Ziegler.
p. cm. — (Interaction of mechanics and mathematics series)
1. Mechanics—Congresses. 2. Mathematics—Congresses.
I. Schneider, Wilhelm, Dipl.-Ing. Dr. techn. II. Troger, H. (Hans),
1943– . III. Ziegler, Franz, 1937– . IV. Title. V. Title:
Eighth Symposium on "Trends in Applications of Mathematics to
Mechanics". VI. Series.
QA801.S94 1989
531—dc20 90-42156
CIP

Printed and bound in Great Britain at the Bath Press, Avon

Foreword

The International Society for the Interaction of Mechanics and Mathematics (ISIMM) was founded in 1977. Its purpose is to promote cooperative research involving the fields of mechanics and pure mathematics.

Its Executive Committee decided that, from time to time, scholarly works relevant to the Society's interests should, by invitation, be published under its auspices. The present volume is one in this series which, it is hoped, will help to advance the objective of the Society.

The Editorial Board

Professor Emeritus Ekkehart Kröner
University of Stuttgart, Germany

In honour of his many contributions to research in Theoretical and Applied Mechanics and Physics on the occasion of his 70th birthday

Ekkehart Kröner

The seventieth birthday of Professor Ekkehart Kröner came to most of us as a surprise. He has been around attending conferences, presenting lectures and organizing science for such a long time that one tends to accept his always desirable presence as obvious. Moreover, his exceptional vitality, good spirits and his sense of humour mean that he simply gets older at a much slower rate than others. Since his retirement (whatever this word means: there are people twenty years younger who are more retired than Ekkehart Kröner) he has been as active as ever before.

This is not the place to go into details of Professor Kröner's outstanding contribution to mechanics, physics of solids and applied mathematics (one hopes that some historian of science will do it), in particular to the theory of defects (dislocations, disclinations, etc.), theory of heterogeneous media (with random properties), the transition from micro- to macro-mechanics. All his papers exhibit originality of thought, an exceptional physical intuition and very sound mathematics; many of his papers opened new chapters in mechanics and physics and for decades have continued to influence science. Although a physicist by nature and education, he contributed to such seemingly abstract topics as non-Riemannian geometry of defects and gauge field theories of solids, for he well understands the importance of the foundations of the continuum theories and their connections with other branches of physics.

Ekkehart Kröner is not only a scientist. He is also an exceptional human being with very high ethical principles. He has numerous friends and followers in many countries. His attitude and help have been very much appreciated and we are all looking forward to future meetings, discussions and collaboration with him.

H. Zorski, Warsaw

The International Society for
Interactions of Mechanics and Mathematics
mourns the death of the Members

R.J. DiPerna

R.H. Krohn

H.H.E. Leipholz

M.A. Sneider

M. Żòrawski

Contents

PART V FREE SECTION: SIGNIFICANT INTERACTIONS OF MATHEMATICS AND MECHANICS

List of contributors

A.L. Afendikov — Keldysh Institute of Applied Mathematics
Miusskaja sq. 4
Moscow 125047
USSR

D.R. Axelrad — McGill University
Micromechanics Research Laboratory
Department of Mechanical Engineering
817 Sherbrooke Street West
Montreal, QC, H3A 2K6
CANADA

A.K. Belyaev — Leningrad Politechnic Institute
Department of Dynamics & Strength of Machines
Politeckhnicheskaya, Dom, 2
Leningrad 195251
USSR

H. Berger — University of Stuttgart
Mathematics Institute A
Lehrstuhl 6
Angewandte Mathematik
Pfaffenwaldring 57
D-7000 Stuttgart 80
GERMANY

J. Brilla — Comenius University
Institute of Applied Mathematics and Computing Technique
Mlynska dolina
842-15 Bratislava
CZECHOSLOVAKIA

P. Chadwick — University of East Anglia
School of Mathematics
Norwich NR4 7TJ
ENGLAND

D. Cioranescu

CNRS-Lab. d'Analyse Numérique
4 Place Jussieu
F-75252 Paris Cedex 05
FRANCE

A. Di Carlo

Università di Roma "La Sapienza",
Dipartimento Ingegneria Strutturale
Via Eudossiana, 18
I-00184 Roma
ITALY

R. Ellinghaus

Technische Universität Berlin
Institut für Theoretische Physik
Sekr. PN 7-1
Hardenbergstraβe 36
D-1000 Berlin 12
GERMANY

F.D. Fischer

University for Mining and Metallurgy
Institute of Mechanics
A-8700 Leoben
AUSTRIA

T.M. Fischer

Institut für Theoretische Stromüngsmechanik
Deutsche Forschungsanstalt für Luft- und
Raumfahrt (DLR)
Bunsenstraβe 10
D-3400 Göttingen
GERMANY

P. Fotiu

Technical University Vienna
Institut für Allgemeine Mechanik (E201)
Wiedner Hauptstraβe 8-10
A-1040 Wien
AUSTRIA

M.A. Goldshtik

Institute of Thermophysics
Novosibirsk 630090
USSR

G. Hackmüller

Technical University Vienna
Institut für Strömungslehre und
Wärmeübertragung (Inst. 322)
Wiedner Hauptstraβe 7
A-1040 Wien
AUSTRIA

H. Hadouaj

Laboratoire de Modelisation en Mécanique
Associé au CNRS, URA 229
Université Pierre et Marie Curie
Tour 66, 4 Place Jussieu
F-75252 Paris Cedex 05
FRANCE

H. Herwig

Ruhr Universität Bochum
Institut für Thermo und Fluiddynamik
Theoretische Strömungsmechanik
Gebäude JB-6/145
Postfach 10 21 48
D-4630 Bochum
GERMANY

G. Iooss

University of Nice
Laboratory of Mathematics
U.A. CNRS 168
Parc Valrose
F-06034 Nice
FRANCE

H. Irschik

Technical University Vienna
Institut für Allgemeine Mechanik (E201)
Wiedner Hauptstraße 8-10
A-1040 Wien
AUSTRIA

A. Kaczyński

Warsaw University of Technology
Institute of Mathematics
Plac Jedności Robotniczej 1
PL-00-661 Warsaw
POLAND

A. Kaveh

Iran University of Science and Technology
Narmak,
Tehran - 16
IRAN

A. Kluwick

Technical University Vienna
Institut für Strömungslehre und
Wärmeübertragung (Inst. 322)
Wiedner Hauptstraße 7
A-1040 Wien
AUSTRIA

N.T. Kovatcheva — Bulgarian Academy of Sciences
Institute of Mechanics
P.O. Box 373
1090 Sofia
BULGARIA

T. Lewinski — Warsaw Technical University
Institute of Structural Mechanics
Armii Ludowej 16,
PL-00-637 Warsaw
POLAND

A. Liolios — Democritus University of Thrace
Department of Civil Engineering
Institute of Structural Mechanics
GR-67100 Xanthi
GREECE

A. Luongo — University of L'Aquila
Dipartimento di Ingegneria delle Strutture,
Acque e del Terreno
I-67040 Monteluco di Roio L'Aquila
ITALY

R. Mahnken — Universität Hannover
Institut für Baumechanik und Numerische
Mechanik
Appelstraße 9A
D-3000 Hannover 1
GERMANY

S.J. Matysiak — University of Warsaw
Institute of Hydrogeology and Engineering
Geology
Al. Zwirki i Wigury 93
PL-02-089 Warsaw
POLAND

G.A. Maugin — Université Pierre et Marie Curie
Laboratoire de Modelisation en Mécanique
Associé au CNRS, URA 229
Tour 66, 4 Place Jussieu
F-75252 Paris Cedex 05
FRANCE

P. Mazilu	Technische Hochschule Darmstadt Petersenstraße 30 D-6100 Darmstadt GERMANY
A. Mielke	University of Stuttgart Mathematics Institute A Pfaffenwaldring 57 D-7000 Stuttgart 80 GERMANY
S. Minagawa	University of Electro-Communications Department of Mechanical and Control Engineering Chofu, Tokyo 182 JAPAN
I. Müller	Technische Universität Berlin Hermann-Föttinger-Institut für Thermo- und Fluiddynamics Str. des 17. Juni 135 D-1000 Berlin 12 GERMANY
W. Muschik	Technische Universität Berlin Institut für Theoretische Physik Sekr. PN 7-1 Hardenbergstraße 36 D-1000 Berlin 12 GERMANY
J.A. Nohel	University of Wisconsin-Madison Center for the Mathematical Sciences 610 Walnut Street Madison, Wisconsin 53705 USA
J.P. Nowacki	Polish Academy of Sciences Institute of Fundamental Technological Research Swietokrzyska 21 PL-00-049 Warsaw POLAND

P.D. Panagiotopoulas	RWTH Aachen Institut für Technische Mechanik D-5100 Aachen GERMANY and Aristotle University Department of Civil Engineering GR-54006 Thessaloniki GREECE
A.D. Pierce	Pennsylvania State University Graduate Program in Acoustics and Department of Mechanical Engineering 157 Hammond Building University Park, Pennsylvania 16802 USA
A.D. Polyanin	USSR Academy of Sciences Institute for Problems in Mechanics Moscow 117526, Vernadsky Avenue 101 USSR
J. Pouget	Université Pierre et Marie Curie Laboratoire de Modélisation en Mécanique Associé au CNRS, Tour 66-4, Place Jussieu F-75252 Paris, Cedex 05 FRANCE
F.G. Rammerstorfer	Technical University Vienna Institute of Lightweight Structures and Aerospace Engineering, A-1040 Vienna AUSTRIA
N. Rizzi	Università di Roma "La Sapienza" Dipartimento di Ingegneria Strutturale e Geotecnica via Eudossiana 18 I-00184 Roma ITALY

J.Saint Jean Paulin	Université de Metz Département de Mathematiques Ile de Saulcy F-57045 Metz, Cedex 01 FRANCE
U. Schaflinger	Technical University Vienna Institut für Strömungslehre und Wärmeübertragung Wiedner Hauptstraβe 7 A-1040 Wien AUSTRIA
M. Schultz	Technical University Vienna Institut für Strömungslehre und Wärmeübertragung (Inst. 322) Wiedner Hauptstraβe 7 A-1040 Wien AUSTRIA
M. Seisl	Technical University Vienna Institute of Mechanics (E325) Wiedner Hauptstraβe 8-10 A-1040 Vienna AUSTRIA
J. Shaw	Michigan State University Department of Mechanical Engineering East Lansing, MI 48824 USA
S.W. Shaw	Department of Mechanical Engineering Michigan State University East Lansing, MI 48824 USA
V.N. Shtern	Institute of Thermophysics Novosibirsk 630090 USSR
H. Sockel	Technical University Vienna Institut für Strömungslehre und Wärmeübertragung (Inst. 322) Wiedner Hauptstraβe 7 A-1040 Wien AUSTRIA

E. Stein

Universität Hannover
Institut für Baumechanik und Numerische Mechanik
Appelstraße 9A
D-3000 Hannover 1
GERMANY

A. Steindl

Technical University Vienna
Institute of Mechanics (E325)
Wiedner Hauptstraße 8-10
A-1040 Vienna
AUSTRIA

A. Tatone

Università di L'Aquila
Dipartimento di Ingegneria delle Strutture
Facolta di Ingegneria
I-67040 Monteluco di Roio (AQ)
ITALY

J.J. Telega

Polish Academy of Sciences
Institute of Fundamental Technological Research
Z TK
Swietokrzyska 21
PL-00-049 Warsaw
POLAND

A. Tiero

Università di Roma "La Sapienza"
Dipartimento Ingegneria Strutturale
Via Eudossiana, 18
I-00184 Roma
ITALY

I. Troch

Technical University Vienna (Inst. 114)
Institut für Technische Mathematik
Wiedner Hauptstraße 8-10
A-1040 Wien
AUSTRIA

H. Troger

Technical University Vienna
Institute of Mechanics (E325)
Wiedner Hauptstraße 8-10
A-1040 Vienna
AUSTRIA

T.S. Vashakmadze	Tbilisi University I.N. Vekua Institute of Applied Mathematics University Str. 2 380043 Tbilisi, Georgia USSR
J. Verhas	Technical University Budapest H-1521 Budapest HUNGARY
F. Vestroni	University of L'Aquila Dipartimento di Ingegneria delle Strutture Acque e del Terreno I-67040 Monteluco di Roio L'Aquila ITALY
G. Warnecke	University of Stuttgart Mathematics Institute A Lehrstuhl 6 Angewandte Mathematik Pfaffenwaldring 57 D-7000 Stuttgart 80 GERMANY
W. Wendland	University of Stuttgart Mathematics Institute A Lehrstuhl 6 Angewandte Mathematik Pfaffenwaldring 57 D-7000 Stuttgart 80 GERMANY
K. Wilmanski	Technische Universität Hamburg-Harburg Arbeitsbereich Meerestechnik II Postfach 90 14 03 D-2100 Hamburg 90 GERMANY
C. Woźniak	Miedzynarodowa 58 m.63 PL-03-922 Warsaw POLAND

F. Ziegler

Technical University Vienna
Institut für Allgemeine Mechanik (E201)
Wiedner Hauptstraße 8–10
A-1040 Wien
AUSTRIA

H. Zorski

Polish Academy of Sciences
Institute of Fundamental Technological Research
ul. Swietokrzyska 21
0-0049 Warsaw
POLAND

Preface

The 8th Symposium on Trends in Applications of Mathematics to Mechanics was the continuation of a series of meetings held every two years under the auspices of the International Society for Interaction of Mechanics and Mathematics. The scientific committee consisted of

G.I. Barenblatt	R.P. Gilbert	G. Herrmann	S. Kaliszky
K. Kirchgässner	J. Kratochvil	S. Leibovich	W. Schneider
H. Troger	W. Wendland	F. Ziegler	H. Zorski.

The symposium was attended by 84 participants from 17 countries and a total of 50 lectures were delivered. The presentations were divided into the following five sections on specific topics:

I. Numerical methods in non-linear continuum mechanics
II. Dynamic systems and bifurcation
III. Asymptotic theory of viscous flow
IV. Material science and thermodynamic aspects of continuum mechanics
V. Free section: Significant interactions of mathematics and mechanics

We wish to thank all participants for their contributions to the lectures and the discussions. We are very grateful to Dr. E. Müller, Director of the Engineering College (HTBLA) Hollabrunn, Lower Austria, and his staff for their kind and generous hospitality, providing a spacious lecture hall and convenient accommodation. Moreover we gratefully acknowledge the support from the following sources:

The Governor of Lower Austria
The Mayor of Hollabrunn
Creditanstalt-Bankverein, branch office Rilkeplatz 8, Vienna
Institut für Allgemeine Mechanik,
Institut für Mechanik and
Institut für Strömungslehre und Wärmeübertragung of the
Technical University of Vienna.

Wien,
March 1990

W. Schneider
H. Troger
F. Ziegler

Part I
Numerical Methods in Nonlinear Continuum Mechanics

F. ZIEGLER, H. IRSCHIK, P. FOTIU

The method of self-stresses in physically nonlinear continuum mechanics[1]

ABSTRACT

Dynamical problems of viscoplasticity are considered by separating the nonlinear strains which are noncompatible in the associated linear elastic solid. The latter is determined by the stiffness distribution of the undeformed body which is kept invariant in time. Even ductile damage when smeared in the yielding zone in that formulation changes the intensity of the plastic sources and renders additional noncompatible strains: equivalence of cracks and dislocations is encountered. The derivation of the integral equations is quite general, and also nonlinear geometric relations are considered. Illustrative applications to bending vibrations are presented in the small-strain limit and a natural definition of the plastic drift is given by the quasistatic response portion.

1. The constitutive relations and ductile damage

Since the solution to dynamic viscoplastic problems is sought by considering the plastic portion of strain as a noncompatible field in the associated linear elastic body the first reference is given to papers on sources of self-stress by Reissner [1] and Nemenyi [2]. Such actual or fictitious eigenstresses from a noncompatible field $\bar{\varepsilon}_{ij}$ are formally defined by the linear elastic constitutive law which has a rate or incremental form

$$\Delta\varepsilon_{ij} - \Delta\bar{\varepsilon}_{ij} = C_{ijkl}\,\Delta\sigma_{kl}. \tag{1}$$

From linearized thermoelasticity the field of noncompatible thermal strains is well known. With a temperature field $\theta(\vec{x}, t)$,

$$\Delta\bar{\varepsilon}_{ij} = \alpha_{ij}\,\Delta\theta. \tag{2}$$

For example, taking Perzyna's law of viscoplasticity, see e.g. Telles [3] for small deformations, the plastic strain rate is given by

[1] Dedicated to Professor E. Kröner, president of ISIMM, on the occasion of his 70th birthday.

$$\dot{\varepsilon}^p_{ij} = \frac{2k}{\tau} \langle(\varphi(F)\rangle \frac{\partial F}{\partial \sigma_{ij}}, \quad F = (\sqrt{J_2}/k - 1), \tag{3a}$$

$$\langle\varphi(F)\rangle = \begin{cases} \varphi(F) = F & \dots\ F > 0 \\ 0 & \dots\ F \leq 0, \end{cases} \tag{3b}$$

where k is the static yield-stress, τ is a characteristic time and J_2 is the second invariant of the deviatoric stress tensor, the noncompatible strain increment is determined by the current stress state, e.g. by the one-step Euler formula. The field $\boldsymbol{\varepsilon}^p$ represents the continuous distribution of the dislocations of the plastic (irreversible) deformations, Kröner [4] and Mura [5]. Thus, the singular dislocation is the elementary source of self-stresses, Kröner [6].

In ductile materials, where the yield stress is below the ultimate strength, the micro-cracks of any material-damage grow approximately into spherical shape. That means the loss of strength to be isotropic. Thus, according to Cocks and Leckie [7] the smeared ductile damage can be described by a scalar internal variable

$$D = \left(\frac{3}{2} f_v\right)^{2/3}, \tag{4}$$

where f_v is the volume porosity of the yielding material. The damage parameter measures the portion of the cross-section which lost the capacity of transferring stress due to the accumulation of micro-cracks, Kachanov [8]. Thus, the effective stress is related to the macroscopically averaged stress σ by

$$\bar{\boldsymbol{\sigma}} = \boldsymbol{\sigma}/(1 - D). \tag{5}$$

The inverse of Eq. (1) is changed to that of a porous elastoplastic body

$$\sigma_{ij} = E_{ijkl}(1 - D)(\varepsilon_{kl} - \varepsilon^p_{kl}) = E_{ijkl}(\varepsilon_{kl} - \bar{\varepsilon}_{kl}) \tag{6}$$

and the nonlinear strain is changed to

$$\bar{\varepsilon}_{ij} = \varepsilon^p_{ij} + D(\varepsilon_{ij} - \varepsilon^p_{ij}) = \varepsilon^p_{ij} + \varepsilon^f_{ij}. \tag{7}$$

Thus, the difference between the total and the plastic strain is the elastic strain of the porous body, due to the actual stress $\bar{\boldsymbol{\sigma}}$,

$$\boldsymbol{\varepsilon}^e = \boldsymbol{\varepsilon} - \boldsymbol{\varepsilon}^p. \tag{8}$$

The additional source of self-stresses due to damage and micro-cracks is thus

$$\dot{\boldsymbol{\varepsilon}}^{\mathrm{f}} = D\dot{\boldsymbol{\varepsilon}}^{\mathrm{e}} + \dot{D}\boldsymbol{\varepsilon}^{\mathrm{e}}. \tag{9}$$

Equivalence of crack and dislocation was pointed out by Bilby and Eshelby [9]. The damage source $\boldsymbol{\varepsilon}^{\mathrm{f}}$ not only is determined by the cross-sectional porosity D but depends on ε^{e} as well and, hence, is related to $\bar{\boldsymbol{\sigma}}$. A crack develops a displacement discontinuity only when a traction is applied. Furthermore, contrary to the plastic source $\dot{\boldsymbol{\varepsilon}}^{\mathrm{p}}$, the damage source $\dot{\boldsymbol{\varepsilon}}^{\mathrm{f}}$ contains a reversible portion $D\dot{\boldsymbol{\varepsilon}}^{\mathrm{e}}$: that is, the additional elastic deformation due to the porosity under the action of the same load. Energy dissipation due to crack propagation and accumulation takes place if $\dot{D} \neq 0$ and always in case of plastic deformation, $\dot{\boldsymbol{\varepsilon}}^{\mathrm{p}} \neq 0$.

Coupling of the given external loading and the loading by the sources due to plasticity and damage is present in general in the constitutive relations:

$$\begin{aligned} \dot{\boldsymbol{\varepsilon}}^{\mathrm{p}} &= \dot{\boldsymbol{\varepsilon}}^{\mathrm{p}}(\bar{\boldsymbol{\sigma}}, \dot{\bar{\boldsymbol{\sigma}}}, D, \alpha_n) \\ \dot{D} &= \dot{D}(\bar{\boldsymbol{\sigma}}, D, \alpha_n) \\ \dot{\alpha}_n &= \dot{\alpha}_n(\bar{\boldsymbol{\sigma}}, D, \alpha_m), \end{aligned} \tag{10}$$

where α_n, $n = 1,\ldots,k$ are internal variables. The rate of the intensity of the generalized plastic sources is to be determined in a time stepping manner substituting the incremental solution into Eqs. (10).

2. The integral equation method of self-stresses

Considering a body with undeformed initial volume V and mass density ρ under the action of body forces b_i and surface tractions f_i as well as loaded by sources of self-stresses $\bar{\varepsilon}_{ij}$ in a Lagrangean description at time instants t and $t + \Delta t$. Second Piola-Kirchhoff stress tensor and the Green's strain tensor in the later instant have increments as well as the displacements have changed to $u_i + \Delta u_i$. D'Alembert's principle requires the sum of work of all forces including the inertia forces done on virtual displacements $\delta(u_i + \Delta u_i)$ when keeping the time $t + \mathrm{d}t$ = constant and on $\delta(u_i)$ at constant t, respectively, to vanish. After some manipulations, at the later time instant the principle renders

$$\oint_{\partial V} (f_i + \Delta f_i)\delta\Delta u_i \,\mathrm{d}S + \int_V (b_i + \Delta b_i)\,\delta\Delta u_i \,\mathrm{d}V$$

$$-\int_V (\sigma_{ij} + \Delta\sigma_{ij})\left[\ \delta\Delta\varepsilon_{ij} + \delta\left(\frac{1}{2}\Delta u_{1,i}\,\Delta u_{1,j}\right)\right]\mathrm{d}V - \int_V \rho\left(\ddot{u}_i + \Delta\ddot{u}_i\right)\delta\Delta u_i \,\mathrm{d}V = 0, \tag{11}$$

where the integration is performed over the initial volume, see Washizu [10] for static loading. The strain increment

$$2\,\Delta\varepsilon_{ij} = (\delta_{\ell j} + u_{\ell,j})\Delta u_{\ell,i} + (\delta_{\ell i} + u_{\ell,i})\,\Delta u_{\ell,j}, \tag{12}$$

is linearized with respect to the small increment Δu_i and the variation of the strain field is performed with respect to Δu_i. Considering the principle applied at t in Eq. (11) gives a form which has been used in deriving discretizations by the Finite Element Method - FEM, Nagtegaal [11].

Workable integral equations are derived by identifying the variations $\delta\Delta u_i$ with the influence displacements $\tilde{u}_{i(k)}$ of the linear elastic body loaded statically by a unit force $\mathbf{F}_{(k)} = 1.\mathbf{e}_k$ at a point x in the direction $\mathbf{e}_k$. In a second step the equilibrium of the auxiliary (~)-problem is considered and the principle of virtual work is applied taking the actual displacement increment Δu_i as the virtual ones. To make the deformations kinematically admissible to the auxiliary problem, kinematic constraints have to be released and reaction forces $\tilde{X}_{i(k)}$ have to be applied,

$$1.\,\Delta u_k + \oint_{\partial V} \tilde{X}_{i(k)}\,\Delta u_i\,\mathrm{d}S - \int_V \tilde{\sigma}_{ij(k)}\,\Delta\varepsilon_{ij}\,\mathrm{d}V = 0. \tag{13}$$

Substituting $\Delta\sigma_{ij}$ for $\Delta\varepsilon_{ij}$ above, using the incremental form of Eq. (6), and further, replacing $\tilde{\varepsilon}_{ij}$ by means of Hooke's law of the auxiliary problem by $\tilde{\sigma}_{ij}$ in the reduced Eq. (11) renders finally

$$1.\;\Delta u_k = \oint_{\partial V} \Delta f_i\,\tilde{u}_{i(k)}\,\mathrm{d}S + \int_V (\Delta b_i - \rho\Delta\ddot{u}_i)\tilde{u}_{i(k)}\,\mathrm{d}V + \int_V \tilde{\sigma}_{ij(k)}\,\Delta\bar{\varepsilon}_{ij}\,\mathrm{d}V + \Delta u_k^{\mathrm{geom}}. \tag{14}$$

Small strains are considered subsequently and $\Delta u_k^{\mathrm{geom}} = 0$. Splitting the increment of displacement into a quasistatic portion and a dynamic part,

$$\Delta u_k = \Delta u_k^{\mathrm{S}} + \Delta u_k^{\mathrm{D}} \tag{15}$$

gives explicitly, in a self-explanatory form of superposition,

$$\Delta u_k^{\mathrm{S}} = \oint_{\partial V} \Delta f_i\,\tilde{u}_{i(k)}\,\mathrm{d}S + \int_V \Delta b_i\,\tilde{u}_{i(k)}\,\mathrm{d}V + \int_V \tilde{\sigma}_{ij(k)}\,\Delta\bar{\varepsilon}_{ij}\,\mathrm{d}V \tag{16}$$

and an integral equation

$$\Delta u_k^{\mathrm{D}} + \int_V \rho\tilde{u}_{i(k)}\,\Delta\ddot{u}_i^{\mathrm{D}}\,\mathrm{d}V = -\int_V \rho\tilde{u}_{i(k)}\,\Delta\ddot{u}_i^{\mathrm{S}}\,\mathrm{d}V. \tag{17}$$

Separating the solution further into the contribution of the external, given loadings and the generalized plastic sources,

$$(\Delta u_i = \Delta u_i^{\mathrm{e}} + \Delta u_i^{*}) \quad (\mathrm{S, D}) \tag{18}$$

Eq. (17) applies individually to the (e) as well as to the (*) loading. Assuming the quasistatic solution to be known, the linear integral equations are readily solved, accuracy and convergence of the dynamic increments are secured due to the splitting; for the linear thermoelastic shock problem of beams see Boley and Barber [12] and for plates see Nowacki [13].

Following Bolotin [14] an eigenfunction expansion is well suited: The Green's tensor $\tilde{u}_{i(k)}$ is developed into a bilinear series,

$$\tilde{u}_{i(k)} = \varphi_{ij}\, \varphi_{kj} / \omega_j^2, \tag{19}$$

where ω_j is the j-th natural frequency. The orthonormalization condition

$$\int_V \rho\, \varphi_{kj}\, \varphi_{kl}\, \mathrm{d}V = \delta_{jl} \tag{20}$$

is understood and

$$\Delta u_k^{\mathrm{D}} = \Delta q_j(t)\, \varphi_{kj}, \quad j = 1,2, \ldots \tag{21}$$

is substituted when Eq. (17) is projected into the space of the function sets, to render formally uncoupled linear oscillators. The time function of the driving agency (incrementally) is assumed to be separable.

$$\Delta \ddot{q}_j + 2\zeta_j\, \omega_j\, \Delta \dot{q}_j + \omega_j^2\, \Delta q_j = -\, \Gamma_j\, \Delta \ddot{f},$$

$$\Gamma_j\, \Delta \ddot{f} = \int_V \rho\, \varphi_{ij}\, \Delta \ddot{u}_i^{\mathrm{S}}\, \mathrm{d}V. \tag{22}$$

Light modal damping has been added to Eq. (22). Γ_j^{e} is a classical participation factor, Γ_j^{*} is generalized and is a weighting factor of the plastic source loading, [15].

The modal expansion is worked out in detail for bending vibrations of beams and plates, [16]–[20].

3. Bending vibrations

In case of physically nonlinear beams, cross-sectional integration is performed, for plates integration over the thickness h reduces the volume integrals in Eq. (22).

With proper static influence functions the quasistatic deflection is

$$w_s = w_s^e + w_s^*, \quad w_s^e = \int_B p(\xi)\tilde{w}(\xi, x)\mathrm{d}B_\xi, \tag{23}$$

and the increment of the plastic drift becomes,

$$\Delta w_s^* = \int_L \tilde{M}(\xi, x)\, \Delta\bar{\kappa}(\xi)\mathrm{d}\xi, \quad \Delta w_s^* = \int_L \tilde{m}_{ij}(\boldsymbol{\xi}, \mathbf{x})\, \Delta\bar{\kappa}_{ij}\,(\boldsymbol{\xi})\mathrm{d}B_\xi \tag{24}$$

respectively. The plastic curvature of a beam is simply

$$\bar{\kappa} = \frac{1}{J}\int_A z\,\bar{\varepsilon}\,\mathrm{d}A. \tag{25}$$

Considering the given lateral loading in the separated form

$$p(x, t) = p_0(x)\, s(t) \tag{26}$$

the set of modal equations, m_n is the modal mass,

$$\ddot{Y}_n^e + 2\zeta_n\,\omega_n\,\dot{Y}_n^e + \omega_n^2\,Y_n^e = -\frac{\Gamma_n^e}{m_n\,\omega_n^2}\,\ddot{s}(t),$$

$$\Delta\ddot{Y}_n^* + 2\zeta_n\,\omega_n\,\Delta\dot{Y}_n^* + \omega_n^2\,\Delta Y_n^* = -\frac{\Delta\Gamma_n^*}{m_n}\,\ddot{g}\,(t, \Delta t),$$

$$\Gamma_n^e = \int_\ell \varphi_n(x)\,p_0(x)\,\mathrm{d}x, \quad \Delta\Gamma_n^* = \int_\ell \psi_n(x)\,\Delta\kappa(x)\,\mathrm{d}x, \tag{27}$$

$$\psi_n(x) = \int_\ell \varphi_n(\xi)\,\tilde{M}(x, \xi)\,\mu\,\mathrm{d}\xi,$$

is readily solved. In addition to the solution of the homogeneous equations the particular integrals are, $\zeta_n \ll 1$,

$$Y_{np}^e(t) = -\frac{\Gamma_n^e}{m_n\,\omega_n^3}\int_0^t \ddot{s}(\tau)\,\mathrm{e}^{-\zeta_n\,\omega_n(t-\tau)}\sin\omega_n(t-\tau)\,\mathrm{d}\tau$$

(28)

$$\Delta Y_{np}^*(t_i) = -\frac{\Delta\Gamma_n^*}{m_n\,\omega_n}\int_0^{\mathrm{t}=\Delta t_i} \ddot{g}(t_{i-1},\tau)\,\mathrm{e}^{-\zeta_n\,\omega_n(\bar{t}-\tau)}\sin\omega_n(\bar{t}-\tau)\,\mathrm{d}\tau.$$

The latter integrations are performed by assuming linear time ramp functions for the plastic source increments. The delta functions at the beginning and at the end of the time interval render simply

$$\Delta Y^*_{np}(t_i) = -\frac{\Delta\Gamma^*_n}{m_n \omega_n \Delta t_i} e^{-\zeta_n \omega_n \Delta t_i} \sin \omega_n \Delta t_i. \tag{29}$$

Recent numerical results of elasto-plastic and deteriorating beams may be found in [2].

Acknowledgement

Support through the grant S30-03 of the Austrian National Science Foundation FWF is gratefully acknowledged.

References

[1] Reissner, H.: Eigenspannungen und Eigenspannungsquellen. *ZAMM* **11** (1931), 1-8.

[2] Nemenyi, P.: Selbstspannungen elastischer Gebilde. *ZAMM* **11** (1931), 59-70.

[3] Telles, J.C.F.: *The boundary element method applied to inelastic problems.* Lecture Notes in Engng. Vol. 1. Springer Verlag, Berlin, 1983.

[4] Kröner, E.: Dislocation: A new concept in continuum theory of plasticity. *J. Math. Phys.* **42** (1963), 27-37.

[5] Mura, T.: *Micromechanics of defects in solids.* (2nd. ed.), Martinus Nijhoff Publ., Dortrecht, 1988.

[6] Kröner, E.: *Kontinuumstheorie der Versetzungen und Eigenspannungen.* Ergeb. Angew. Math. 5., Springer Verlag, Berlin, 1958.

[7] Cocks, A.C.F., Leckie, F.A.: Creep constitutive equations for damaged materials. *Adv. Appl. Mech.* **25** (1987), 239-294. Academic Press, Orlando 1987.

[8] Kachanov, L.M.: *Introduction to continuum damage mechanics*, Martinus Nijhoff Publ., Dortrecht, 1986.

[9] Bilby, B.A., Eshelby, J.D.: Dislocations and the theory of fracture. In: *Fracture 1* (Ed. H. Liebowitz). Academic Press, New York, 1968, 99-182.

[10] Washizu, K.: *Variational methods in elasticity and plasticity.* (2nd ed.), Pergamon Press, Oxford, 1975.

[11] Nagtegaal, J.C.: Some recent developments in combined geometric and non-linear finite element analysis. In: *Recent Advances in Nonlinear Comput. Mechanics* (Ed.: Hinton, E., Owen, D.R.J.; Taylor, C.) Pineridge Press, Swansea, 1982, 87-118.

[12] Boley, B., Barber, A.D.: Dynamic response of beams and plates to rapid heating. *J. Appl. Mech.* **24** (1957), 413-420.

[13] Nowacki, W.: *Thermoelasticity*. Pergamon Press, Oxford, 1962.

[14] Bolotin, V.V.: *The dynamic stability of elastic systems*. Holden-Day, San Francisco, 1964.

[15] Irschik, H., Ziegler, F.: Dynamics of linear elastic structures with selfstress: A unified treatment for linear and nonlinear problems. *ZAMM* **68** (1988), 199-205.

[16] Ziegler, F., Irschik, H.: Dynamic analysis of elastic plastic beams by means of thermoelastic solutions. *Int. J. Solids Struct.* **21** (1985), 819-829.

[17] Irschik, H., Ziegler, F.: Thermal shock loading of elastoplastic beams, *J. Thermal Stresses* **8** (1985), 53-69.

[18] Irschik, H.: Biaxial dynamic bending of elastoplastic beams. *Acta Mechanica* **62** (1986), 332-342.

[19] Fotiu, P., Irschik, H., Ziegler, F.: Forced vibrations of an elastic-plastic and deteriorating beam. *Acta Mechanica* **69** (1987), 193-203.

[20] Irschik, H.: Berechnung inelastischer Platten mittels Einflußfunktionen. *Ing. Archiv* **56** (1986), 332-342.

[21] Fotiu, P., Irschik, H., Ziegler, F.: Dynamic Plasticity: Structural drift and modal projections. In: *Proc. IUTAM-Symp. on Nonlinear Dynamics in Engng. Systems*, Stuttgart 1989 (Ed. W. Schiehlen), Springer-Verlag, Berlin, 1990, 75-82.

F. Ziegler, H. Irschik and P. Fotiu
Institut für Allgemeine Mechanik (E201)
Technical University of Vienna
Wiedner Hauptstraβe 8-10
A-1040 Wien
AUSTRIA

F.D. FISCHER AND F.G. RAMMERSTORFER

The thermally loaded heavy beam on a rough surface

1. Introduction

The paper deals with the analysis of an initially straight elastic homogeneous beam lying freely on a plane surface and loaded actively only by its deadweight and a temperature moment being constant over the beam length. The thermally activated transversal deformations of the beam parallel to the supporting surface are restrained by friction between the beam and this surface.

The technical background for this investigation is the cooling process of very long railway rails on a cooling bed down from approximately 900°C after rolling. Due to the nonhomogeneously developing temperature field (the rail head is hotter than the rail foot) the rail shows a curved configuration soon after the beginning of this cooling process accompanied by the development of a significant residual stress field [1].

An ideal rail for use in a straight railway track is perfectly straight and has a residual stress state with axial compression stresses both at the head and at the foot. Currently sufficiently straight rails can only be obtained by a straightening procedure after cooling. Considerable costs could be saved if this production step could be omitted.

In the case of very long rails (100m) one can observe that the end portions deflect significantly while the main middle portion remains totally straight on the bed. The obvious reason for the straight middle region is the fact that after a certain distance from the ends the moment due to the friction forces is equal to the 'temperature moment', which varies with time. If one cuts the long rail into two (~ 50m) parts, immediately, both new end portions show a similar deformation to the former end portions. Also if one lifts the long rail a constant curvature can be observed immediately in the middle region which was originally totally straight.

The detailed investigation of the described phenomena requires a thermo-viscoplastic consideration [1]. However, to understand this mechanism principally a thermally loaded, elastic heavy beam with a weight q per unit length on a rough surface parallel to the $x - z$ plane, friction coefficient μ, is investigated (see Figure 1). For the sake of simplicity we assume that a cross-section is only bent in the $x - z$ plane. The temperature field $T(z, t)$ does not vary with y and leads to a temperature moment as similarly defined in [2] by

$$m_{\Theta}(t) = \int_A E \alpha T(z, t) z \,\mathrm{d}y\,\mathrm{d}z, \tag{1}$$

where α is the linear coefficient of thermal expansion. E is Young's modulus.

2. Principal features

Let us consider possible friction force distributions along the beam axis. In slip regions, i.e. in intervals in which the transverse deflection $w(x)$ varies with time (due to $m_\Theta(t) \neq$ constant) the transverse load per unit length, $p(x)$, is governed by Coulomb's friction law

$$|p(x)| = \mu q \qquad \forall x : \frac{\partial w(x)}{\partial t} \neq 0. \tag{2}$$

$p(x)$ acts in the $\pm z$-direction and changes its sign in those points $x_i(t)$ where the velocity just vanishes:

$$\frac{\partial w}{\partial t}(x = x_i(t)) := \dot{w}(x_i) = 0, \quad i = 1,2,...,N(t). \tag{3}$$

The question arises if regions $[a, b]_j$ along the beam axis exist in which adhesive friction appears, i.e.

$$\frac{\partial w}{\partial t} \equiv 0 \quad \forall x \in [a, b]_j \tag{4}$$

within intervals of finite length.

From the requirement (4) and the equation for thermoelastic beam deflections [3]

$$\frac{\partial^2 w}{\partial x^2} = -\frac{1}{EJ}(M + m_\Theta) \tag{5}$$

(M being the bending moment of mechanical loads which in our case are simply the distributed forces $p(x)$) we can state that in adhesive friction regions

$$\frac{\partial M}{\partial t} = -\frac{\partial m_\Theta}{\partial t} \qquad \forall x \in [a, b]_j \tag{6}$$

must be met.

The temperature field and, therefore, m_Θ are assumed to be independent of x. Hence, the condition (6) requires that within adhesive friction regions the time variation of the bending moment due to the mechanical loads $p(x)$ must be independent of x, too:

$$\frac{\partial^2 M}{\partial x \, \partial t} \equiv 0 \quad \forall\, x \in [a, b]_j. \tag{7}$$

Changing the sequence of differentiation we can conlcude from (7):

$$\frac{\partial^2 M}{\partial t \, \partial x} = \frac{\partial}{\partial t} Q \equiv 0 \quad \forall\, x \in [a, b]_j \tag{8}$$

(Q being the transverse shear resultant).

If we assume a monotonic development of the process, starting with $m_\Theta(t = 0) \equiv 0$, $w(x, t = 0) \equiv 0$ and, hence, $Q(t = 0) \equiv 0$, condition (8) leads to

$$Q \equiv 0 \qquad \forall\, x \in [a, b]_j \tag{9}$$

and, hence

$$\frac{\partial Q}{\partial x} = - p(x) \equiv 0 \quad \forall\, x \in [a, b]_j. \tag{10}$$

From (10) we can conclude that regions which are at rest are not held by adhesive friction forces $p(x) \neq 0$ but are subject to the following conditions:

$$M(x) \equiv - m_\Theta, \tag{11a}$$

$$Q(x) \equiv 0, \tag{11b}$$

$$p(x) \equiv 0 \tag{11c}$$

for all $x \in [a, b]_j$.

Let us now consider where such regions might appear for $\partial m_\Theta / \partial t \neq 0$. For this purpose consider Fig. 1. Because of (11a) the end portions must not be at rest for $t > 0$. One inner region - due to symmetry requirements in the middle of the beam - could be at rest since (11a-c) can be fulfilled by any number $N > 1$ of tightly connected load intervals according to Eqs. (2) and (3) as long as $L(t) := x_N(t)$ is smaller than half the complete length of the beam.

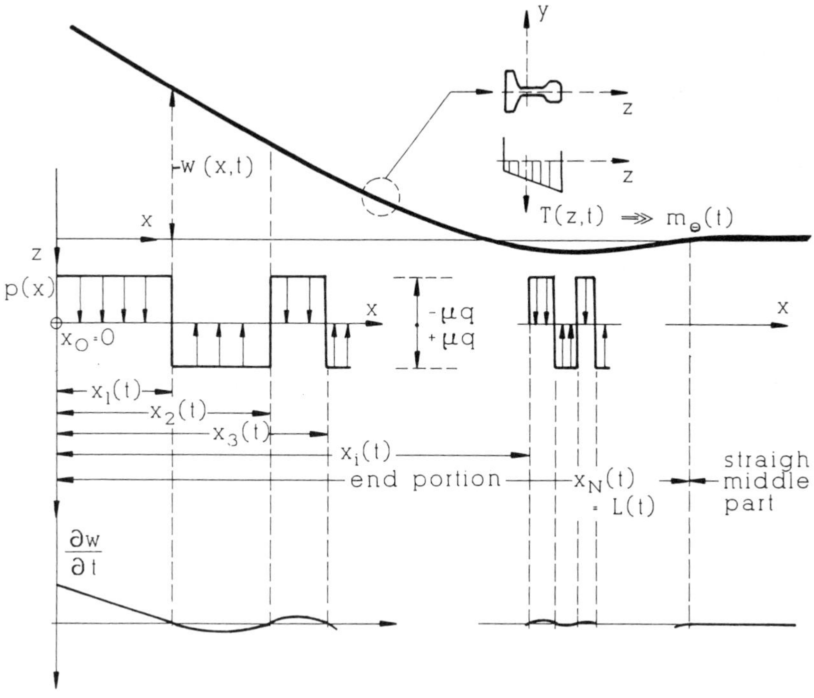

Figure 1. Friction force system

If besides this straight middle region at rest other regions at rest would appear, the conditions (11a–c) would require that in the slip regions between those regions self-equilibrated friction force systems would have to act:

$$\mu\, q \sum_{k=1}^{n} (\xi_k - \xi_{k-1})\, (-1)^{k-1} = 0, \tag{12}$$

$$\mu\, q \sum_{k=1}^{n} (\xi_k - \xi_{k-1})\, \left(a - \frac{\xi_k + \xi_{k-1}}{2}\right) (-1)^{k-1} = 0, \tag{13}$$

with ξ_k being the interval boundaries of the n tightly connected load intervals according to (2) and (3) and a being the starting coordinate for the next stick interval on the right. Equations (12) and (13) are independent of $m_\Theta(t)$, and no 'driving force' exists for the activation of such self-equilibrated force systems. Therefore, the existence of stick regions in addition to the straight middle region can, in accordance with the experimental observations, be excluded.

3. Formulation of the initial value problem

One can conclude from the above consideration that the sliding end portion of the beam can be divided into $N(t)$ tightly connected intervals $[x_{i-1}(t),\ x_i(t)]$, $i = 1,\dots,N(t)$, $x_0 = 0$, $x_{N_1} = L(t)$, with an alternatingly constant transverse load $p(x) = (-1)^i \mu q$ (see Fig. 1).

$$\int_0^{x_N} p(x)\,dx = \mu q \sum_{i=1}^{N} (-1)^{i-1} (x_i - x_{i-1}) = 0, \tag{14}$$

$$\int_0^{x_N} p(x)(x_N - x)\,dx = \mu q \sum_{i=1}^{N} (-1)^{i-1} (x_i - x_{i-1}) \left(x_N - \frac{x_i + x_{i-1}}{2}\right) = m_\Theta. \tag{15}$$

Eq. (14) can be rewritten as

$$2 \sum_{i=1}^{N-1} (-1)^{i-1} x_i = (-1)^N x_N. \tag{16}$$

Using (14) or (16) Eq. (15) can be reformulated as

$$2 \sum_{i=1}^{N-1} (-1)^{i-1} x_i^2 = (-1)^N x_N^2 - 2m_\Theta/(\mu q). \tag{17}$$

Additionally we have the N velocity conditions (3). To study the character of the differential equations for $x_i(t)$, $i = 1,\dots,N(t)$, the deflection w is given as the following sum:

$$w(x, t) = \sum_{i=1}^{N} (-1)^{i-1}\, {}_iw(x_{i-1}, x_i, x_N, x) - \frac{m_\Theta}{2EJ_y}(x_N - x)^2. \tag{18}$$

${}_iw(x_{i-1}, x_i, x_N, x)$ is the deflection of a beam clamped at

$$x_N \ \left({}_iw\,|_{x_N} = 0,\ \frac{\partial\, {}_iw}{\partial x}\Big|_{x_N} = 0\right)$$

and only loaded by μq in $[x_{i-1}, x_i]$.

With the definition

$${}_i\tilde{w} := 24\, EJ_y\, {}_iw/(\mu q), \tag{19}$$

${}_i\tilde{w}$ is given as

$$[0, x_{i-1}]: \; {}_i\tilde{w} = (x_N - x_{i-1})^3\,(-4x + 3x_N + x_{i-1})$$

$$- (x_N - x_i)^3\,(-4x + 3x_N + x_i), \tag{20a}$$

$$[x_{i-1}, x_i]: \; {}_i\tilde{w} = (x - x_{i-1})^4 + (x_N - x_{i-1})^3\,(-4x + 3x_N + x_{i-1})$$

$$-(x_N - x_i)^3\,(-4x + 3x_N + x_i), \tag{20b}$$

$$[x_i, x_N]: \; {}_i\tilde{w} = (x - x_{i-1})^4 + (x_N - x_{i-1})^3\,(-4x + 3x_N + x_{i-1})$$

$$- (x_N - x_i)^3\,(-4x + 3x_N + x_i) - (x - x_i)^4. \tag{20c}$$

With the logical function $\Lambda(i)$,

$$\Lambda(i) = 1 \;\text{ for }\; i \le k, \;\; \Lambda(i) = 0 \;\text{ for }\; i > k, \tag{21}$$

we obtain for $\dot{w}(x_k) = 0$, i.e. Eq. (3),

$$\sum_{i=1}^{N} (-1)^{i-1} \{ [-\Lambda(i-1)\,(x_k - x_{i-1})^3 + (x_N - x_{i-1})^2\,(3x_k - 2x_N - x_{i-1})]\,\dot{x}_{i-1}$$

$$+ [\Lambda(i)\,(x_k - x_i)^3 - (x_N - x_i)^2\,(3x_k - 2x_N - x_i)\,]\,\dot{x}_i$$

$$+ 3(x_N - x_k)\,[(x_N - x_{i-1})^2 - (x_N - x_i)^2]\,\dot{x}_N\}$$

$$-6\,(x_N - x_k)\,\frac{m_\Theta}{\mu\, q}\dot{x}_N - \frac{3}{\mu\, q}\dot{m}_\Theta\,(x_N - x_k)^2 = 0, \quad k = 1,2,\ldots,N. \tag{22}$$

This equation can be reformulated and results in the following relation by use of the equilibrium equations:

$$2\sum_{i=1}^{N-1} (-1)^{i-1}\,[\Lambda(i)\,(x_k - x_i)^3 - (x_N - x_i)^2\,(3x_k - 2x_N - x_i)\,]\,x_i$$

$$= \frac{3}{\mu\, q}\dot{m}_\Theta\,(x_N - x_k)^2\,, \quad k = 1,2,\ldots,N-1. \tag{23}$$

It can immediately be seen that for $k = N$ the above equation leads to a trivial identity!

The conclusion is that $(N - 1)$ differential equations plus two algebraic equations (the equilibrium equations) exist. Therefore, $N + 1$ equations can be written for N unknowns $x_1(t),\ldots,x_N(t)$ with the initial conditions $x_1,\ldots,x_N = 0$ at $t = 0$ and an additional unknown quantity viz. the integer variable $N(t)$ itself. This differential/algebraic system of an unknown number of equations has not been solved up to now.

With respect to the experimental observations it can be argued that N is an infinite number with a convergent sequence $x_1, x_2,...,x_N$, $N \to \infty$. In the opinion of the authors further mathematical research is necessary, particularly with respect to the existence of a unique solution.

4. Numerical Solutions

4.1. Discretization by finite elements

In order to obtain a reference solution a proper spatial discretization using finite elements has been performed. Because of the known friction force the frictional behaviour is modelled by transversally oriented truss elements with ideal-plastic material. The yield stress and the cross-section of these trusses are chosen so that the maximum truss forces represent the distributed load μq. The flow rule automatically controls the sign of the transverse friction force with respect to the orientation of the sliding velocity.

Results of the incremental-iterative solution of this nonlinear problem are shown in Figs. 2 and 3. They confirm the tightly connected sliding intervals and the development of a convergent sequence $x_k(t)$ with

$$\dot{w}(t)\,|_{x = x_k} = 0.$$

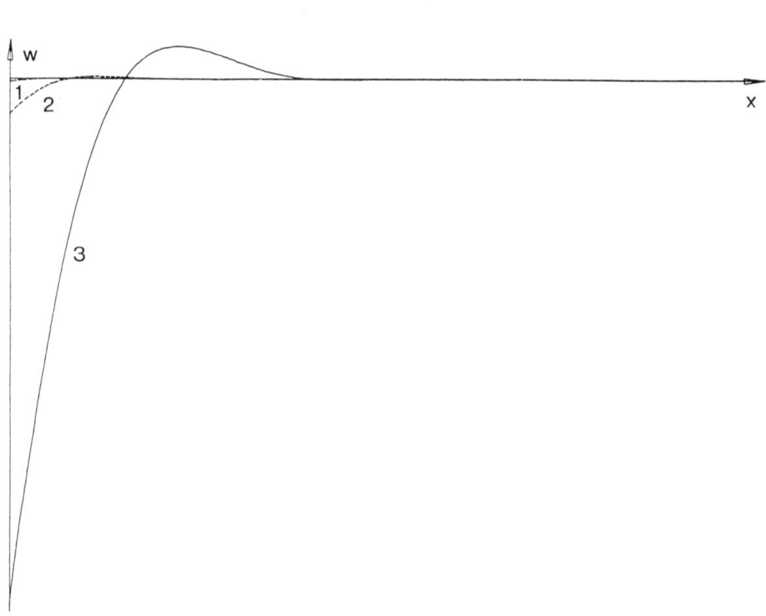

Figure 2. Displacements for three different increased temperature moments

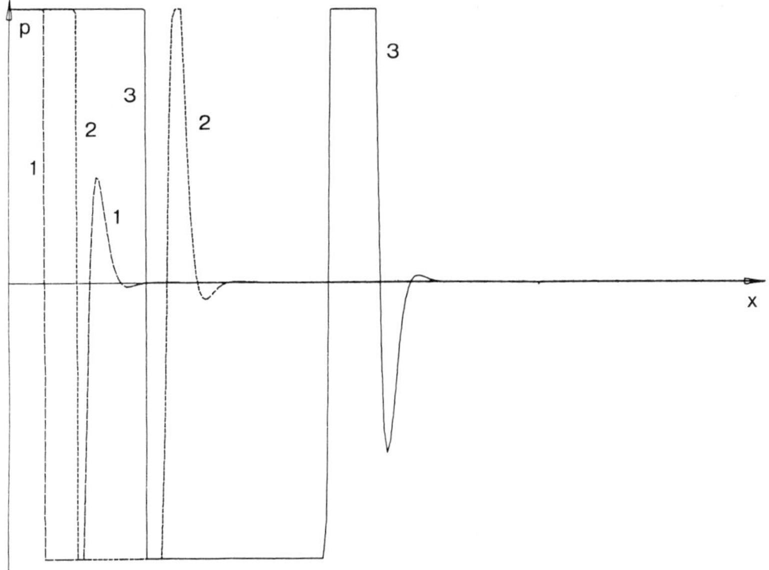

Figure 3. Friction force distributions for temperature moment corresponding to Figure 2.

Of course, the behaviour cannot be resolved completely due to the limited fineness of the model. For example the elasticity in the trusses (even if chosen to be nearly rigid within the yield surface) leads to a garbled force distribution in the developing intervals.

4.2. Discretization of the supporting conditions

In reality the rail acts as an equal-span continuous beam with an infinite (to be precise, very large) number of spans each of length l. The Clapeyron equation [3] can be used – as long as no slip appears – to find the supporting reactions ${}_{\Theta}A_i$, ${}_{q}A_i$ and moments ${}_{\Theta}M_i$, ${}_{q}M_i$, (the indices Θ and q refer to loading by m_{Θ} or q, respectively). With the following defintions we obtain

$$ {}_{q}M_i = -\frac{q\,l^2}{F_i}, \quad {}_{\Theta}M_i = -\frac{m_{\Theta}}{G_i} \tag{24a} $$

$$ \frac{1}{F_i} = \frac{1}{2} - \frac{1}{F_{i-2}} - \frac{4}{F_{i-1}}, \quad \frac{1}{G_i} = 6 - \frac{1}{G_{i-2}} - \frac{4}{G_{i-1}}, \quad i \geq 2 \tag{24b} $$

$$ {}_{\Theta}M_0 = {}_{q}M_0 = 0, \quad F_1 = 2(3+\sqrt{3}), \quad G_i = \frac{F_i}{12}. $$

The following relations can be formulated for the reaction forces

$${}_qA_i = a_i\, ql, \qquad i = 0 : a_0 = \frac{(3+\sqrt{3})}{12},$$

$$i \geq 1 : a_i = (\frac{1}{2} + \frac{6}{F_i}); \tag{24c}$$

$${}_\Theta A_i = b_i\, \frac{m_\Theta}{l}, \quad i = 0 : b_0 = -\,(3 - \sqrt{3}),$$

$$i \geq 1 : b_i = -\,6(1 - \frac{1}{G_i}). \tag{24d}$$

Data for a_i, G_i, can be taken from Table 1.

Table 1

i	1	2	3	4	5	6	7	8
G_i	0.78868	1.07735	0.98113	1.00518	0.99862	1.00037	0.99991	1.00001
a_i	1.13397	0.96410	1.00762	0.99742	1.00069	0.99982	1.00005	0.99999

The deflection in a field $x \in [\hat{x}_{i-1}, \hat{x}_i]$, $\hat{x}_j$ representing support positions, due to m_Θ is described by

$$-\,EJ\, w''_\Theta = m_\Theta \,[\, 1 - \frac{1}{G_{i-1}} - (\frac{1}{G_{i-1}} - \frac{1}{G_i})\, \frac{(x - x_{i-1})}{(x_i - x_{i-1})} \,]. \tag{25}$$

One can immediately see from Table 1 that for $i \geq 5$ $w''_\Theta \approx 0$ and, therefore, $w_\Theta \approx 0$. This agrees with the observation of a straight middle part of the rail.

If we assume that sliding occurs in the end portion at the supports $0,1,\ldots,(k-1)$ the corresponding friction forces R_i (positive in the positive z-direction) are defined by

$$|R_i| = \mu \, _qA_i = \mu \, ql \, a_i \tag{26}$$

and lead to the following reaction force $\tilde{R}_k$ at support k:

$$\tilde{R}_k = \sum_{i=0}^{k-1} R_i + \beta \sum_{i=0}^{k-1} R_i \, (k - i). \tag{27}$$

Since the reaction force at the first support of an infinite equal-span beam due to a moment $\tilde{M}$ at this support is $-(3 - \sqrt{3}\,)\tilde{M}/\ell$ the coefficient β is found to be $\beta = 3 - \sqrt{3}$. It follows that

$$\tilde{R}_k = \mu \, q\ell \, [\sum_{i=0}^{k-1} \text{sign}\,(R_i)\, a_i + \beta \sum_{i=0}^{k-1} (k - i) \, \text{sign}\,(R_i)\, a_i]. \tag{28}$$

The reaction force due to m_Θ at support k corresponds to that one at the first support of an infinite equal-span beam and is $-\beta m_\Theta/l$. Let us define $_km_\Theta$ to be that temperature moment which just leads to sliding at support k. Now $m_\Theta(t)$ is assumed to be a monotonically increasing positive function of time. Sliding occurs at support k at the same instant as at support k-1, if

$$|\tilde{R}_k - \beta_{k-1} \, m_\Theta/l\,| \geq \mu \, ql \, a_k, \tag{29}$$

$$_km_\Theta = {}_{k-1}\, m_\Theta.$$

If this multiple support sliding does not appear, sliding occurs under an increased m_Θ:

$$_km_\Theta = \ell \, (\tilde{R}_k + \mu \, ql \, \text{sign}\,(R_k)\, a_k)/\beta \tag{30}$$

with sign (R_k) selected so that $_km_\Theta$ is the nearest positive and greater value to $_{k-1}m_\Theta$. This algorithm leads to the following result:

* Starting with sign $(R_0) = +1$ always a positive pair of signs is followed by a negative pair,
  sign $(R_0) = +1$, sign $(R_1) = +1$, sign $(R_2) = -1$, sign $(R_3) = -1$,
  sign $(R_4) = +1$, sign $(R_5) = +1$, sign $(R_6) = -1$, ...

* Table 2 for $_k\tilde{m}_\Theta = {}_km_\Theta/(\mu q l^2)$ can be given.

Table 2

k	0	1,2,3	4	5,6,7	8	9,10,11
${}_k\tilde{m}_\Theta$	0.3110	1.59969	2.47683	3.81807	4.68793	6.02946

Using the constant $a = -0.44718$ very simple relations can be obtained referring to the resultant force.

For $k = 12, 16, 20, 24,...$ we obtain

$$R = \sum_{i=0}^{k-1} R_i = a\,\mu\,ql \tag{31}$$

$$ {}_k\tilde{m}_\Theta = {}_{k-4}\tilde{m}_\Theta + 4(1+a) = {}_{k-4}\tilde{m}_\Theta + 2.21128. \tag{32}$$

For $k = 13, 17, 21$ and so on, we obtain

$$ {}_k\tilde{m}_\Theta = {}_{k-1}\tilde{m}_\Theta + \frac{1}{3-\sqrt{3}} + (1+a) = {}_{k-1}\tilde{m}_\Theta + 1.34150. \tag{33}$$

The analytically derived solution for the rail on discrete slip-stick supports shows the principal features of the original problem.

5. Conclusion

A hot rail lying on a cooling bed and being free to deform under friction conditions is modelled as a beam with unit weight q. The analysis of the deformation state of the end portions leads to a differential/algebraic system of an unknown number of equations for the unknown interval boundaries $x_i(t)$, $i = 1,...,N$. The friction force $\pm\mu q$ changes its sign at the boundaries of the tightly connected length intervals.

A twofold discretization, once by finite elements and once by an infinite number of discrete supports, leads to physically meaningful results. These solutions may serve as reference solutions for the direct solution of the formulated set of equations of the original problem by a sophisticated mathematical treatment. Since the friction force distribution represents a 'bang-bang' function with respect to x methods of optimum control [4] seem to be the most promising tool.

References

[1] Fischer, F.D., Hinteregger, E. and Rammerstorfer, F.G., The influence of different geometrical and thermal boundary conditions and the phase transformation on the residual stress state in railroad rails after heat treatment, *Proc. ICRS*-2, Nancy, France, 1988.

[2] Parkus, H., *Thermoelasticity*, Springer, Vienna and New York, 1976.

[3] Ziegler, F., *Technische Mechanik der festen und flüssigen Körper*, Springer, Vienna and New York, 1985.

[4] Bell, D.J., *Recent Mathematical Developments in Control*, Academic Press, 1979.

F.D. Fischer
Institute of Mechanics
University for Mining and Metallurgy
Leoben
AUSTRIA

F.G. Rammerstorfer
Institute of Lightweight Structures
and Aerospace Engineering
Vienna Technical University
Vienna
AUSTRIA

R. MAHNKEN AND E. STEIN

Error analysis and adaptive time-step control for finite-element computations of creep-problems

1. Introduction

The problem of long term prediction of creep is very important in industry, especially if high temperatures are present. Criteria to tackle this problem are twofold: Firstly, there is a great need for reliable evolution laws modelling the creep, classically consisting of three stages such as the primary, secondary and tertiary stage. Secondly, the numerical integration process has to be reliable for the integration in both time and space.

In the present paper we briefly give some basic ideas for the development of evolution laws modelling creep. Then, the main purpose is to discuss the integration in time of constitutive equations. As an extension of the error analysis in [1] for Norton type creep problems, we present an error analysis for time-dependent creep problems including internal variables. Since the choice of the step size can crucially affect the numerical results, an adaptive step-size control based on error estimators is described. The effects are demonstrated in a simple numerical example of a cylindrical specimen.

2. Problems and modelling of creep deformations

2.1. Concepts for modelling

Creep is very sensitive to several factors such as *external conditions* (type of material (salt, carbon steel, copper,...), shape and processing of the specimen, temperature, stress, testing or equipment) or *internal conditions* (dislocation density (lattice distortions, edge/screw type), chemical composition, phase structure, type of crystals (e.g. face- or body-centred cube), damage (voids, cracks), agglomoration or diffusion of foreign atoms (e.g. alloys)). There is a great variety in modelling creep. Usually a phenomenological strategy describes creep in the form

$$\varepsilon^{in} = \sum_i \Big(f_1(\Theta)\ \ f_2(\sigma)\ \ f_3(t)\Big)_i\,, \qquad (1)$$

where ε^{in} are the inelastic strains, σ are stresses, t denotes the time and Θ the temperature. The temperature dependence in f_1 can be described in most cases by an *Arrhenius term*

$$f_1 = A \exp\left(\frac{-Q}{R\,\Theta}\right) \tag{2}$$

and is well known in statistical physics. In Eq. (2) A is a constant, Q is the activation energy, and R is the Boltzman constant. Some of the stress functions f_2 can be physically explained (hyperbolic creep, and in some cases power creep) but are still not completely understood. The time functions f_3 are based purely phenomenologically and are of no value if conditions change. Thus, in modern developments they are omitted and replaced by creep functions $f_3(\varepsilon)$ or, even better, by incorporation of *internal variables*, called η in the sequel, considering the microstructure of the material [2].

What makes a solid material, e.g. a crystal, deforming? Even in virgin crystals there are always some defects, classified as point defects (intercrystalline or missing atoms), line defects (dislocations), plane (cracks) and volume defects (holes). Deformation is carried by diffusion of point defects (usually connected with high-temperature creep and low loading) and gliding (high stresses) or climbing (high temperature) of dislocations. Furthermore, the dislocations can multiply and interact because of their accompanying stress fields which greatly influences the material behaviour and so makes the dislocation density one of the most favoured internal variables (but by no means the only one).

A three-dimensional generalization is carried out by observing the experimental fact that in most cases creep is dependent only on the deviatoric stresses $\mathbf{s}$ of the stress tensor $\boldsymbol{\sigma}$ and thus the hydrostatic stresses play no role. As an example a Norton type creep law often used to model salt creep behaviour and which, incidently, is a special case of Eq. (1) in rate form, is presented below [3]:

$$\boldsymbol{\varepsilon}^{in} = \kappa F^{m}\, \mathbf{s} \tag{3a}$$

$$\mathbf{s} = \mathbf{D}\boldsymbol{\sigma} \tag{3b}$$

$$\mathbf{D} = \mathbf{I} - \frac{1}{3}\mathbf{1} \otimes \mathbf{1} \tag{3c}$$

$$F = \sqrt{\frac{3}{2}\mathbf{s}\mathbf{s}} \tag{3d}$$

$$\kappa = \frac{3}{2}\left(A_p n \exp(-\,nt) \exp\left(-\frac{Q_p}{R\,\Theta}\right) + A_s \exp\left(-\frac{Q_s}{R\,\Theta}\right)\right). \tag{3e}$$

In the above equations F is a von Mises function depending only on the deviatoric stress tensor $\mathbf{s}$, $\mathbf{D}$ is a projection tensor relating $\mathbf{s}$ to the Cauchy stress tensor $\boldsymbol{\sigma}$, m is a constant, usually between 3 and 5, and κ is a function dependent on time and temperature, thus accounting for primary and secondary creep.

2.2. Initial-boundary-value creep problem

Using standard notation, we now state the set of equations for the initial-boundary value problem with small strains for a material body, denoted by Ω, subjected to external and internal loads:

$$\mathrm{div}\,\boldsymbol{\sigma} + \rho\mathbf{b} = 0 \tag{4a}$$

$$\boldsymbol{\sigma} = \mathbf{E}\boldsymbol{\varepsilon}^{el} \tag{4b}$$

$$\boldsymbol{\varepsilon} = \frac{1}{2}(\nabla\mathbf{u} + (\nabla\mathbf{u})^T) \tag{4c}$$

$$\boldsymbol{\varepsilon} = \boldsymbol{\varepsilon}^{in} + \boldsymbol{\varepsilon}^{el}. \tag{4d}$$

$$\dot{\boldsymbol{\varepsilon}}^{in} = \mathbf{f}(\boldsymbol{\sigma}, \boldsymbol{\eta}, \Theta, t), \quad \dot{\boldsymbol{\eta}} = \mathbf{f}(\boldsymbol{\sigma}, \boldsymbol{\eta}, \Theta, t). \tag{4e, f}$$

Eq. (4a) is the equilibrium condition and Eq. (4b) is Hooke's law, with $\mathbf{E}$ being the fourth-order elasticity tensor. Eq. (4c) describes the linear kinematic relation between displacements $\mathbf{u}$ and the strains $\boldsymbol{\varepsilon}$, while Eq. (4d) denotes the additive decomposition of the total strains into elastic strains $\boldsymbol{\varepsilon}^{el}$ and the inelastic strains $\boldsymbol{\varepsilon}^{in}$. Finally, Eq. (4e) and Eq. (4f) denote the evolution equations for the inelastic strains $\boldsymbol{\varepsilon}^{in}$ and the internal parameters $\boldsymbol{\eta}$. Additionally, boundary conditions such as $\mathbf{u}(\mathbf{x}, t) = \bar{\mathbf{u}}(\mathbf{x}, t)$, $\mathbf{x} \in \partial\Omega_1$, $\boldsymbol{\sigma}(\mathbf{x}, t)\mathbf{n} = \bar{\mathbf{t}}(\mathbf{x}, t)$, $\mathbf{x} \in \partial\Omega_2$, $t \in [0, T]$ and initial conditions $\mathbf{u}(\mathbf{x}, 0) = \mathbf{u}_0(\mathbf{x})$, $\boldsymbol{\sigma}(\mathbf{x}, 0) = \boldsymbol{\sigma}_0(\mathbf{x})$, $\boldsymbol{\varepsilon}^{in}(\mathbf{x}, 0) = \boldsymbol{\varepsilon}_0^{in}(\mathbf{x})$ and $\boldsymbol{\eta}(\mathbf{x}, 0) = \boldsymbol{\eta}_0(\mathbf{x})$, $\mathbf{x} \in \Omega$, $t = 0$ are given. T denotes the total time of interest.

3. Discretization of the initial-boundary value problem

3.1. Finite element formulation

Using a finite-element-discretization with n_e elements the approximation in space is carried out in the context of a Galerkin method using test functions $\mathbf{u}^h$ and trial fucntions $\mathbf{q}^h$. Thus the discrete approximation of the weak form of equilibrium reads

$$g(\mathbf{u}_e, \mathbf{q}_e) = \bigcup_{e=1}^{n_e} \mathbf{q}_e^T \left[\int_{\Omega_e} \mathbf{B}^T \sigma(\mathbf{u}_e)\, d\Omega - \int_{\Omega_e} \mathbf{N}^T \rho_0 \mathbf{b}\, d\Omega - \int_{\partial\Omega_e} \mathbf{N}^T \bar{\mathbf{t}}\, ds \right] = 0, \tag{5}$$

where the matrix $\mathbf{B}$ is defined by the strain–displacement relation $\boldsymbol{\varepsilon} = \mathbf{B}\mathbf{u}_e$, and $\mathbf{N}$ are the shape functions. The resulting system of equations (FEM–residual vector) is then given by

$$\mathbf{G}^h = \bigcup_{(e)} \int_{\Omega_e} \mathbf{B}^T \sigma(\mathbf{u}_e)\, d\Omega - \mathbf{P} = \mathbf{0} \quad \forall\ t \in [0, T],. \tag{6}$$

where the vector $\mathbf{P}$ is made up of the inner and outer forces.

3.2. Numerical single-step time integration

Let $[0, T]$ be the time interval of interest discretized in I steps $h_{i+1} = t_{i+1} - t_i$, $i = 0,\ldots,I - 1$. Then for each time-step we formulate locally, i.e. for each Gaussian quadrature point the basic problem as follows: Let $\boldsymbol{\varepsilon}_i, \boldsymbol{\varepsilon}_i^{in}, \boldsymbol{\varepsilon}_i^{el}, \boldsymbol{\sigma}_i, \boldsymbol{\eta}_i$ be the given data at time t_i. Furthermore, in the context of a finite element method an increment of the displacements $\Delta\mathbf{u}_{i+1} = \mathbf{u}_{i+1} - \mathbf{u}_i$ is known from the global iteration such that the increment of total strains can be obtained from

$$\Delta\boldsymbol{\varepsilon}_{i+1} = \boldsymbol{\varepsilon}_{i+1} - \boldsymbol{\varepsilon}_i = \mathbf{B}\Delta\mathbf{u}_{i+1} \tag{7}$$

(strain driven algorithm). Thus, the unknowns at time t_{i+1} are $\boldsymbol{\varepsilon}_{i+1}^{in}, \boldsymbol{\varepsilon}_{i+1}^{el}, \boldsymbol{\sigma}_{i+1}, \boldsymbol{\eta}_{i+1}$.

A generalized integration scheme for the increment of inelastic strains and the increment of internal variables, respectively, can be written as

$$\Delta\boldsymbol{\varepsilon}_{i+1}^{in} = \boldsymbol{\varepsilon}_{i+1}^{in} - \boldsymbol{\varepsilon}_i^{in} = h_{i+1} \sum_{j=1}^{l} w_j\, \dot{\boldsymbol{\varepsilon}}^{in}(\zeta_j); \quad \Delta\boldsymbol{\eta}_{i+1} = \boldsymbol{\eta}_{i+1} - \boldsymbol{\eta}_i = h_{i+1} \sum_{j=1}^{l} w_j\, \dot{\boldsymbol{\eta}}(\zeta_j), \tag{8}$$

where ζ_j are quadrature points, w_j are weights and l is the number of quadrature points. In Eq. (8) $\mathbf{f}(\zeta_j)$ means

$$\mathbf{f}(\zeta_j) = \mathbf{f}(\boldsymbol{\sigma}_i + \zeta_j\, \Delta\boldsymbol{\sigma}_{i+1},\ \boldsymbol{\eta}_i + \zeta_j\, \Delta\boldsymbol{\eta}_{i+1}, \ldots), \tag{9}$$

i.e. $\dot{\boldsymbol{\varepsilon}}^{in}(\zeta_j)$ and $\dot{\boldsymbol{\eta}}(\zeta_j)$ are obtained from linear interpolation of the stresses and the internal variables. Thus, the resulting difference equation for the stresses is

$$\Delta\boldsymbol{\sigma}_{i+1} = \mathbf{E}\ (\Delta\boldsymbol{\varepsilon}_{i+1} - h_{i+1} \sum_{j=1}^{l} w_j\, \dot{\boldsymbol{\varepsilon}}^{in}(\zeta_j)). \tag{10}$$

Some well-known special cases of the above integration scheme are firstly, the *generalized trapezoidal rule*, where $l = 2,\ \zeta_1 = i,\ \zeta_2 = i+1,\ w_1 = 1-\theta,\ w_2 = \theta$ and secondly the *generalized midpoint rule*, where $l = 1,\ \zeta = i+\theta,\ w = 1$ and where θ is a variable between 0 and 1. Note, that for $\theta = 0$ and $\theta = 1$ both integration schemes coincide and are known as the *Euler (expl.)* forward rule and the *Euler (impl.)* backward rule, respectively.

The determination of $\boldsymbol{\varepsilon}^{in}_{i+1}, \boldsymbol{\varepsilon}^{el}_{i+1}, \boldsymbol{\sigma}_{i+1}, \boldsymbol{\eta}_{i+1}$ at time t_{i+1} for given $\boldsymbol{\varepsilon}_{i+1}$ and time step h_{i+1} leads to a nonlinear system of, generally, low dimension. The solution can be accomplished in a *local iteration* procedure most effectively with a Newton iteration scheme. Additionally, the stresses $\Delta\boldsymbol{\sigma}_{i+1}$ have to satisfy the discretized FEM-system of equations, i.e. an equivalent of Eq. (6) in incremental form. The solution of the nonlinear equation, which dimension can be very high, is accomplished in a *global iteration* procedure, where several methods such as Newton-, quasi-Newton- or multi-grid methods can be applied.

4. Error analysis

We mentioned before that the accuracy of the numerical results may significantly depend on the step size. Thus a strategy would be beneficial to select the time step adaptively based on error estimators. The purpose of this chapter is to introduce the basic mathematical concepts for an error analysis and to describe briefly how error estimators can be obtained.

4.1. Local truncation error

Let $\boldsymbol{\sigma}(t_{i+1}), \boldsymbol{\varepsilon}(t_{i+1}), \boldsymbol{\varepsilon}^{in}(t_{i+1}), \boldsymbol{\eta}(t_{i+1})$ be the exact solution of the initial value problem (in the sense that the spatial discretization plays no role) and let $\boldsymbol{\sigma}_{i+1}, \boldsymbol{\varepsilon}_{i+1}, \boldsymbol{\varepsilon}^{in}_{i+1}, \boldsymbol{\eta}_{i+1}$ be the solution of the difference equations. For a strain-driven algorithm we assume that $\boldsymbol{\varepsilon}_{i+1} = \boldsymbol{\varepsilon}(t_{i+1})$. Then, a *local truncation error* for the inelastic strains, the internal variables and the stresses, respectively, is defined as

$$\mathbf{L}(\boldsymbol{\varepsilon}^{in}) = \boldsymbol{\varepsilon}^{in}(t_{i+1}) - \boldsymbol{\varepsilon}^{in}(t_i) - h_{i+1} \sum_{j=1}^{l} w_j\, \dot{\boldsymbol{\varepsilon}}^{in}(t_{i+\zeta_j})$$

$$\mathbf{L}(\boldsymbol{\eta}) = \boldsymbol{\eta}(t_{i+1}) - \boldsymbol{\varepsilon}(t_i) - h_{i+1} \sum_{j=1}^{l} w_j\, \dot{\boldsymbol{\eta}}(t_{i+\zeta_j}) \tag{11}$$

$$\mathbf{L}(\boldsymbol{\sigma}) = \boldsymbol{\sigma}(t_{i+1}) - \boldsymbol{\sigma}(t_i) - \mathbf{E}\left(\boldsymbol{\varepsilon}(t_{i+1}) - \boldsymbol{\varepsilon}(t_i) - h_{i+1}\sum_{j=1}^{l} w_j\, \dot{\boldsymbol{\varepsilon}}^{in}(t_{i+\zeta_j})\right).$$

Let $(\boldsymbol{\varepsilon}^{in}, \boldsymbol{\eta}, \boldsymbol{\sigma}) \in \mathbb{R}^p$ be replaced by $\mathbf{y}$. Thus a Taylor series expansion gives

$$\mathbf{L}(\mathbf{y}_i) = \sum_{j=0}^{n+1} \mathbf{C}_j\, h_{i+1}^j\, \mathbf{y}^{(j)}(t_i) + \mathrm{O}(h^{n+2}), \tag{12}$$

where $\mathbf{C}_j$ is a diagonal matrix with elements $C_{r,j}$, $r = 1,\ldots,p$. The method is said to be of order n if $C_{r,0} = 0,\ldots,C_{r,n} = 0$, $C_{r,n+1} \neq 0$. It is well known that, for both the *generalized trapezoidal rule* and the *generalized midpoint rule*, second-order accuracy ($n = 2$) is achieved if $\theta = 1/2$ while otherwise both methods are only of first-order accuracy ($n = 1$). For methods of order n the term $\mathbf{C}_{n+1}\, h^{n+1}\, \mathbf{y}^{(n+1)}(t_i)$ is frequently called the *principal local truncation error* [4].

4.2. Error estimators

In this section we describe, how error estimators for methods of order n with the property

$$l(\mathbf{y}) = \mathbf{C}_{n+1}\, h^{n+1}\, \mathbf{y}^{(n+1)}((t_i) + \mathrm{O}(h^{n+2}) \tag{13}$$

can be obtained.

4.2.1. Error estimators by means of higher-order methods

Let $\mathbf{y}_{i+1}$ be the solution of the method with order n. Furthermore, let $\tilde{\mathbf{y}}_{i+1}$ be the solution of the method with order $\tilde{n} \geq n + 1$. Thus, it can be shown that the difference

$$\mathbf{l}(\mathbf{y}) = \tilde{\mathbf{y}}_{i+1} - \mathbf{y}_{i+1} \tag{14}$$

is an error estimator with the property given in Eq. (13).

4.2.2. Error estimators by means of extrapolation method

Let $\mathbf{y}_{i+1}$ be the solution of the nth order scheme. Now the same scheme may be used twice, but with step-size $h_{i+1}/2$. Denote its solution by $\tilde{\mathbf{y}}_{i+1}$. Accordingly, the error estimator is

$$\mathbf{l}(\mathbf{y}) = \frac{2^n}{2^n - 1}(\tilde{\mathbf{y}}_{i+1} - \mathbf{y}_{i+1}). \tag{15}$$

The extrapolation method can be combined very efficiently with a subincrementation technique for second-order integration schemes. For this purpose, both the time step and the total inelastic strains are divided, such that

$$h_{i+\frac{1}{2}} = \frac{h_{i+1}}{2} \tag{16}$$

$$\boldsymbol{\varepsilon}_{i+\frac{1}{2}} = \frac{1}{2}(\boldsymbol{\varepsilon}_i + \boldsymbol{\varepsilon}_{i+1}). \tag{17}$$

Accordingly both the total strains and, because of Eq. (4), the stresses are only of first-order accuracy. Therefore no error estimators are available for the stresses by use of this technique. However, substitution of the linearized stresses into Eq. (8) preserves

$$L(\boldsymbol{\varepsilon}^{in}) = \mathrm{O}(h^3); \quad L(\boldsymbol{\eta}) = \mathrm{O}(h^3), \tag{18}$$

which means that error estimators are available for the inelastic strains and the internal variables by use of the subincrementation technique.

5. Adaptive time-step control

An adaptive step-control policy can be applied in practice as follows. The step size is chosen according to two criteria: firstly a stability criterion and secondly an accuracy criterion. The first criterion is automatically satisfied for $\theta \geq 1/2$ (A-stability, unconditional stability). Otherwise an eigenvalue analysis is necessary (see References [4-7] for further considerations). Concerning the accuracy criterion the step size is determined according to

$$\epsilon_{min} \leq \| \mathbf{l}^{(m)}(\mathbf{y}) \| \leq \epsilon_{max}. \tag{19}$$

In Eq. (19) $\mathbf{l}^{(m)}(\mathbf{y})$ is the error estimator computed by one of the devices presented in Section 4.2. ϵ_{min} and ϵ_{max} are prescribed tolerances, thus requiring a bounded error and allowing the increase of the step size, m refers to the iteration index of the step-size control. If Eq. (19) is not satisfied for $h_{i+1}^{(m)}$ a correction can be carried out in the iteration process according to

$$h_{i+1}^{(m+1)} = h_{i+1}^{(m)} \left(\frac{\in}{\| \mathbf{l}^{(m)}(\mathbf{y}) \|} \right)^{\frac{1}{n+1}}, \tag{20}$$

where $\in = \in_{min}$ or $\in = \in_{max}$, respectively. Additionally, bounds such as $h_{i+1}^{(m+1)} \leq 5\, h_{i+1}^{(m)}$ and $h_{i+1}^{(m+1)} \geq 0.1\, h_{i+1}^{(m)}$ should be used to avoid excessive differences in the iteration process.

If attention is paid to the accuracy criterion Eq. (19) at each time-step an estimator for the *global truncation error* at time T can be obtained by summing up all error estimators from each step, i.e.

$$\| \tilde{\mathbf{G}}(\mathbf{y}) \| \simeq \sum_{i=1}^{l} \| \mathbf{l}_i(\mathbf{y}) \| . \tag{21}$$

It should be pointed out that the estimator in Eq. (21) has no bound character.

Remarks

(i) In the context of a finite element method, the step-control policy described in this section seems to be, at first sight, necessary at all Gaussian points within the whole structure wherever a local iteration is carried out. However, this procedure would be very expensive, and therefore we suggest restricting it to only a small number of points, e.g. where high stresses (or stress gradients) occur.

(ii) The step-control policy has to be carried out (at least) twice for each time step: firstly, at the beginning of the global iteration to initialize h_{i+1} (for predicted total strains $\Delta\varepsilon_{i+1}^{pred}$). After finishing the global FEM-system iteration the time step is controlled (for the final total strains $\Delta\varepsilon_{i+1}^{final}$) and, in the case that Eq. (19) does not hold, it may become necessary to repeat the global iteration with a new time step.

6. Example: cylinder under pressure

The example is concerned with a rock-salt cylinder discretized with three elements. The creep behaviour is modelled with the three-dimensional generalization of the Norton type creep law (see Eq. (3)). Parameters used in the computation are given as follows: $\mathbf{A}_p = 0.21$ $[\mathrm{MN/m^2}]^{-5}$, $Q_s = 12.9$ kcal/mol, $Q_p = 10.7$ kcal/mol, $R = 1.986 \times 10^{-3}$ kcal/mod K, $\Theta = 26°\mathrm{C}$, $A_s = 0.18$ $[\mathrm{MN/m^2}]^{-5}$, $n = 0.35$. The exponent for Eq. (3a) is $m = 4$. Young's modulus is $E = 14.000$ MPa and Poisson's ratio is $\nu = 0.1$. The total time is $T = 25$ days.

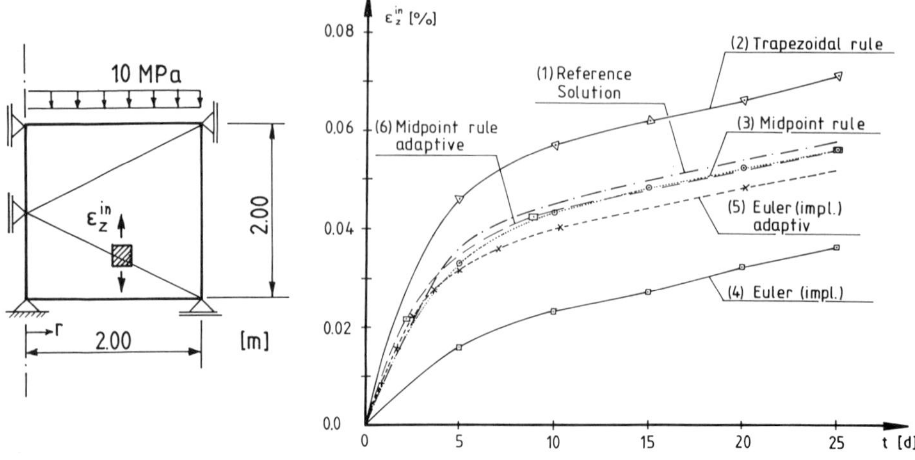

Figure 1. Comparison of solutions for a specific inelastic strain for different integration schemes

In Fig. 1 and Table 1 the results of several calculations for the inelastic strain ε_z^{in} in a specific point are compared. Curve (1) was obtained by dividing the total time T = 25 days into 75 equal spaced increments and thus it may be regarded as the reference solution. Curve (2) was obtained by use of the trapezoidal rule (θ = 0.5), curve (3) by the midpoint rule (θ = 0.5) and curve (4) by Euler's backward rule (θ = 1.0). For each, the total time T was equally spaced into 5 steps per 5 days. It is obvious that the choice of the time step crucially affects the final results. It can also be seen that, although there is no significant difference in CPU-time consumed, second-order methods give better results than the first-order Euler method and that, in this example, the midpoint rule gives better results than the trapezoidal rule. Curves (5) and (6) were obtained by use of adaptive procedures, based on Eq. (19) where $\varepsilon_{min} = 0.9\, tol$, $\varepsilon_{max} = 1.1\, tol$ and $tol = 0.001$. Curve (5) is the result of an implicit Euler integration where the higher-order midpoint rule was used to calculate error estimators. It can be seen that this curve is more accurate than curve (4). Curve (6) was generated with the midpoint scheme via error estimators of the extrapolation method combined with subincrements. Although only 3 time steps were used, the result is slightly more accurate in the primary phase than curve (3) where the number of iterations and the CPU-time consumed, demonstrates the superiority of the second adaptive technique.

Table 1. Comparison of solutions, 'exact' and estimated errors, number of iterations and CPU time for different integration schemes

	ε_z^{in}	Error estimator	'Exact' error	Number of time-steps	CPU time
(1) Reference solution	0.057	-	-	75	5552
(2) Trapezoidal rule	0.071	0.007	0.014	6	707
(3) Midpoint rule	0.056	0.004	0.001	5	677
(4) Euler (impl.)	0.036	0.012	0.021	5	676
(5) Euler (impl.) (adaptive)	0.054	0.006	0.003	11	2011
(6) Midpoint rule (adaptive)	0.056	0.002	0.001	3	800

7. Conclusions

As demonstrated in the numerical examples the choice of the time step in a step-by-step integration algorithm may significantly affect the accuracy of the results. Thus it is necessary to determine the time step adaptively based on error estimators. The cost for this procedure can be kept low, if it is restricted to only few Gaussian points of the FEM structure, e.g. where high stresses are present. For second-order methods the extrapolation method combined with a subincrementation technique is very efficient for calculating error estimators for inealstic strains and internal variables.

Concerning further investigations we intend to apply adaptive techniques to finite visco-plastic deformation problems. Furthermore, it seems to be beneficial to combine A-stable and L-stable methods to achieve optimal stability and convergence.

Acknowledgement

We should like to thank Dr-Ing. U. Heemann for his careful reading of the first part of this paper.

References

[1] Mahnken, R. and Stein, E. Adaptive time-step control in creep analysis, *Int. J. Num. Meth. Eng.* **28** (1989), 1619-1633.

[2] Heeman, U. Transientes Kriechen und Kriechbruch in Steinsalz, *Forschungs- und Seminarberichte aus dem Bereich der Mechanik der Universität Hannover*, Report-Nr. F89/2, Doctoral Thesis, 1989.

[3] Langer, M. and Hunsche, U. Das Verformungs- und Bruchverhalten von Steinsalz, *Zusammenfassende Darstellung einiger Forschungsergebnisse der BGR zur Salzmechanik*, Hannover, Okt. 1980, Salzmechanik II.

[4] Lambert, J.D. *Computational Methods in Ordinary Differential Equations*, Wiley, Chichester, New York, Brisbane, Toronto (1983).

[5] Hughes, T.J.R. and Taylor, R.L. Unconditionally stable algorithms for quasi-static elasto/viscoplastic finite element analysis, *Comp. & Struct.* **8** (1978), 169-173.

[6] Ortiz, M. and Popov, E.P. Accuracy and stability of integration algorithms for elasto plastic constitutive equations, *Int. J. Num. Meth. Eng.* **21** (1985), 1561-1576.

[7] Cormeau, F. Numerical stability in quasi-static elasto/visco-plasticity, *Int. J. Num. Math. Eng.* **9** (1975) 109-127.

R. Mahnken and E. Stein
University of Hannover
Institute of Structure Mechanics
and Numerical Mechanics
Appelstraße 9A
D-3000 Hannover 1
GERMANY

M. SCHULTZ AND H. SOCKEL

Pressure transients in railway tunnels

Nomenclature

A_T tunnel cross-section area
A_Z train cross-section area
L_T tunnel length
L_Z train length
P_Z work
R friction force
T temperature
V_Z train speed

a speed of sound
c_p specific heat
p pressure
s entropy
t time
w fluid velocity
x coordinate

ε path line
η Mach line
ξ Mach line
ρ density
Δt time stop
Δx mesh size

1. Introduction

To make travelling by train more attractive new high-speed lines have been introduced in several countries. One problem with high-speed trains is caused by pressure transients due to the passage of a train through a tunnel. Because these transients may cause discomfort to the ears of passengers, it is necessary to study the air flow in railway tunnels and to check if the pressure fluctuations exceed some critical value.

2. Mathematical model

2.1. Simplifying assumptions

For long tunnels, where the diameter is much smaller than the tunnel length, one-dimensional flow may be assumed [1-5]. This assumption is experimentally verified for the pressure, the quantity we are mainly interested in. So all quantities are averaged values over the tunnel cross section. The remaining independent variables are time t and the coordinate x along the tunnel.

The air inside the tunnel is treated as a perfect gas. Furthermore constant cross sections of the tunnel and the trains are assumed and it is assumed that there is no heat transfer across the walls. So in the regions 1, 2 and 3 of Fig. 1 there is a compressible one-dimensional unsteady air flow.

At discontinuities of flow area, that is at the tunnel portals and at the train ends (Fig. 1) the fundamental equations for quasisteady one-dimensional flow are applied.

This gives boundary conditions at the tunnel portals and conditions at the moving train ends linking the flow in the annulus between train and tunnel wall to the flow in the empty tunnel. Furthermore initial conditions have to be prescribed. This could either be the air at rest inside the tunnel, or a fully developed flow [1].

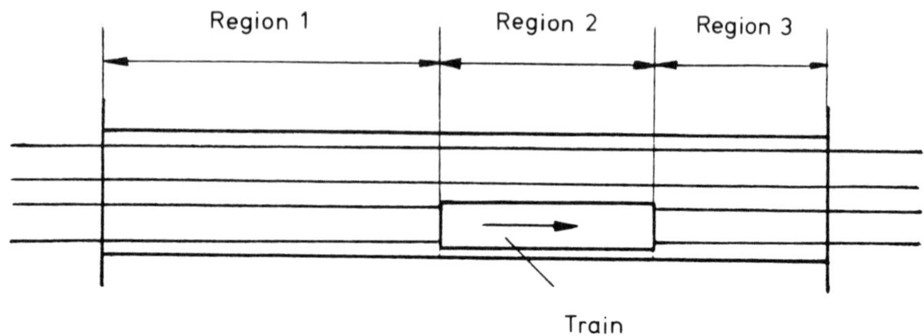

Figure 1. Partition of the integration domain

2.2. Basic equations for unsteady one-dimensional flow

These equations are the conservation of

mass
$$\frac{\partial \rho}{\partial t} + w\frac{\partial \rho}{\partial x} + \rho\frac{\partial w}{\partial x} = 0, \tag{1}$$

momentum
$$\frac{\partial w}{\partial t} + w\frac{\partial w}{\partial x} = -\frac{1}{\rho}\frac{\partial p}{\partial x} + R \tag{2}$$

and energy
$$T\left[\frac{\partial s}{\partial t} + w\frac{\partial s}{\partial x}\right] = P_Z + wR \tag{3}$$

and they are applied to the flow in the regions 1, 2 and 3 of Fig. 1. A derivation of these equations is given in [1]. It is necessary to find an expression for the friction force R acting at the tunnel and train walls. Usually a quasisteady friction model is used by setting the wall shear stress proportional to the square of the relative velocity between the fluid and the wall. In the energy equation P_Z is the work done by the train walls.

It is well known that the equations for unsteady compressible flow form a hyperbolic system. The transformation of this system to the characteristic curves leads first to the direction conditions:

$$\left.\begin{aligned} \left(\frac{dx}{dt}\right)_\xi &= w + a; \quad \xi = const \\ \left(\frac{dx}{dt}\right)_\eta &= w - a; \quad \eta = const \end{aligned}\right\} Mach\ lines \tag{4}$$

$$\left(\frac{dx}{dt}\right)_\varepsilon = w; \qquad \varepsilon = const \Big\} \ path\ lines$$

There are two families of Mach lines, the propagation curves of weak pressure disturbances, and the family of path lines which are the propagation curves of entropy disturbances. Along the characteristic curves the compatibility conditions are valid.

$$\left(\frac{dw}{dt}\right)_\xi + \frac{1}{\rho a}\left(\frac{dp}{dt}\right)_\xi + R - \frac{a}{c_p}\frac{wR + P_Z}{T} = 0; \quad \xi = const$$

$$-\left(\frac{dw}{dt}\right)_\eta + \frac{1}{\rho a}\left(\frac{dp}{dt}\right)_\eta - R - \frac{a}{c_p}\frac{wR + P_Z}{T} = 0; \quad \eta = const \tag{5}$$

$$\left(\frac{ds}{dt}\right)_\varepsilon - \frac{wR + P_Z}{T} = 0; \qquad \varepsilon = const$$

3. Numerical integration procedure

Using these compatibility equations a very effective numerical integration procedure is possible. For the numerical integration a rectangular grid in the $x - t$ plane is used. The grid size Δx is constant, but the time step Δt is variable. The differential equations are replaced by finite difference equations.

To calculate the flow parameters in a general grid point 3 at time $t = t_{j+1}$ (Fig. 2) it is first necessary to find on the characteristic curves passing through the point 3 the flow parameters at time $t = t_j$. Since the footpoints of the characteristic lines passing through the grid point 3 in general do not coincide with grid points at time $t = t_j$, interpolation is necessary. But the positions of the footpoints of the characteristics depend also on the flow parameters in the grid point 3. Therefore an iteraive computation using a predictor corrector method is applied [1].

The flow parameters at the tunnel portals and at the train ends are found using the compatibility conditions and the equations for steady compressible one-dimensional flow linking the flow in the corresponding domains (Fig. 1).

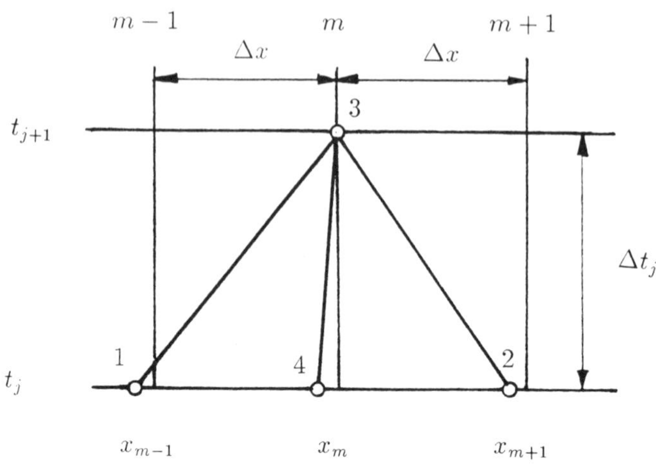

Figure 2. Mesh for numerical method of characteristics

A variable time step was introduced by Waclawiczek [2] and Steinrück [1] in order to minimize interpolation errors. Furthermore the right choice of the time step is very important for the correct calculation of the speed of wave propagation. Steinrück [1] proposed that the time step Δt should be found so that $\Delta x/\Delta t$ equals the propagation speed of a weak shock wave. By using Pfriem's bisector rule he finds for one single mesh element

$$\Delta t_{m,j} = \frac{2\,\Delta x}{(w_{m-1} + a_{m-1,j} + w_{m+1,j} + a_{m+1,j})} \tag{6}$$

However, many waves are running simultaneously and an overall time step Δt_j has to be chosen. So for every mesh it is checked whether a left or right running wave prevails and the optimum time step is calculated with the prevailing wave. Then the overall time step Δt_j is evaluated as the average of those steps weighted with the strength of the wave which passes the corresponding mesh element [1].

4. Results

In the lower part of Fig. 3 the pressure on a point 20 m behind the train head during the passage of a tunnel is shown. The upper part of Fig. 3 is the $x - t$ plane where the paths of the train ends and the observed point are shown. Furthermore the propagation lines of the waves due to the train ends entering the tunnel are included.

When the train nose enters the tunnel a compression wave is generated which runs through the tunnel. This wave is reflected on the far side tunnel portal and returns as a rarefaction wave. Due to the entrance of the train tail an expansion wave is generated.

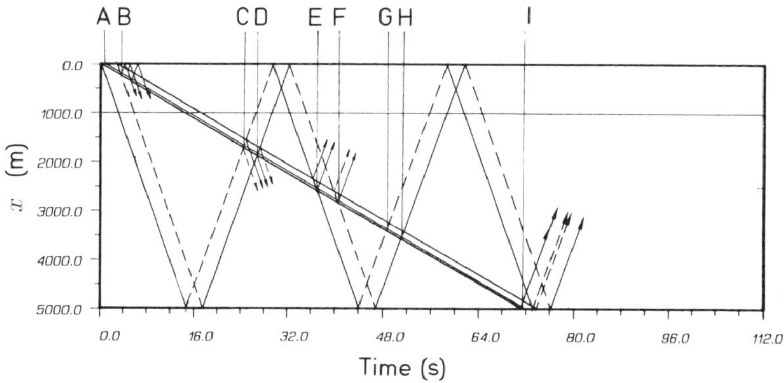

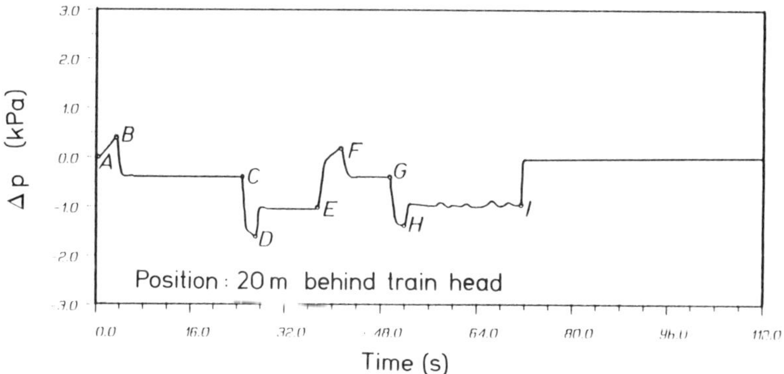

Figure 3. $x - t$ diagram and pressure transient on a train during the passage of a tunnel
Tunnel length: $L_T = 5000\text{m}$, Train length: $L_Z = 185\text{m}$
Train speed: $V_2 = 70\ \text{m/s}$, Blockage ratio: $A_Z/A_T = 0.12$

The pressure increase between A and B is due to the wall friction between the train and the tunnel wall. The pressure drop at B is caused by the rarefaction wave due to the train tail entering the tunnel. All further main pressure changes (CDEFGH) are caused by the two main waves passing the measuring point. There are also minor wave reflections at the train ends. The pressure increase at I is caused by the passage of the train head at the tunnel portal.

In Fig. 4 a comparison is made between calculated and measured pressure transients. The measurements were performed by British Railways in the Patchway tunnel [6]. In the beginning the agreement between calculation and measurement is very good and therefore the largest pressure fluctuations are predicted correctly. But after long times the attenuation of the pressure waves is underestimated by the computation.

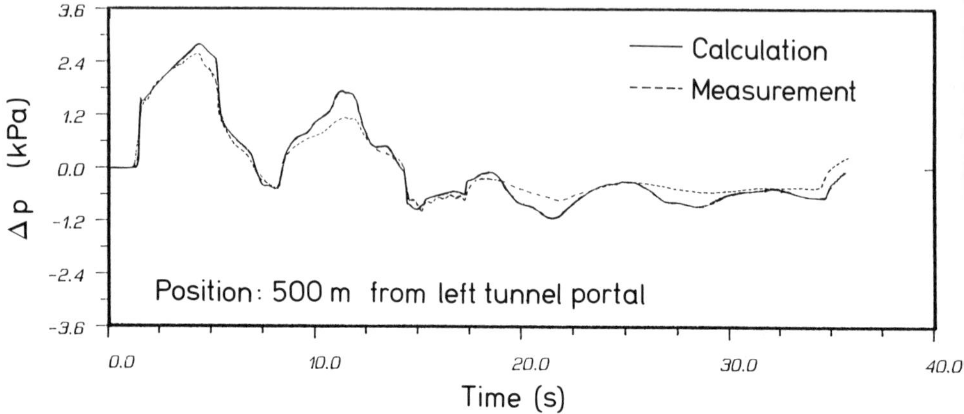

Figure 4. Pressure transients inside the Patchway tunnel during train passage. Comparison between calculation and measurement.
Tunnel length: $L_T = 1140$ m, Train length: $L_Z = 100.3$ m
Train speed: $V_Z = 34.7$ m/s, Blockage ratio: $A_Z/A_T = 0./363$

Possible reasons for the discrepancies between calculations and measurements could be, cumulative effects due to nonlinearity, minor effects of unsteady friction at the walls [7], the assumption of quasisteady flow for the boundary conditions, the neglect of heat transfer or general limitations of the simple one-dimensional theory.

Conclusions

The accuracy of todays computer programs evaluating the air flow in railway tunnels is sufficient for engineering purposes. The differences occurring between calculations and measurements are not only due to nonlinear effects.

References

[1] Steinrück, P. *Ein Verfahren zur Berechnung instationärer Strömungsvorgänge bei Fahrten von mehreren Zügen durch Tunnels mit und ohne Portalvorbauten.* Dissertation, Technical University of Vienna, 1984.

[2] Waclawiczek, M. and Sockel, H. Pressure transients and aerodynamic power in railway tunnels with special reference to entropy and airshafts. *4th Int. Symp. in Aerodynamics and Ventilation of Vehicle Tunnels, BHRA Fluid Engineering*, York, 1982, p. 337-351.

[3] Woods, W.A. and Pope, C. On the range of validity of simplified one-dimensional theories for calculating unsteady flows in railway tunnels. Paper D2, Proc. *3rd Int. Symp. on the Aerodynamics and Ventilation of Vehicle Tunnels, BHRA Fluid Engineering*, Sheffield, 1979.

[4] Fox, J.A. and Vardy, A.E. The generation and alleviation of air pressure transients caused by the high speed passage of vehicles through tunnels. Paper G3, *Proc. 1st Int. Symp. on the Aerodynamics and Ventilation of Vehicle Tunnels, BHRA Fluid Engineering*, Canterbury, 1973.

[5] Steinrück, P. and Sockel, H. Further calculations on transient pressure alleviation and simplified formulae for initial tunnel design. *5th Int. Symp. on the Aerodynamics and Ventilation of Vehicle Tunnels, BHRA Fluid Engineering*, Lille 1985, p. 317-341.

[6] Gawthorpe, R.G. and Pope, C.W. The measurement and interpretation of transient pressures generated by trains in tunnels. *Proc. 2nd Int. Symp. on the Aerodynamics and Ventilation of Vehicle Tunnels, BHRA,* Cambridge, March 1976.

[7] Schultz, M. and Sockel, H. The influence of unsteady friction on the propagation of pressure waves in tunnels. Paper B3, *Proc. 6th Int. Symp. on the Aerodynamics and Ventilation of Vehicle Tunnels, BHRA*, Durham September, 1988.

M. Schultz and H. Sockel
Institut für Strömungslehre und
Wärmeübertragung (Inst. 322)
Technische Universität Wien
Wiedner Hauptstraße 7
A-1040 Wien
AUSTRIA

H. BERGER, G. WARNECKE AND W. WENDLAND

Finite element approximation of transonic flows

The presented paper gives a review on recent work on the finite element approximation of entropy solutions to the transonic full potential equation ([1]-[4]). Our method is based on a minimization problem using a least squares concept whose functional contains an additional penalty term modelling an appropriate selection principle. This is a modification of the method by Bristeau *et al.* [5], and it is based on results by Feistauer and Nečas [6].

Here, we consider a steady, plane, adiabatic, isoenergetic and homentropic flow of a non-viscous compressible gas in the exterior of a given profile, which travels with a given subsonic travelling speed $\vec{v}_\infty$. If $|\vec{v}_\infty|$ is about 0.8 of the speed of sound or larger, then on top and bottom of the profile, the flow develops domains with supersonic speed and with shocks. We meet the following difficulties:

(i) The full potential equation is nonlinear, which implies discontinuous solutions. Without an additional selection principle, uniqueness is lost for weak solutions.

(ii) The differential equation changes type and the boundaries between subsonic (elliptic) and supersonic (hyperbolic) regions are free boundaries.

(iii) There exists a maximum speed at which the local density of the gas vanishes and the potential equation degenerates.

(iv) The exterior domain is infinite.

(v) The potential model needs to be extended by a circulation flow for providing lift.

For (i) and (ii) we use Bateman's variational principle and an additional selection principle by limiting accelerations to prevent expansion shocks. The corresponding penalty term can be seen as an artificial annihilation of momentum for large acceleration. This concept allows a corresponding finite element formulation which can be treated with methods of optimal control yielding a conjugate gradient method where a sequence of simple Poisson problems is to be solved. To prevent (iii), we modify the speed-density relation by requiring a positive minimum density for all speeds. For (iv), we introduce an artificial boundary sufficiently far from the profile.

There we either impose a Neumann condition with the parallel flow $\vec{v}_\infty$ or we use a coupling via a boundary element method with a Prandtl-Glauert flow in the exterior. The latter gives far better computational results. For (v), we allow a non-periodic potential on a Riemann surface and require the Kutta-Joukowski condition at the trailing edge.

For flows in bounded domains, Berger showed in [1] and [2] convergence of the finite element method by extending a compactness result by Mandel and Nečas [7], provided a weak entropy solution to the boundary value problem with the full potential equation exists. For our much more complex model of exterior flows around profiles, however, only partial results are so far justified.

References

[1] Berger, H. *Finite-Element-Approximation für transsonische Strömungen*, Dissertation, University of Stuttgart, 1989.

[2] Berger, H. A convergent finite element formulation for transonic flow. *Numer. Math* **56** (1989), 425–447.

[3] Berger, H., Warnecke, G. and Wendland, W. Finite elements for transonic potential flows. *Numer. Meth. Partial Differential Equs.*, **6** (1990) 17–42.

[4] Berger, H., Warnecke, G. and Wendland, W. On the coupling of finite elements and boundary elements for transonic flows. In: *Finite Element Analysis in Fluids* (eds. T.J. Chung, G.R. Karr). University of Alabama Press, Huntsville, Alabama, 1989, pp. 763–768.

[5] Bristeau, M.O., Glowinski, R., Periaux, J., Perrier, P., Pironneau, O. and Poirier, G. Application of optimal control and finite element methods to the calculation of transonic flows and incompressible flows. In: *Numerical Methods in Applied Fluid Dynamics* (B. Hunt ed.), Academic Press, New York, 1980, pp. 203–312.

[6] Feistauer, M. and Nečas, J. On the solvability of transonic potential flow problems. *Z. Anal. Anwend.* **4** (1985), 305–329.

[7] Mandel, J. and Nečas, J. Convergence of finite elements for transonic potential flows. *SIAM J. Numer. Anal.*, **24** (1990) 985–996.

H. Berger, G. Warnecke and W. Wendland
University of Stuttgart
Mathematics Institute A
Lehrstuhl 6
Angewandte Mathematik
Pfaffenwaldring 57
D-7000 Stuttgart 80
GERMANY

Part II
Dynamic Systems and Bifurcation

G. IOOSS AND A. MIELKE

Reduction and normalization of hydrodynamical stability problems in infinitely long cylinders

This paper is dedicated to E. Kröner on the occasion of his 70th birthday

1. Introduction

Hydrodynamic stability problems in extended domains lead to continuous spectra for the linearized operators. The classical way for studying such problems in infinitely long cylinders is to assume spatial periodicity. This assumption discretizes the spectrum and makes possible the use of techniques such as *centre manifolds* and *normal forms*, for studying the evolution of the dynamics in the neighbourhood of a basic solution which is weakly unstable. There are in fact other possible things to do which consist in considering only *steady* solutions or *time-periodic* ones, and not imposing any specific spatial behaviour in space. We can make this type of analysis, starting with either a steady basic solution translation-invariant in the direction of the generators of the cylinder, or even with a steady spatially periodic basic solution. We also can manage problems perturbing such situations. The advantage of this is that one can find solutions of type different from spatially periodic ones, for instance, spatially quasi-periodic, or solutions joining two different types of 'classical' solutions at both infinities. The disadvantage of such an analysis is that one does not know (for an exception see Collet and Eckmann (1986, 1987)) anything on the stability of these new solutions with respect to small initial perturbations. It should be noticed that the method of reduction of elliptic problems in cylindrical domains, using centre manifold theory, comes back to Kirchgässner (1982), and is extensively used for water wave problems (see Mielke (1986b); Amick and Kirchgässner (1989); and Kirchgässner (1988), and in elasticity (see Mielke (1988b).

Fixing the time behaviour leads to a problem where the space variable $x \in \mathbb{R}$ is required to play the role of usual time. The difficulty is that ellipticity prevents us from considering this as an evolution problem (the initial value problem is ill-posed). However, we can show that it is again possible to use the centre manifold theorem and the normal form technique to reduce the problem into a small dimensional ordinary differential equation where we are interested in solutions bounded for $x \in \mathbb{R}$, bifurcating from the basic one. It appears to be often possible to solve completely the reduced system expressed in normal form. The results are unfortunately only valid up to a given order N, due to the polynomial character of the normal form. The next difficulty is then to prove that the solutions we find persist in some sense when we consider the full equations.

We restrict our presentation to showing the reduction process for several typical situations such as reflection symmetric systems with a translationally invariant steady solution having a steady bifurcation, or systems displaying a Hopf bifurcation, with or without a reflection symmetry. We also give all the solutions of the reduced problems. For the full system, we are able in particular to prove the persistence of the spatially quasi-periodic solutions (see Iooss and Los (1989) and Pluschke (1989)), to prove the existence of static fronts of travelling waves for the standard Hopf bifurcation case (see Iooss and Mielke (1989) and, for the Hopf bifurcation with reflection symmetry, to prove the existence of co-existing symmetric travelling waves on two different sides of the system. These results apply for instance to classical hydrodynamical stability problems such as the Taylor-Couette problem between very long cylinders or the Bénard convection problem in a very long box, or even to the Poiseuille flow in a pipe.

2. Stability of steady translation-invariant solutions

Let us denote by $Q = \Omega \times \mathbb{R}$ the domain of the flow, satisfying the Navier-Stokes equations, where Ω is a bounded regular domain of $\mathbb{R}$ or $\mathbb{R}^2$. One of the important points is that the boundary conditions on $\partial\Omega \times \mathbb{R}$ are steady and independent of x. It is known that, after subtracting a fully symmetric solenoidal vector field satisfying the boundary conditions, the equations for the perturbation can be put into the form of a differential equation lying in a suitable function space:

$$\frac{\mathrm{d}U}{\mathrm{d}t} = L_\mu U + N(\mu, U). \tag{1}$$

In (1) U is, in most cases, the velocity vector field in Q, and $\mu \in \mathbb{R}$ represents a distinguished parameter among the set of parameters of the problem.

Now, there is a very important symmetry property of the system: the translational invariance $(x \to x + a)$. It is expressed by the property that L_μ and $N(\mu, .)$ commute with a one parameter group of linear operators τ_a, $a \in \mathbb{R}$.

We start with (1) and study the stability of the maximally symmetric solution $U = 0$. The usual classical linear theory of hydrodynamical stability looks for perturbations of the form $\hat{U}_k e^{ikx}$, where $\hat{U}_k$ is a function of variables lying in Ω. The corresponding eigenvalues of L_μ are denoted by $\sigma(\mu, ik)$:

$$L_\mu(\hat{U}_k e^{ikx}) = \sigma(\mu, ik))\hat{U}_k e^{ikx}. \tag{2}$$

For each k, there is a denumerable set of eigenvalues $\{\sigma_m; m \in \mathbb{N}\}$ and since the only restriction on the behaviour in x is the boundedness of vector fields, it is clear that the set of all eigenvalues $\{\sigma_m(\mu, ik)\}$ is not discrete (since k can vary continuously). If an h-periodicity is assumed in x, one only allows k to take multiple values of $2\pi/h$. It is a classical result that the full spectrum of L_μ is then discrete.

Another important point, which appears in many cases, is the effect of the reflection symmetry $x \to -x$. This symmetry property is expressed by the commutativity of operators L_μ and $N(\mu, .)$ with a symmetry operator S ($S^2 = \text{Id}$) such that we have the property

$$\tau_a S = S\tau_{-a}. \tag{3}$$

Now, $(S\hat{U}_k)\,e^{-ikx}$ is an eigenvector belonging to the same eigenvalue $\sigma(\mu, ik)$, hence we have

$$\sigma(\mu, -ik) = \sigma(\mu, ik), \;\; \hat{U}_{-k} = S\hat{U}_k. \tag{4}$$

We mainly consider in the following two cases: either the eigenvalue σ_0 with largest real part is real, and we have the reflection symmetry $x \to -x$, or the eigenvalue with largest real part is complex (with or without reflection symmetry) and the eigenvector $\bar{\hat{U}}_k$ belongs to $\bar{\sigma}(\mu, ik) = \sigma(\mu, -ik)$. In both cases, the neutral stability curve $\mu = \mu_c(k)$ is defined by $\text{Re}\,\sigma_0(\mu, ik) = 0$, and μ is *an even function* of k.

Let us assume that $\mu_c(k)$ passes through a minimum $\mu = 0$ at $k = k_c$. We arrange notations in such a way that

(i) for $\mu < 0$, $\text{Re}\,\sigma_0 < 0$ for any k, and the 0 solution is exponentially stable, while

(ii) for $\mu > 0$, $\text{Re}\,\sigma_0 > 0$ for some k, and 0 is linearly unstable.

We sum up the situation on Fig. 1(a). In the (k, μ) plane, if we look at a fixed value of $\mu > 0$, then the values of $|k|$ giving points inside the parabolic region $\mu > \mu_c(k)$ lead to instability (see Fig. 1(b)), while outside of this region perturbations $\hat{U}_k e^{ikx}$ are damped.

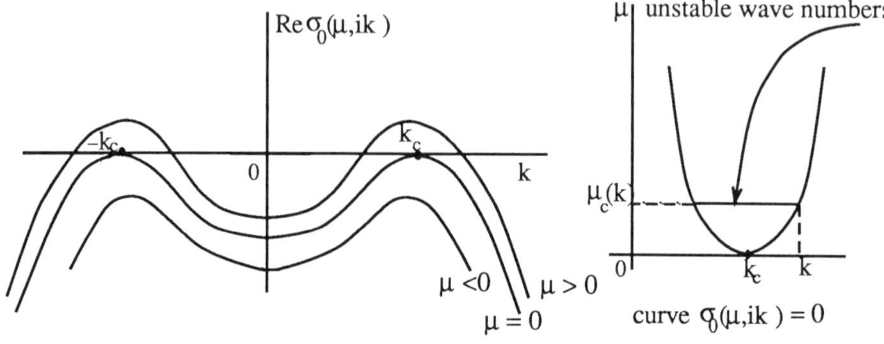

Figure 1(a) Figure 1(b)

3. The reduction process for steady or time-periodic solutions

Now we can show that these situations lead to a specific property for the spectrum of a linear operator associated with steady or time periodic solutions of the Navier–Stokes equations.

It is shown in Iooss *et al.* (1989)] that the *steady* Navier–Stokes equations can be written in the following form:

$$\frac{d\mathbf{V}}{dx} = \mathcal{A}_\mu \mathbf{V} + \mathcal{B}_\mu(\mathbf{V}, \mathbf{V}), \tag{5}$$

where the three first components of $\mathbf{V}$ give the previous vector field U of (1). There are no longer differentiations in x on the right-hand side. Here $\mathcal{A}_\mu$ respresents the linear part, while $\mathcal{B}_\mu$ represents the quadratic terms. A suitable functional frame is introduced in Iooss *et al*. (1989), and it is shown that $\mathcal{B}_\mu$ is a continuous quadratic map in the domain $\mathcal{D}$ of $\mathcal{A}_\mu$.

The reflection invariance of the problem in this case is now expressed by a reversibility property. Introducing the corresponding symmetry operator S leads to the identities:

$$S\mathcal{A}_\mu = -\mathcal{A}_\mu S, \quad \mathcal{B}_\mu \circ S = -S\mathcal{B}_\mu. \tag{6}$$

For the time periodic case, one introduces the variable $s = \omega t$ where ω is an unknown frequency which plays the role of an additional parameter. The Navier-Stokes equations become, in this case

$$\frac{d\mathbf{V}}{dx} = -\mathcal{A}_\mu \mathbf{V} + \omega\mathcal{E}_\mu \mathbf{V} + \mathcal{B}_\mu(\mathbf{V}, \mathbf{V}), \tag{7}$$

where $\mathbf{V}$ lies in a space where functions are 2π-periodic in s, and where the linear operator $\mathcal{E}_\mu$ contains the derivatives with respect to s. In the case when we have the reflection symmetry we have properties (6) and $S\mathcal{E}_\mu = -\mathcal{E}_\mu S$.

Now, the properties we mentioned in Section 2 for the operator L_0 lead to very specific properties for the linear operator $\mathcal{A}_\mu + \omega\mathcal{E}_\mu$ at $\mu = 0$. In fact, for the steady case with symmetry, we have, at $\mu = 0$, *two double non semi-simple eigenvalues* $\pm ik_c$ for $\mathcal{A}_0$, which split into two pairs of pure imaginary simple eigenvalues for $\mu > 0$ (as can be seen in making $\sigma_0 = 0$ on Fig. 1(a)), and two pairs of complex eigenvalues symmetric with respect to the imaginary axis for $\mu < 0$, the remainder of the spectrum of $\mathcal{A}_\mu$ staying 'far' from the imaginary axis. It should be noticed that the spectrum of $\mathcal{A}_\mu$ is discrete and symmetric with respect to the imaginary axis, situated in a sector centred on the real axis (Iooss *et al.*, 1989). The linear operator $\mathcal{A}_0$, reduced to the invariant subspace belonging to the pure imaginary eigenvalues, can be written as follows, in a suitable basis:

$$\mathcal{T}_0 = \begin{pmatrix} ik_c & 1 & 0 & 0 \\ 0 & ik_c & 0 & 0 \\ 0 & 0 & -ik_c & 1 \\ 0 & 0 & 0 & -ik_c \end{pmatrix}, \tag{8}$$

where an element of the invariant subspace is denoted by

$$Y_0 = A\mathbf{V}_0 + B\mathbf{V}_1 + \bar{A}\bar{\mathbf{V}}_0 + \bar{B}\bar{\mathbf{V}}_1 \tag{9}$$

and reversibility is represented here by the following action:

$$(A, B, \bar{A}, \bar{B}) \to (\bar{A}, -\bar{B}, A, -B). \tag{10}$$

In the time periodic case, if we denote the critical eigenvalues of L_0 by $\pm i\omega_0$, it is clear that $\pm\, ik_c$ are eigenvalues of $\mathcal{A}_0 + \omega_0\mathcal{E}_0$ with 2π-periodic eigenvectors $\mathbf{V}_0\, e^{is}$ $(\in ik_c)$ and $\bar{\mathbf{V}}_0\, e^{-is}$ $(\in -ik_c)$ (for this convention, ω_0 might be < 0). If, in addition, we have the reflection symmetry, these eigenvalues are now double semi-simple, the other eigenvectors being $S\mathbf{V}_0\, e^{is}$ $(\in -ik_c)$, and $S\bar{\mathbf{V}}_0\, e^{-is}$ $(\in ik_c)$. Using a suitable functional frame (see Iooss and Mielke, 1989), the linear operator $\mathcal{A}_0 + \omega_0\mathcal{E}_0$, reduced to the invariant subspace belonging to the pure imaginary eigenvalues, can then be written as follows:

$$\mathcal{T}_0 = \begin{pmatrix} ik_c & 0 \\ 0 & -ik_c \end{pmatrix}, \text{ in the case with no reflection symmetry,} \tag{11}$$

and, in the case with reflection symmetry where an element of the four-dimensional invariant subspace is denoted by

$$Y_0 = A\mathbf{V}_0\, e^{is} + B\mathbf{V}_1\, e^{is} + \bar{A}\bar{\mathbf{V}}_0\, e^{is} + \bar{B}\bar{\mathbf{V}}_1\, e^{-is}, \tag{12}$$

we have

$$\mathcal{T}_0 = \begin{pmatrix} ik_c & 0 & 0 & 0 \\ 0 & -ik_c & 0 & 0 \\ 0 & 0 & -ik_c & 0 \\ 0 & 0 & 0 & ik_c \end{pmatrix} \tag{13}$$

and reversibility is then represented by the action:

$$(A, B, \bar{A}, \bar{B}) \rightarrow (B, A, \bar{B}, \bar{A}). \tag{14}$$

It should be noticed that in the time-periodic case the action of time shift corresponds to a non-trivial SO(2) action, contrary to the steady case:

$$(A, B, \bar{A}, \bar{B}) \rightarrow (Ae^{i\theta}, Be^{i\theta}, \bar{A}e^{-i\theta}, \bar{B}e^{-i\theta}) \quad \text{for} \quad \theta \in \mathbb{R}. \tag{15}$$

It is shown in Iooss *et al.* 1989 (for the steady case) and in Iooss and Mielke, 1989 (for the time-periodic case) that we can apply the centre manifold reduction using methods of Kirchgässmer (1982) and Mielke (1986a; 1988a). This means that, since we want 'small' bounded solutions on $\mathbb{R}$, these solutions lie on a finite-dimensional centre manifold (here of dimension 4, or 2 in the non-symmetric Hopf case). In addition, the centre manifold reduction respects the invariances of the problem, so we can arrange the reduced ordinary differential equation into a normal form, equivariant under the relevant symmetries.

For the steady case with reflection symmetry (typically the Taylor-Couette problem with co-rotating cylinders, or the Bénard convection in a long box), it is shown in Iooss *et al.* (1989) that the reversible normal form, corresponding to the choice of coordinates satisfying (8-10), reads

$$\begin{cases} \dfrac{dA}{dx} = ik_c A + B + iAP[\mu ; |A|^2; \frac{i}{2}(A\bar{B} - \bar{A}B)] \\ \dfrac{dB}{dx} = ik_c B + iBP[\mu ; |A|^2; \frac{i}{2}(A\bar{B} - \bar{A}B)] + AQ[\mu ; |A|^2; \frac{i}{2}(A\bar{B} - \bar{A}B)]. \end{cases} \tag{16}$$

Here P and Q are real polynomials in their two last arguments, with μ dependent coefficients, and such that $P(0, 0, 0) = Q(0, 0, 0) = 0$. We have to understand that the full system on the centre manifold contains in fact terms $O(|A|+|B|)^N]$ in the right-hand side of (16) (N being fixed arbitrarily).

For the Hopf bifurcation case with no reflection symmetry (typically the Poiseuille flow case), the $SO(2)$ action of the time shift coincides with the group action of normalization (see Elphick *et al.* (1987)). It results that there is no need of normalization, and the full system has the following form on the centre manifold:

$$\frac{dA}{dx} = ik_c A + Af[\mu, \omega, |A|], \tag{17}$$

where f is a smooth function, even in its third argument, taking complex values, such that $f[0, \omega_0, 0] = 0$.

For the Hopf bifurcation case with reflection symmetry (typically the Taylor-Couette flow instability in the counter-rotating case, or the convection problem in a binary mixture) the reversible normal form corresponding to the choice of coordinates satisfying (12-15), is as follows:

$$\begin{cases} \dfrac{dA}{dx} = ik_c A + AP[\mu, \omega, |A|^2, |B|^2], \\ \dfrac{dB}{dx} = -ik_c B - BP[\mu, \omega, |B|^2, |A|^2], \end{cases} \tag{18}$$

where P is a polynomial in its two last arguments taking complex values, such that $P[0, \omega_0, 0, 0] = 0$, and where, as for (16), we have to understand that the full system on the centre manifold contains in fact terms $O[|A|+|B|)^N]$ in the right-hand side of (18).

4. Integration of the truncated normal forms

4.1. Steady case with reflection symmetry

Let us sum up below the complete integration of system (16) (details can be found in Iooss *et al*. (1989)). First we set:

$$A = r_0 e^{i(k_c x + \psi_0)}, \quad B = r_1 e^{i(k_c x + \psi_1)}. \tag{19}$$

Then, the system (16) has the two integrals:

$$\begin{cases} r_0 r_1 \sin(\psi_1 - \psi_0) = K, \\ r_1^2 - G(\mu, r_0^2, K) \ = H, \end{cases} \tag{20}$$

where $G(\mu, r_0^2, K) = \int_0^{r_0^2} Q(\mu, s, K)\, ds.$

If we set $u_0 = r_0^2$, $u_1 = r_1^2$, taking account of (20), system (16) reduces to

$$\left(\frac{d u_0}{d x}\right)^2 = 4\left\{u_0 [G(\mu, u_0, K) + H] - K^2\right\}, \tag{21}$$

$$\begin{cases} \dfrac{d\psi_0}{dx} = P(\mu, u_0, K) + K / u_0, \\ \dfrac{d\psi_1}{dx} = P(\mu, u_0, K) + K/u_0 - K(u_0 u_1)^{-1} [u_0 Q(\mu, u_0, K) + u_1]. \end{cases} \tag{22}$$

More precisely, let us define the principal part of the polynomial Q as

$$Q(\mu, u_0, K) = -q_1\mu + q_3 K + q_2 u_0 + \dots, \tag{23}$$

where $q_1 > 0$ by convention, and $q_2 > 0$ to obtain a supercritical bifurcation, and let us set, for $\mu > 0$,

$$K = \mu^{3/2} k, \quad H = \mu^2 h, \quad h_m = \frac{q_1^2}{2 q_2}. \tag{24}$$

Then, it is easy to show that, for $0 < h < (4/3)(h_m)$ and

$$(1+v)^2 (1-2v) < \frac{27\, q_2^2}{4 q_1^3} k^2 < (1-v)^2 (1+2v) \tag{25}$$

where $v = \left(1 - \frac{3\,h}{4\,h_m}\right)^{\frac{1}{2}}$ (see shaded region on Fig. 2), we obtain, for $|\mu|$ small enough, a two-paramter family of quasi-periodic soltuions of the form:

$$\begin{cases} A_0(x) = r_0(\omega_1 x)\, e^{i[\omega_0 x + \Psi_0(\omega_1 x)]}, \\ B_0(x) = r_1(\omega_1 x)\, e^{i[\omega_0 x + \Psi_1(\omega_1 x) + \pi/2]}, \end{cases} \tag{26}$$

where r_0, r_1, Ψ_0, Ψ_1 are 2π-periodic functions, and r_0, r_1 are even while Ψ_0, Ψ_1 are odd. Functions r_0, r_1, Ψ_0, Ψ_1 depend on μ, h, k, as well as ω_0 and ω_1, and in addition r_0 and r_1 have the size $O(\mu^{1/2})$. More precisely, if we define the function $f(u_0)$ by

$$f(u_0) = u_0\,[G(\mu, u_0, K) + H\,] - K^2 \tag{27}$$

we then have $f(\mu v_0) = \mu^3 f_0(v_0) + O(\mu^{7/2})$, and r_0^2/μ lies between the two smallest positive roots $v_0^{(1)}$, $v_0^{(2)}$ of f_0 (cubic polynomial) (see Fig. 2).

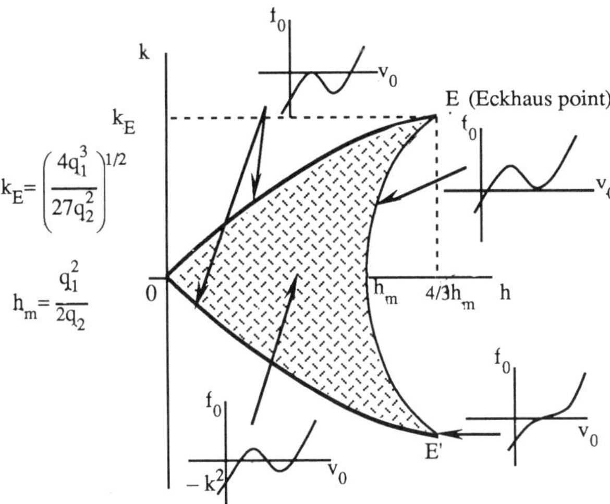

Figure 2. Different types of bounded solutions (bounded positive part of f_0)

It is easy to show that we have for the frequencies:

$$\begin{cases} \omega_0 = k_c + \sqrt{\mu}\ \alpha(h,k) + O(\mu), \\ \omega_1 = \sqrt{\mu}\ \beta(h,k) + O(\mu). \end{cases} \tag{28}$$

On the boundary of the shaded region of Figure 2, we obtain steady solutions in u_0, u_1, i.e. spatially periodic solutions for the truncated system. These are in fact the solutions obtained in the classical analysis, when spatial periodicity is assumed for the velocity vector field. On the curve EE′, we have in addition homoclinic solutions, such that at infinity they tend towards periodic solutions (differing by a phase shift). One of these 'homoclinic' solutions is such that the amplitude A cancels in $x = 0$, so it looks like a 'defect solution'. It can be shown that at least two of these homoclinic solutions persists when one considers the full system of equations (i.e. taking account of the high-order terms in the 4th-order differential equation on the centre manifold) [see Iooss and Pérovème (1990)]. In addition, for the quasi-periodic solutions, it is shown in Iooss and Los (1989) that they mostly persist for the full system, i.e. for small enough $\mu > 0$, there is a two-parameter set where they persist, corresponding to the shaded region of Fig. 2, but replaced locally by the product of a line by a Cantor set.

4.2. Hopf bifurcation with no reflection symmetry

Let us start with equation (17), and set $A = re^{i(k_c x + \Psi)}$, then we have:

$$\begin{cases} \dfrac{dr}{dx} = rf_r(\mu, \omega, r), \\ \dfrac{d\psi}{dx} = f_i(\mu, \omega, r), \end{cases} \tag{29}$$

where f_r and f_i denote the real and imaginary parts of f. This system is easily integrable. We notice that a non-zero solution of $f_r(\mu, \omega, r) = 0$ gives a travelling wave solution for the original problem (don't forget that the eigenvector in factor of A has the form $\mathbf{V}_0 e^{is}$, and that $s = \omega t$). More precisely, the principal part of f may be written as

$$f(\mu, \omega, r) = a\mu + ib(\omega - \omega_0) + cr^2 + O\left([\,|\mu| + |\omega - \omega_0| + r^2\,]^2 \right) \tag{30}$$

where b is real, while a and c are complex, with $a_r > 0$ to fix the idea, and where $c_r < 0$ corresponds to the supercritical Hopf bifurcation case. We then obtain immediately for any fixed positive value of μ if $c_r < 0$ (resp. for any negative value of μ if $c_r > 0$) a set of travelling waves, corresponding to the spatial wave numbers giving unstable (resp. stable) modes in Fig. 1(b). These solutions are the classical Hopf bifurcated solutions obtained in the usual way when one fixes the spatial period. Now, we also have solutions of (29) such that (for $c_r < 0$) they tend towards 0 at $-\infty$ and to one of the above travelling waves solutions at $+\infty$ (the converse holds if $c_r > 0$). Since equation (17) is complete, we then obtain a new solution of the Navier–Stokes equation into the form of a *static front of travelling waves*.

4.3. Hopf bifurcation with reflection symmetry

We consider the system (18) and set

$$A = r_0 e^{i(k_c x + \Psi_0)}, \quad B = r_1 e^{i(-k_c x + \Psi_1)}, \tag{31}$$

then we obtain a system in r_0 and r_1 uncoupled from the phases:

$$\begin{cases} \dfrac{dr_0}{dx} = r_0 P_r(\mu, \omega, r_0^2, r_1^2), \\[2ex] \dfrac{dr_1}{dx} = -r_1 P_r(\mu, \omega, r_1^2, r_0^2), \end{cases} \tag{32}$$

$$\frac{d\psi_0}{dx} - P_i(\mu, \omega, r_0^2, r_1^2), \quad \frac{d\psi_1}{dx} = -P_i(\mu, \omega, r_1^2, r_0^2). \tag{33}$$

A remarkable fact is that we can completely build the phase portraits of the system (32), and that there is an explicit integral for this system truncated at cubic order (see Iooss and Mielke, 1989). Let us make the principal part of P_r precise:

$$P_r(\mu, \omega, r_0^2, r_1^2) = \nu + br_0^2 + cr_1^2 + \text{higher-order terms} \tag{34}$$

where $\nu = a_r\mu + O(|\mu| + |\omega_0|)^2$, as in (30). The different phase portraits are described on Fig. 3.

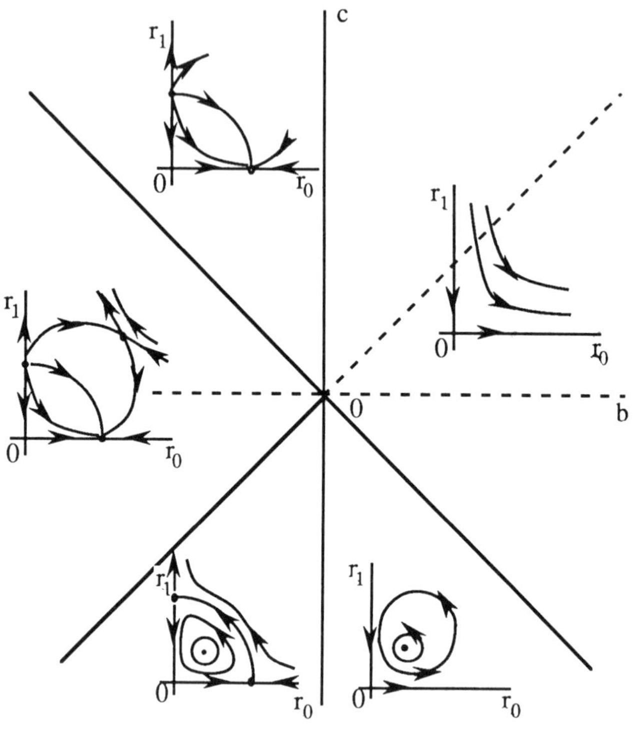

Figure 3. Phase portraits ($\nu > 0$) depending on coefficients of P_r

Equilibrium points on each axis correspond to travelling waves, as in Section 4.2. They are exchanged by the symmetry S. The equilibrium points on the diagonal correspond to standing waves, again periodic in time and space. The size of solutions depend on ν, i.e. on the couple (μ, ω), and (r_0, r_1) tends towards 0 when μ tends to the neutral stability curve of Fig. 1(b). All these solutions are the well-known classical solutions when one deals with Hopf bifurcation with $O(2)$ symmetry (the spatial periodicity is then imposed), and there is a one-parameter family of such solutions corresponding to different spatial wave numbers. Now, we observe that there are other types of bounded solutions. In particular, there are families of closed trajectories corresponding to spatially quasi-periodic solutions, there are heteroclinic types of solutions joining for instance two symmetric travelling waves at both infinities, and there are plenty of other bounded solutions joining from $-\infty$, either one of the travelling waves, or the standing wave, or the basic solution (translation invariant and steady), to one of these solutions at $+\infty$. It is clear that the 'defect' solution such as the one joining two symmetric travelling waves is physically observed, in particular in the Taylor-Couette problem. This type of solution is shown to exist not only for the normal form (18), but also for the full equation (with 'flat terms'), using symmetry arguments and the fact that the full equation is still rotationally invariant (see Iooss and Mielke, 1989), and many other connections of Fig. 3 are also shown to persist for the full equation. One can also prove the persistence of 'most' of the spatially quasi-periodic, time periodic solutions (see the method in Iooss and Los, 1989, or Pluschke, 1989) as in the steady case of Section 4.1.

References

Amick C.J. and K. Kirchgässner (1989). A theory of solitary water-waves in the presence of surface tension. *Arch. Rat. Mech. Anal.* **105**, 1-50.

Collet P. and Eckmann J.P. (1986). The existence of dendritic fronts. *Comm. Math. Phys.* **107**, 39-92.

Collet P. and Eckmann J.P. (1987). The stability of modulated fronts. *Helvetica Phys. Acta* **60**, 969.

Elphick C., Tirapegui E., Brachet M.E., Coullet P. and Iooss G. (1987). A simple global characterization for normal forms of singular vector fields. *Physica* **29D**, 95-127.

Iooss G., Mielke A. and Demay Y. (1989). Theory of steady Ginzburg-Landau equation, in hydrodynamic stability problems. *Europ. J. Mech. B/Fluids*, **8** (3), 229-268.

Iooss G. and Los J. (1990). Bifurcation of spatially quasi-periodic solutions in hydrodynamic stability problems. Nonlinearity **3**, 851-871.

Iooss G. and Mielke A. (1989). On the time-periodic solutions of Navier-Stokes equations in an infinite cylinder to appear in J. Nonlinear Science 1991.

Iooss G. and Pérovème H.C. (1990). Perturbed homoclinic solutions in reversible 1:1 resonance vector fields. Preprint Univ. Nice 1990.

Kirchgässner K. (1982). Wave solutions of reversible systems and applications. *J. Diff. Eq.* **45** 113-127.

Kirchgässner K. (1988). Nonlinearly resonant surface waves and homoclinic bifurcation. *Adv. in Applied Mech.* **26** 135-181.

Mielke A. (1986a). A reduction principle for nonautonomous systems in infinite dimensional spaces. *J. Diff. Eq.* **65**, 68-88.

Mielke A. (1986b). Steady flows of inviscid fluids under localized perturbations. *J. Diff. Eq.* **65**, 89-116.
Mielke A. (1988a). Reduction of quasilinear elliptic equations in cylindrical domains with applications. *Math. Meth. Appl. Sci.* **10**, 51-66.
Mielke A. (1988b). Saint Venant's problem and semi-inverse solutions in nonlinear elasticity. *Arch. Rat. Mech. Anal.* **102**, 205-229.
Pluschke W. (1989). Invariant tori bifurcating from fixed points on non-analytical reversible systems. . *EEC meeting in Poigny-La-Forêt*, May 1989 .

G. Iooss
Laboratoire de Mathematiques
U.A. CNRS 168
University of Nice
Parc Valrose
F-06034 Nice
FRANCE

A. Mielke
Mathematisches Institut A,
Universität Stuttgart
Pfaffenwaldring 57
D-7000 Stuttgart 80
GERMANY

A.L. AFENDIKOV

Bifurcation on the slow cycle in some hydrodynamic problems

In the recent years a viewpoint has become widespread that the transition to turbulence in viscous incompressible fluid flow is connected with the appearance of 'chaotic' dynamics on the strange attractor of the boundary initial value problem for the Navier–Stokes equation. If we accept this viewpoint then our first step in studying turbulence and dealing with localization of attracting sets and investigation of their structures is impossible without the use of computers.

On the other hand, it is now well known that direct simulation of turbulence by solving numerically the initial boundary value problem for the Navier–Stokes equations may lead to 'turbulence' of pure finite-dimensional nature, which completely disappears by improving the approximation of the original infinite-dimensional problem (see survey [1]).

Therefore, theoretical facts about elementary attracting sets such as stationary points or limit cycles play an essential role in planning the numerical experiment. These sets as well as their rearrangements must be implemented in the numerical experiment with immaculate acuracy.

For the computer simulation of a 'turbulent' attractor it is very important to know a characteristic length of time interval on which we should consider the solution. This is not an easy question, because of the existence of some cases where a flow with an arbitrarily large period can be realized for the stationary external force.

One of the possible examples is connected with an elementary consideration. We consider the problem

$$\dot{x} = f(x, \varepsilon), \;\; f(0, \varepsilon) = 0, \;\; f: R^2 \times R_+ \to R^2 \tag{1}$$

where for the sake of simplicity $f(x, \varepsilon)$ is assumed to be real-analytical. If equation (1) is equivariant under the action of the group $SO(2)$

$$T^n_\theta f(x, \varepsilon) = f(T^n_\theta x, \varepsilon),$$

where T^n_θ is a matrtix element of an irreducible representation $SO(2)$

$$T^n_\theta = \begin{pmatrix} \sin n\theta, \; \cos n\theta \\ -\cos n\theta, \; \sin n\theta \end{pmatrix}$$

then in a suitable coordinate system (1) may be written in the form

$$\dot{z} = zA(|z|^2, \varepsilon).$$

In a small neighbourhood of the point $z = 0$ all the periodic solutions of the equation are of the form

$$z(t) = \rho(\varepsilon)\exp(\mathrm{i}\Omega(\varepsilon)t),$$

where $\rho(\varepsilon)$ is determined from the equation

$$\mathrm{Re}\,(A(\rho^2, \varepsilon)) = 0,$$

and $\Omega(\varepsilon) = \mathrm{Im}(A(\rho^2(\varepsilon), \varepsilon))$. Let

$$\mathrm{Re}\,(A(\rho^2, \varepsilon)) = \sum_{j,k=0}^{\infty} a_{jk}\, \varepsilon^j \rho^{2k}, \; a_{00} = 0.$$

In order to guarantee the uniqueness of a branching cycle it is usually assumed that

$$a_{10} > 0, \quad a_{01} \neq 0. \tag{3}$$

If $\mathrm{Im}(A(0,\ 0)) \neq 0$, the conditions (3) lead to the classical Andronov–Hopf bifurcation. But if $\mathrm{Im}(A(0, 0)) = 0$, the slow cycle

$$\Omega(0) = 0, \; \Omega(\varepsilon) \neq 0$$

may appear. The condition $\mathrm{Im}(A(0, 0)) = 0$ is evidently equivalent to the one when the martrix $f'(0,0)$ has double semi-simple eigenvalue $\lambda_{1,2} = 0$.

It is exactly this situation that occurs in the case of the generalized Kolmogorov flow [2].

We consider the flow of viscous incompressible fluid flow on the two-dimensional torus

$$T^2 = \{(x, y) : 0 \leq x \leq 2\pi/\alpha,\; 0 \leq y \leq 2\pi\};$$

if u and p are the velocity vector and the pressure, the motion is described by the Navier–Stokes system

$$\partial_t u + (u \cdot \nabla)u + \mathrm{grad}\, p = \nu\Delta u + f, \;\; \mathrm{div}\, u = 0 \tag{4}$$

where f is external force and ν is kinematic viscosity. The density drops out, since it may be assumed that $\rho = 1$. As boundary conditions we shall take the periodicity conditions

$$u(t, x, y) = u(t, x, y + 2\pi) = u(t, x + 2\pi/\alpha, y),$$

with the additional condition that the velocity average over $T - Q$,

$$Q = \alpha/4\pi^2 \int_{T^2} u \, dx \, dy,$$

is the constant vector $Q = (c_0, v)$. If we assume that $f = (F(y), 0), \langle f \rangle = 0$, then one of the possible stationary solutions of the system (5) will have the form

$$u = (U(y), \nu) \quad p = \text{const.} \quad U(y) = \gamma^{-1} \sum_{k=-\infty}^{\infty} c_k e^{iky}, \quad 2 \sum_{k=1}^{\infty} | c_k |^2 = 0$$

where $U(y)$ and $F(y)$ are connected by the relation

$$R^{-1} U'' - BU' + F(y) = 0$$

where $R = \gamma/\nu$, $B = v/\gamma$ are assumed to be dimensionless constants. In the original Kolmogorov problem $F(y) = \gamma/\nu \sin y$, $Q = (0, 0)$. In order to study the stability and bifurcation of these steady-state solutions it is convenient to introudce the stream function of the perturbed flow

$$\Psi(t, x, y) = \Psi_0(x, y) + \Psi(t, x, y)$$

where $\Psi_0(x, y)$ is the stream function corresponding to the steady flow $U(y)$. Then we obtain the problem

$$\begin{cases} \dfrac{\partial \Delta \Psi}{\partial t} + U \dfrac{\partial \Delta \Psi}{\partial x} - U'' \dfrac{\partial \Psi}{\partial x} + \dfrac{\partial \Delta \Psi}{\partial x} \dfrac{\partial \Psi}{\partial y} - \dfrac{\partial \Delta \Psi}{\partial y} \dfrac{\partial \Psi}{\partial x} = R^{-1} \Delta^2 \Psi - v \dfrac{\partial \Delta \Psi}{\partial y} \\ \Psi(t, x, y) = \Psi(t, x + 2\pi/\alpha, y) = \Psi(t, x, y + 2\pi). \end{cases} \tag{5}$$

By rejecting nonlinear terms in (5) and searching for the solution in the form $\psi = \chi(y) \exp(\lambda t \pm i\alpha x)$, we obtain the spectral problem of stability:

$$\begin{cases} R^{-1} \ell_\alpha^2 \chi \pm i\alpha (U \ell_\alpha \chi - U'' \chi) - v \dfrac{d}{dy} \ell_\alpha \chi - \lambda \ell_\alpha \chi = 0 \\ \chi(y) = \chi(y + 2\pi), \ \ell_\alpha = \dfrac{d^2}{dy^2} - \alpha^2. \end{cases} \tag{6}$$

If we denote the eigenvalue of (6) with maximal real part by $\lambda_0(R, \alpha)$ we obtain the function $A(\alpha) = \mathrm{Im}(\lambda_0(R, \alpha))$ for the R-values such that $\mathrm{Re}(\lambda_0(R, \alpha)) = 0$. Now let us take

$$F(y) = R(2 \sin y + 4\sqrt{2} \cos 2y)/\sqrt{3}, \quad Q = (0, 0),$$

then we have for $A(\alpha)$ the picture shown in Fig. 1.

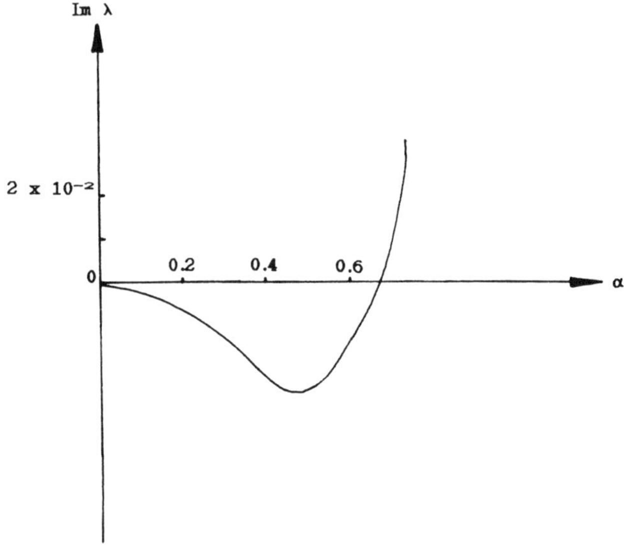

Figure 1

The problem (5) allows a group of the translations $x \to x + \beta \pmod{2\pi/\alpha}$ as a group of transformations, and for $\alpha_0 \simeq 0.676628$, $R_0 \simeq 2.11322$ the bifurcation of the slow cycle occurs. This time a periodic solution of the problem (5) has the form $\psi(x - c(\varepsilon)t, y)$, where $\varepsilon = |R - R_0|$. For $c(\varepsilon)$ we have the expansion

$$c(\varepsilon) = \sum_{k=0}^{\infty} c_k \varepsilon^k .$$

For the above values of parametes we have, with four significant digits, $c_0 = 0$, $c_1 = 0.0127$, $c_2 = 0.0025$, $c_3 = 0.0001$. All these computations were done in collaboration with Dr V.P. Varin. The same results are valid for large class of external forces.

This result may be obtained by the reduction of the problem on the centre manifold, however in [3] the same result was obtained in the framework of the Lyapunov-Schmidt method, developed for hydrodynamic problems by Iooss, Judovich, Joseph and Sattinger.

This example led us to the idea that for each hydrodynamic problem there is the characteristic value of time interval on which the problem must be considered. It must be larger than possible periods of the time-periodic solutions and large enough so that we have dynamics on the attractor but not a transition part of the non-stationary solution.

References

[1] Afendikov A.L. and Babenko K.I. On mathematical simulation of transition to turbulence in viscous incompressible fluid flows. *Mathematical Simulation* **1** (n8), 1989, Moscow, USSR (in Russian).

[2] Afendikov A.L. and Babenko K.I. Bifurcations in the presence of a symmetry group and loss of stability of some plane flows of viscous fluid. *Soviet Math. Dokl.* **33** (3), 1986.

[3] Afendikov A.L., Babenko K.I. and Varin V.P. On the loss of stability and bifurcation of some plane viscous incompressible fluid flow. Preprint of the Keldysh Inst. of Appl. Math., N 76, 1987, Moscow (in Russian).

A.L. Afendikov
Keldysh Institute of Applied Mathematics
Miusskaja sq. 4
Moscow 125047
USSR

S.W. SHAW AND J. SHAW

Nonlinear dynamics of a rotating shaft

Abstract

The interaction between the destabilizing forces of internal damping and those due to unbalance are studied in the context of a symmetric rotor. A nonlinear single-mode model is considered and a complete picture of the rotor response near resonance and the onset of instability is given for different levels of unbalance. It is shown that constant amplitude synchronous motions and amplitude-varying, nearly synchronous motions are possible, and a complete picture of their interaction is given.

1. Introduction

It is well known that a perfectly balanced symmetric rotor will undergo an instability if the impressed rotational speed exceeds a certain value. This instability gives rise to a constant amplitude whirling motion which is, in general, subsynchronous and occurs at speeds above the instability speed. On the other hand, any amount of unbalance has the effect of causing synchronous whirl and results in a resonance near the fundamental flexural vibration frequency of the shaft. The interaction between these effects is rarely considered and, to the authors' knowledge, has never been thoroughly examined. Bolotin [1963] gives a qualitative description of the possible responses in this situation, but does not provide a detailed account of them. This problem is a specific case of a periodically perturbed Hopf bifurcation; see Gambaudo [1985].

The present work does so for a restricted range of parameters which includes the most interesting cases. This provides a guide as to what happens for more general parameter values. The tools employed in this analytical study are from the qualitative theory of dynamical systems, in particular, the centre manifold and bifurcation theory.

In this paper we emphasize the model and the results obtained from that model. The mathematical and numerical procedures used to obtain these results are described in Shaw and Shaw [1989b]. The model considered is obtained by truncating all modes but the first for a symmetric rotating shaft. The shaft is assumed to be slender, straight, with a uniform circular cross-section, hinged ends, made a Voigt viscoelastic material, rotating at a constant speed Ω, subject to external frictional forces proportional to the absolute velocity of shaft elements, with a structural nonlinearity due to midline stretching, and with a centre of mass offset from the axis of rotation by a distance e. This model, while highly idealized, provides the essential ingredients required to study the behaviour of interest. The derivation of the shaft model and the modal truncation are relatively standard (see Shaw and Shaw [1989a]) and result in a pair of equations of motion which govern the first mode vibrations in terms of coordinates which rotate with speed Ω about the axis of shaft rotation. After some rescaling the equations are:

$$\ddot{u} + (\mu_e + \mu_i \Omega_1^2)\,\dot{u} - 2\Omega\dot{v} + (\Omega_1^2 - \Omega^2)\,u - \mu_e \Omega v + \alpha\Omega_1^2 (u^2 + v^2)\,u$$

$$+ 2\alpha\mu_i\,\Omega_1^2(u\dot{u} + v\dot{v})\,u = e\Omega^2$$

$$(1)$$

$$\ddot{v} + (\mu_e + \mu_i \Omega_1^2)\dot{v} + 2\Omega\dot{u} + (\Omega_1^2 - \Omega^2)\,v + \mu_e\,\Omega u + \alpha\Omega_1^2 (u^2 + v^2)\,v$$

$$+ 2\alpha\mu_i\,\Omega_1^2\,(u\dot{u} + v\dot{v})\,v = 0.$$

The coordinates u and v represent transverse vibration amplitudes of the shaft in the rotating frame and overdots correspond to derivatives which respect to a dimensionless time. The parameters μ_i and μ_e represent damping mechanisms: μ_i for the internal damping from the shaft material and μ_e for the external damping. The fundamental vibrational frequency is Ω_1, and α is a parameter which models the structural nonlinearities; $\alpha > 0$ so that the nonlinerity is of stiffening type.

Eqs. (1) have a special symmetry when $e = 0$: u and v can be interchanged without affecting the equations of motion. The straight configuration of a perfectly unbalanced shaft is known (see Bolotin [1963]) to undergo an instability at the critical speed

$$\Omega_1^* = \Omega_1 + \mu_e/\mu_i\Omega_1.$$

This is a Hopf bifurcation which, for $\alpha > 0$, throws off a constant amplitude nonsynchronous whirling motion. In fact, for $\mu_e/\mu_i\Omega_1$ small, the motion is nearly synchronous and becomes synchronous as $\mu_e/\mu_i\Omega_1 \to 0$. In terms of the (u, v) coordinates, this motion is seen as a circular limit cycle with frequency approximately equal to $\mu_e/\mu_i\Omega_1$ (see Shaw and Shaw [1989a]).

For $e \neq 0$ models of this type are typically studied with no internal damping, $\mu_i = 0$, which hides the instability. This leads to a standard nonlinear response curve for synchronous shaft motions in terms of vibration amplitude versus frequency. An analysis of this type yields the same response curves, but different stability types, for $\mu_i \neq 0$.

For a constant amplitude synchronous response, u and v will attain constant values which correspond to equilibria of Eqs. (1). Writing these as $\bar{u} = R\,\cos\theta$ and $\bar{v} = R\,\sin\theta$ and setting all time derivatives in Eqs. (1) to zero yields, after some manipulations, two equations for the amplitude and relative phase of synchronous response; these are analogous to similarly obtained amplitude-phase equations for the steady-state response of a forced nonlinear oscillator, typically obtained by harmonic balance or a perturbation method. Here the analysis is exact, no mathematical approximations are required due to the symmetry in u and v and the fact that the time periodic nature of the disturbance (i.e. the unblance) is eliminated by the use of rotating coordinates. Also, the presence of internal damping has no effect on the synchronous response amplitude or phase, but has a decided effect on their stability.

Internal and external damping mechanisms are present in all rotor systems, thus the value of $\mu_e/\mu_i\,\Omega_1$ will be taken to be nonzero, but will be assumed to be small. This restriction forces the point of instability, Ω_1^*, to be near to the first resonance, Ω_1. Also, speeds near Ω_1^* (and Ω_1) are to be considered, so the parameter $\varepsilon = \Omega - \Omega_1^*$ will be assumed to be small also. Finally, since the transition between resonance and instability is of interest, the unbalance, e will also be restricted to small values. In this parameter range the linearized system has two eigenvalues which lie well into the left half of the complex plane. The dynamics associated with these eigenvalues is relatively fast and rapidly decaying, and the overall dynamics are dominated by the slow behaviour associated with the two near-critical eigenvalues. There exists a mathematically rigorous method which provides the dynamic equations associated with these eigenvalues and correctly accounts for all essential nonlinear coupling between the fast and slow components. This method is the centre manifold method; see Guckenheimer and Holmes [1986] or Carr [1981] for details. The method is constructive and provides the equations which govern the slow dynamics. This process requires substantial work that is rewarded by a pair of autonomous nonlinear differential equations which govern the local dynamics near instability. They can be analysed using standard phase plane methods in conjunction with simulation studies.

Equilibria in the centre manifold correspond to equilibria for (u, v), i.e. $(\bar{u}, \bar{v})$. Since the centre manifold is attractive in phase space, the stabilities of centre manifold equilibria corresponds to those for $(\bar{u}, \bar{v})$. These correspond to constant amplitude synchronous whirling of the shaft. Limit cycles in the centre manifold correspond to limit cycles in the full (u, v) space. In all cases studied, limit cycles were locally stable solutions in which (u, v) oscillate about a synchronous whirling solution $(\bar{u}, \bar{v})$. In terms of the polar coordinates (R, θ), these indicate that the amplitude and relative phase of the motion undergo oscillations. The frequency of these oscillations is small and is approximately equal to $\mu_e/\mu_i\Omega_1$. In terms of the shaft these slow oscillations, when viewed from a laboratory reference frame, are associated with motions in which the shaft undergoes very slow amplitude and phase modulations about a synchronous motion. These arise as a combination of the synchronous whirl caused by the unbalance and the nearly synchronous whirl caused by the instability: when two oscillations of nearly equal frequency are added, the result is long-period beats, i.e. slow amplitude and phase modulations. These are not unexpected, and can be predicted by classical perturbation methods, but the complete picture of how they interact with, and arise from, synchronous motions requires a detailed study of the centre manifold dynamics, the results of which are now described.

2. Results

The main results are summarized in Figs. 1 and 2. Fig. 1 shows a series of curves in (e, ε) space which distinguish different types of shaft behaviours. The curves represent various bifurcations which are described below. Fig. 1 also shows frequency response curves for different levels of unbalance. Fig. 2 depicts the

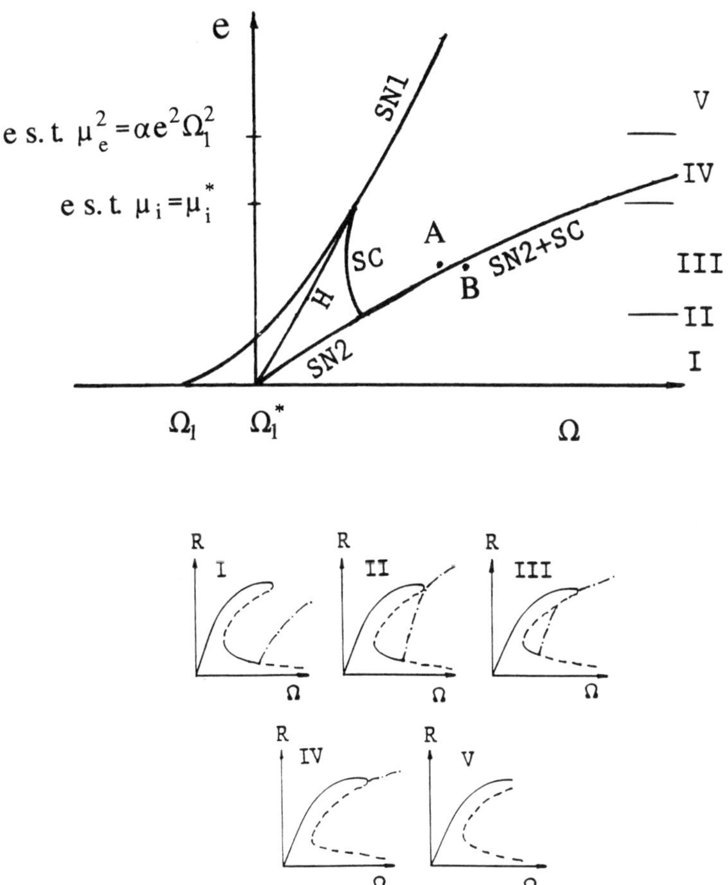

Figure 1. Bifurcation diagram in e - ε space and response amplitude vs. frequency curves; —— stable synchronous response, ------ unstable synchronous response, -.-.-.- . stable amplitude modulated response

Figure 2. Qualitative nature of the phase portraits obtained from the centre manifold dynamic equations

qualitative features of the phase portraits obtained from the centre manifold dynamics in various parameter regions. By correlating the relationships between the equilibria and limit cycles and the shaft response, a complete picture of the shaft rsponse near resonance for varying levels of unbalance can be obtained.

In the amplitude vs. frequnecy diagrams of Fig. 1 stable synchronous motions are shown as solid lines, unstable synchronous motions as dashed lines, and amplitude modulated (or quasiperiodic) responses are shown as broken lines.

In Fig. 1 the curves marked SN1 are associated with the conditions for the first saddle node bifurcation in which the lower two branches of the synchronous response curve are born as Ω increases. The curves SN2 and SN2 + SC correspond to the saddle-node bifurcation in which the upper and middle synchronous response branches merge and disappear. Note that this curve is asymptotic to a finite value of e (e such that $\mu_e^2 = \alpha e^2 \Omega_1^2$) as Ω increases, indicating that for sufficiently large unbalance this bifurcation will not occur. The SN2 curve corresponds to a simple saddle-node, whereas the SN2 + SC curve corresponds to a more complicated bifurcation in which the saddle-node bifurcation interacts with a saddle connection resulting in the appearance of a limit cycle, corresponding to amplitude modulated motion (this is depicted in Fig. 2, portraits 7-9, with 8 showing the condition at SN2 + SC). Schecter [1987] gives a complete account of this bifurcation. The curve H corresponds to the point at which the lowest synchronous response branch becomes unstable via a Hopf bifurcation, at which point a quasiperidoic motion is born. The curve marked SC corresponds to the condition in which the limit cycle amplitude has grown to the point where it touches the saddle point corresponding to the unstable middle branch of the synchronous response curve, forming a simple saddle connection (portrait 5 in Fig. 2). This bifurcation annihilates the limit cycle. The points where these curves meet correspond to codimension two bifurcations which are, in themselves, of interest. (This is not pursued here since the required information is obtained in a direct fashion using simulations.) The various curves fit together in a consistent manner, providing a complete description of the shaft response near resonance.

Simulations of the full equations (Eqs. (1)) were carried out in order to verify the results obtained from the above analysis. The simulation results are shown in terms of the nonrotating laboratory reference frame in order to aid in the visualization of the shaft response. Two-parameter conditions of particular interest were chosen for presentation; these are labelled A and B in Fig. 1, and the results for displacement vs. time are given in Fig. 3. Corresponding to point A, a constant amplitude synchronous motion is shwn in Fig. 3(a). By increasing Ω slightly, just moving across the curve SN2 + SC in Fig. 1, the result in Fig. 3(b) is obtained, associated with point B in Fig. 1. Note that, as predicted, a low-frequency amplitude modulation occurs in the shaft response. Other simulations further verify the response predicted in Fig. 1.

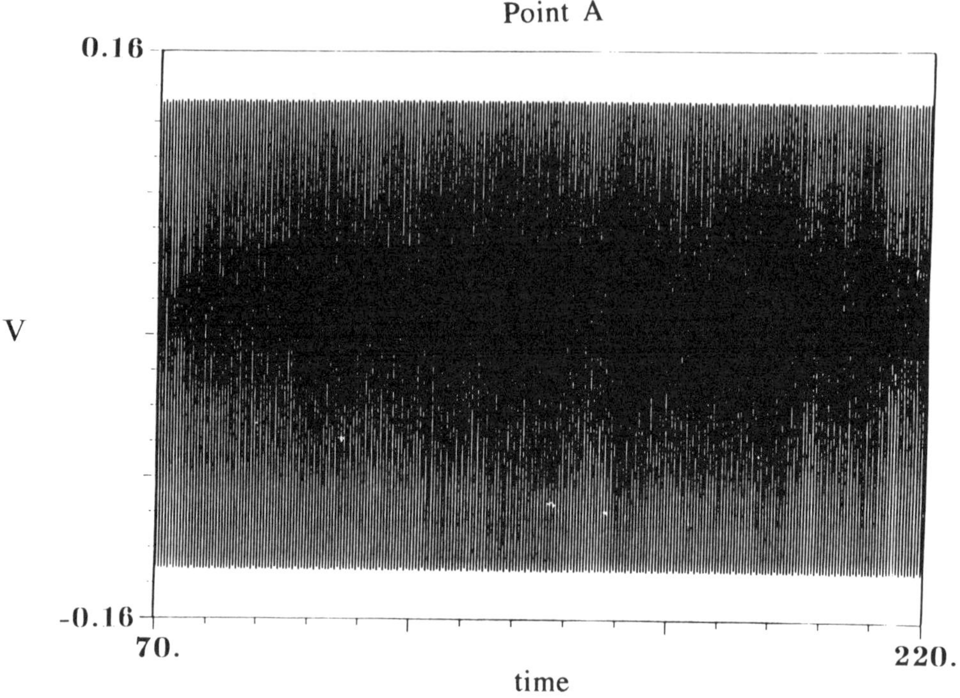

Figure 3(a). Simulations of Equations (1); $\mu_e = 0.08$, $\mu_i = 4.5/\Omega_1^2$, $\alpha = 10.0$, $e = 0.001$.
$\Omega = 10.7$.

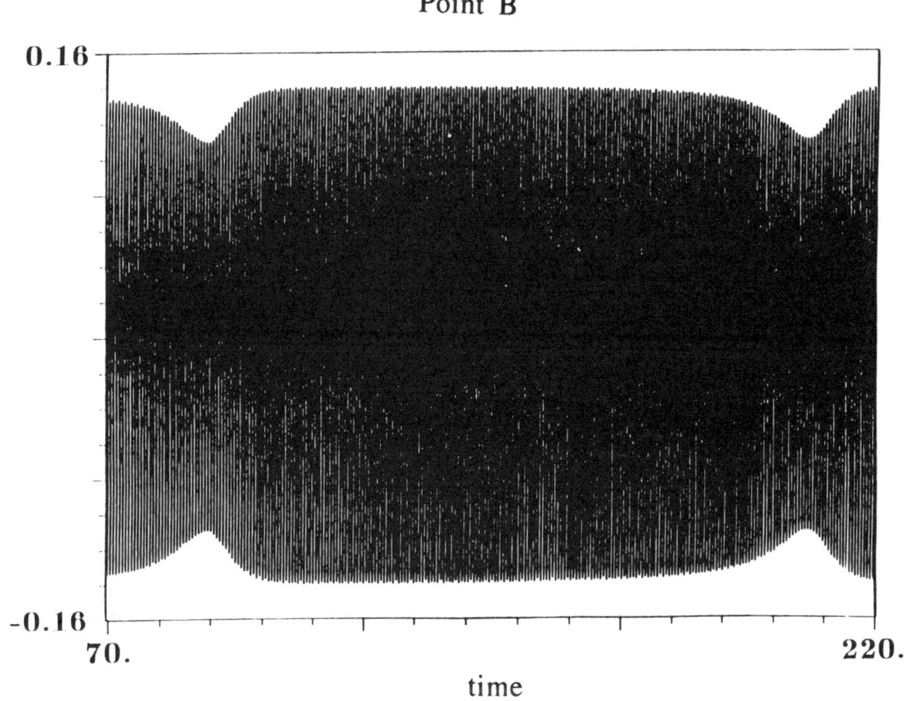

Figure 3(b). Simulations of Equations (1); $\mu_e = 0.08$, $\mu_i = 4.5/\Omega_1^2$,
$\alpha = 10.0$, $e = 0.001$.
$\Omega = 10.8$.

References

Bolotin, V.V. (1963). *Nonconservative Problems of the Theory of Elastic Stability*, Pergamon Press.

Carr, J. (1981). *Applications of Center Manifold Theory*. Springer-Verlag.

Gambaudo, J.M. (1985). Perturbation of a Hopf bifurcation by an external time-periodic forcing, *Journal of Differential Equations* **57**, 172-199.

Guckenheimer, J. and Holmes, P. (1986). *Nonlinear Oscillations, Dynamical Systems and Bifurcations of Vector Fields*, Springer-Verlag, 2nd Edition.

Schecter, S. (1987). *The saddle-note separatrix-loop bifurcation*, North Carolina State University preprint.

Shaw, J. and Shaw, S.W. (1989a). Instabilities and bifurcations in a rotating shaft, *Journal of Sound and Vibration* **132**, 227-244.

Shaw, J. and Shaw, S.W. (1986). Nonlinear resonance of an unbalanced rotating shaft with internal damping, *Journal of Sound and Vibration*, to appear.

S.W. Shaw and J. Shaw
Department of Mechanical Engineering
Michigan State University
East Lansing, MI 48824
USA

F. VESTRONI AND A. LUONGO

Perturbation analysis of finite oscillations of an orbiting string

Introduction

Some years ago, Colombo et al. [1] suggested running various experiments in space, using a satellite tethered to an orbiting station. This idea will be realized in future shuttle missions. The shuttle rotates in the *XY* plane along the circular orbit of radius *a* from the centroid of the earth in equilibrium under centrifugal and gravitational force (Fig. 1). A satellite deployed in the outward (or inward) direction from the shuttle is not in equilibrium; on the external (internal) side the centrifugal (gravitational) force is greater than the gravitational (centrifugal) force and the satellite puts the string in tension.

Due to several disturbance and control forces the system oscillates; dynamic phenomena mainly concern the string and the satellite. The recent space research programme has stimulated many investigations on the dynamic behaviour of the tethered satellite system (TSS). They have been mainly restricted to linear dynamics [2-5] and have shown that the frequencies of oscillations both within the plane of orbit and outside it are very close. Notwithstanding the fact that the nonlinearities which establish a coupling between the motion in the two planes are weak, it appears interesting to analyse the stability of the motion as the system operates almost in the condition of internal resonance.

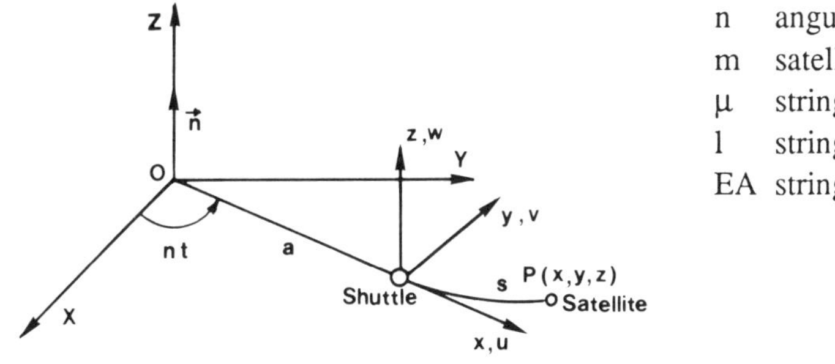

n	angular velocity
m	satellite mass
μ	string mass density
l	string length
EA	string axial stiffness

Figure 1. Tethered satellite system. Reference frames and characteristics.

Nonlinear equations of motion

Two reference systems are considered: one (OXYZ) is a fixed frame with the origin in the mass centre of the earth; the other is connected with the shuttle and rotates with constant angular velocity n. It is assumed that the distance of the shuttle remains unchanged because the mass of the shuttle is much greater than that of the satellite. The exact equations of motion are obtained through the Hamilton principle [4]; they admit a static solution $x_0(s)$ according to which the string lies along the x axis. To obtain an analytical asymptotic solution, the equations are expanded in Taylor series up to third order. After introducing nondimensionalized quantities:

$$\alpha^2 = \frac{\mu n^2 l^2}{EA}, \quad \gamma = \frac{\mu l}{m}, \quad \xi = \frac{l}{a},$$

$$\tilde{x} = x/l, \; \tilde{y} = y/l, \; \tilde{z} = z/l, \; \tilde{s} = s/l, \; \tilde{t} = nt, \; \tilde{u} = u/l, \; \tilde{v} = v/l, \; \tilde{w} = w/l \tag{1}$$

they read:

$$\alpha^2\ddot{u} - 2\alpha^2\dot{v} - 3\alpha^2 u\,(1 - 2\xi x_0) - \left[\, u' + \frac{v'^2 + w'^2}{2} - \left(u'^2 - \frac{v'^2 + w'^2}{2}\right) u' \,\right]' = 0$$

$$\alpha^2\ddot{v} + 2\alpha^2\dot{u} - \alpha^2\,[f(s)v']' - 3\alpha^2\xi x_0 v - \left[\left(u' + \frac{v'^2 + w'^2}{2}\right) v' - u'^2 v'\right]' = 0 \tag{2}$$

$$\alpha^2\ddot{w} + \alpha^2 w(1 - 3\xi x_0) - \alpha^2\,[f(s)w']' - \left[\left(u' + \frac{v'^2 + w'^2}{2}\right) w' - u'^2 w'\right]' = 0$$

with the related boundary conditions [5], where $f(s)$ is associated with the stress gradient of the static solution. It is possible to recognize the different force terms; inertial, gyroscopic, a combination of centrifugal and gravitational, and elastic. The nonlinear equations are all coupled, but plane motion can exist, while out-of-plane components always involve a spatial motion.

In the study of transversal oscillations a consistent ordering of the terms in Eq. (2_1) allows omission of inertial and centrifugal-gravitational forces while obtaining an integro-differential relation among u, v, w.

By using this relation to eliminate $u(s, t)$, the transversal oscillation of TSS are governed by the two equations:

$$\ddot{v} - [f(s)v']' - 3\xi x_0 v - \int_0^s (v'^2 + w'^2)^{\cdot} ds - \left[\frac{2}{\gamma}\dot{v}_1 v' - v' \int_1^s 2\dot{v}\, ds\right]' = 0$$

$$\ddot{w} - [f(s)w']' + (1 - 3\xi x_0)w - \left[\frac{2}{\gamma}\dot{v}_1 w' - w' \int_1^s 2\,\dot{v}\, ds\right]' = 0 \tag{3}$$

and the related boundary conditions.

The system exhibits only quadratic weak nonlinearities associated with gyroscopic force. The nature of the nonlinear terms ensures the existence of plane monofrequent oscillations, while the other is spatial.

With some consistent approximations the linear uncoupled oscillations are described by:

in-plane	out-of-plane	
$v(s, t) = A_v \sin ps \sin \omega_v t$	$w(s, t) = A_w \sin qs \sin \omega_w t$	
$\tan p = \gamma/p$	$\tan q = \gamma/q$	(4)
$\omega_v^2 = p^2 f_0$	$\omega_w^2 = q^2 f_0 + 1.$	

The characteristic equation is the same, while the frequencies squared differ by a unit; this is of some relevance only for the first frequencies.

Perturbational solution

To present the solution in more concise form, the equations of motion along with the boundary conditions are rewritten:

$$M(\ddot{v}) + L_v(v) + B(\dot{v}, v) + C(\dot{w}, w) = 0$$

$$M(\ddot{w}) + L_w(w) + D(\dot{v}, w) = 0, \tag{5}$$

where the linear M, L and bilinear B, C, D operators are deduced by comparison with Eq. (3). Following the method of Lindstedt–Poincaré, time and displacement variables are expanded in series of small parameter ε:

$$\tau = \Omega t = \omega_k (1 + \varepsilon^2 \alpha_2)t, \quad v = \varepsilon v_1 + \varepsilon^2 v_2 + \varepsilon^3 v_3, \quad w = \varepsilon w_1 + \varepsilon^2 w_2 + \varepsilon^3 w_3 \tag{6}$$

and linear systems at ε, ε^2 and ε^3 orders are obtained:

$$\begin{cases} \omega_k^2 M(\ddot{v}_1) + L_v(v_1) = 0 \\ \omega_k^2 M(\ddot{w}_1) + L_w(w_1) = 0 \end{cases}$$

$$\begin{cases} \omega_k^2 M(\ddot{v}_2) + L_v(v_2) = -\omega_k B(\dot{v}_1, v_1) - \omega_k C(\dot{w}_1, w_1) \\ \omega_k^2 M(\ddot{w}_2) + L_w(w_2) = -\omega_k D(\dot{v}_1, w_1) \end{cases} \tag{7}$$

$$\begin{cases} \omega_k^2 M(\ddot{v}_3) + L_v(v_3) = -2\alpha_2\, \omega_k^2 M(\ddot{v}_1) - \omega_k\{B(\dot{v}_1, v_2) + B(\dot{v}_2, v_1)\} \\ \qquad\qquad - \omega_k\{C(\dot{w}_1, w_2) + C(\dot{w}_2, w_1)\} \\ \omega_k^2 M(\ddot{w}_3) + L_w(w_3) = -2\alpha_2\, \omega_k^2 M(\ddot{w}_1) - \omega_k\{B(\dot{v}_1, w_2) + B(\dot{v}_2, w_1)\}. \end{cases}$$

By assuming one component generating solution (Eq. (8)), at ε^2 order a correction v_2 is found (Eqs. (9), (10)) which depends on all the eigenfunctions φ_j; in practice only modes with ω_j close to $2\omega_k$ are involved. The imposition of Fredholm conditions in Eq. (7_3), where use has been made of v_1 and v_2 furnishes the correction of frequency α_2 (Eq. (11)), which is a function of the oscillation amplitude.

$$v_1 = A_k\, \varphi_k(s) \sin\tau \qquad\qquad w_1 = A_k\, \psi_k \sin\tau \tag{8}$$

$$v_2 = A_k^2\, \Sigma_j\, c_j\, \varphi_j \sin 2\tau \qquad\qquad v_2 = A_k^2\, \Sigma_j\, c_j\, \varphi_j \sin 2\tau \tag{9}$$

$$c_j = \frac{\omega_k}{2(\omega_j^2 - 4\omega_k^2)} \int_0^1 \varphi_j\, B(\varphi_k, \varphi_k)\, ds \qquad c_j = \frac{\omega_k}{2(\omega_j^2 - 4\omega_k^2)} \int_0^1 \varphi_j\, C(\psi_k, \psi_k)\, ds \tag{10}$$

$$\alpha_2 = \frac{A_k^2}{2\omega_k}\, c_j \int_0^1 \varphi_k\, B(\varphi_k, \varphi_j)\, ds \qquad \alpha_2 = \frac{A_k^2}{2\omega_k}\, c_j \int_0^1 \psi_k\, C(\varphi_j, \psi_k)\, ds. \tag{11}$$

Stability analysis

Let us consider a disturbance of the steady oscillation $v_0(s, t)$, $w_0(s, t)$:

$$v(s, t) = v_0(s, t) + \eta(s, t), \quad w(s, t) = w_0(s, t) + \zeta(s, t), \tag{12}$$

where $\eta(s, t)$, $\zeta(s, t)$ are described by one mode suitably selected:

$$\eta(s, t) = \varphi_i(s)\ \eta(t), \quad \zeta(s, t) = \psi_\ell(s)\ \zeta(t). \tag{13}$$

Eqs. (12) are introduced in Eqs. (5); the system has been linearized in $\eta(s, t)$, $\zeta(s, t)$. Making use of Eqs. (13) and following a Galerkin technique, then

$$\begin{bmatrix} 1 & 0 \\ 0 & 1 \end{bmatrix} \begin{Bmatrix} \ddot{\eta} \\ \ddot{\zeta} \end{Bmatrix} + \left(\begin{bmatrix} \omega_{vi}^2 & 0 \\ 0 & \omega_{wl}^2 \end{bmatrix} + \begin{bmatrix} f_{11} & f_{12} \\ 0 & f_{22} \end{bmatrix} \right) \begin{Bmatrix} \eta \\ \zeta \end{Bmatrix} + \begin{bmatrix} 0 & g_{12} \\ -g_{12} & 0 \end{bmatrix} \begin{Bmatrix} \dot{\eta} \\ \dot{\zeta} \end{Bmatrix} = \begin{Bmatrix} 0 \\ 0 \end{Bmatrix} \tag{14}$$

The off-diagonal terms establish a coupling between $\eta(t)$ and $\zeta(t)$; the nature of f_{12} and g_{12} is such that they vanish when $w_0(s, t) \equiv 0$.

In the study of *stability of in-plane oscillation* for an out-of-plane disturbance the system (14) is uncoupled and the stability is governed by the second equation. By introducing the nondimensionalized quantities

$$2\tau = \Omega t, \quad \delta = \left(\frac{2\omega_j}{\Omega} \right)^2, \quad \varepsilon = c\frac{A_k}{\Omega}, \tag{15}$$

this equation can be reduced to the classical Mathieu equation:

$$\ddot{\zeta}(\tau) + (\delta + \varepsilon \cos 2\tau)\ \zeta(\tau) = 0 \tag{16}$$

The instability region emanates from $\Omega = 2\omega_j$; this occurs when an in-plane nonlinear frequency is twice an out-of-plane frequency. This condition can be easily verified between two modes that are higher than the first one. As expected, due to weak nonlinearities, the instability region is very narrow; but the ratio ω_{vk}/ω_{wj} is just less than 2 and the frequency-amplitude curve is hardening; this curve then enters into the instability region for a certain amplitude.

The *stability of the monofrequenct out-of-plane* oscillation is governed by the coupled system (14) with periodic coefficient f_{11}, f_{22} of frequency 2Ω and f_{12}, g_{12} of frequency Ω.

After introducing the normalized quantities:

$$\tau = \Omega t, \quad \delta = \left(\frac{\omega_p}{\Omega} \right)^2, \quad \varepsilon_{1,2} = 2A_k^2\, c_{1,2}, \quad \varepsilon_3 = A_k\, c_3, \quad \nu = \omega_{w_1}/\omega_{v_1}, \tag{17}$$

system (14) reads:

$$\ddot{\eta} + (\delta^2 + \varepsilon_1 \cos 2\tau)\eta + \varepsilon_3\, \delta(\zeta \sin \tau)^{\cdot} = 0$$
$$\ddot{\zeta} + (\nu^2\delta^2 + \varepsilon_2 \cos 2\tau)\zeta - \varepsilon_3\, \delta\, \dot{\eta} \sin \tau = 0. \tag{18}$$

According to the Floquet theory, the boundaries of the unstable regions are associated with the periodic solution of Eqs. (18) with frequency 1/2 or 1. The first condition does not furnish any region, since the boundary curves coincide. The second condition gives two distinct regions, one emanating from $\Omega = \omega_\nu$ and the other from $\Omega = \omega_\nu$.

The ith out-of-plane mode disturbed by the ith in-plane mode is always stable. Indeed the frequency-amplitude curve always remains on the right side of the first instability region because $\Omega \cong \omega_{wi} > \omega_{\nu i}$ and the nonlinearity is hardening. Moreover the curve coincides with a boundary of the second region. This circumstance is related to the fact that at this approximation order the coupling term does not play any role in the second region and the problem is formally reduced to the stability of the free motion of a single oscillator.

Conclusions

Nonlinear motion of an orbiting tethered satellite system is studied. Monofrequent oscillations in the orbit plane exist, while in the motion with a prevailing out-of-plane component the in-plane one is always forced due to nonlinear coupling of gyroscopic forces. Although the nonlinearities are weak, the frequencies of the oscillations in the two planes are so close that unstable phenomena can occur. The analysis reveals that the in-plane oscillations are unstable in a certain range of amplitudes while the out-of-plane oscillations are always stable.

Acknowledgements

This work is partially supported by funds from Italian Ministry of Education (MPI-60%, 1988).

References

[1] Colombo, G., Gaposchkin, E.M., Grossi, M.D. and Weiffenbach, G.C., The 'Skyhook': a Shuttle born tool for low-orbital-altitude research, *Meccanica*, (March 1975), 3-20.

[2] Banerjee, A.K. and Kane, T.R., Tether Deployment Dynamics, *Journal of the Astronautical Sciences*, vol. **30** (4) 1982.

[3] Misra, A.K., Xu, D.M. and Modi, V.J., On vibrations of orbiting tethers, *Acta Astronautica*, 1987.

[4] Pasca, M., Luongo, A., Pignataro, M. and Vestroni, F., *Free Dynamics of the Shuttle-Tethered-Satellite System*, Studi e Ricerche, Dept. Struct. Geot. Eng., University Roma 'La Sapienza', No. 3, 1987.

[5] Luongo, A. and Vestroni, F., *Nonlinear Analysis of TSS Plane Oscillations* (in Italian), DISAT, University of L'Aquila, Rep. N. 9, 1988.

F. Vestroni and A. Luongo
University of L'Aquila
Dipartimento di Ingegneria delle Strutture,
Acque e Terreno
I-67040 Monteluco di Roio, L'Aquila
ITALY

G.A. MAUGIN AND H. HADOUAJ

Solitary surface acoustic waves on elastic structures

1. Introduction

Nonlinearity and dispersion (or scale effect, or microstructure) are two necessary ingredients for the existence of solitary waves, the latter existing when strict compensation between the two antagonist effects occurs. In so far as elastic surface waves are concerned, a purely mechanical means of realizing such a situation consists in superimposing a thin soft elastic layer on a nonlinear elastic substrate [1]. That is, nonlinearity will be generated by the substrate while the superimposed layer, which exhibits a characteristic length, its thickness, yields dispersion, as is well known from Love SH (shear horizontal) surface acoustic waves [2], [3]. But a superimposed layer of finite thickness is a wave guide, and thus, not only is it dispersive, but it is also multimode. In the present work the multimode process is eliminated by considering the superimposed layer (Fig. 1) as a linear elastic thermodynamical interface of vanishingly small thickness and perfectly bonded to the substrate. Then in the linear approximation Murdoch SH surface acoustic waves [4] exist which are dispersive but monomode. Using the Whitham–Newell approach to nonlinear dispersive, small amplitude, almost monochromatic waves, we present a complete nonlinear analysis which proves the existence of stable envelope-solitary SH surface wave solutions. These are very much like the electromagnetic dark and bright solitons that can propagate in nonlinear optical fibres. The problem is governed by a nonlinear Schrödinger equation at the interface. The existence conditions obtained allow one to select a couple of materials from which we may effectively build such a surface soliton acoustic guide. The present analysis neglects the nonlinear coupling of the SH mode with the Rayleigh component. The study of this coupling can be carried out resulting in the coupling between a nonlinear Schrödinger equation for the slowly varying amplitude of the SH mode and two d'Alembert equations for the components of the Rayleigh mode which are nonlinearly driven by this amplitude. This is examined in further works as well as the numerical simulation of the present analysis, that of the interaction of two solitary waves showing the practically solitonic collision of the SH modes in spite of the non exact integrability of the original system, and the more tedious one of the coupled SH–Rayleigh system.

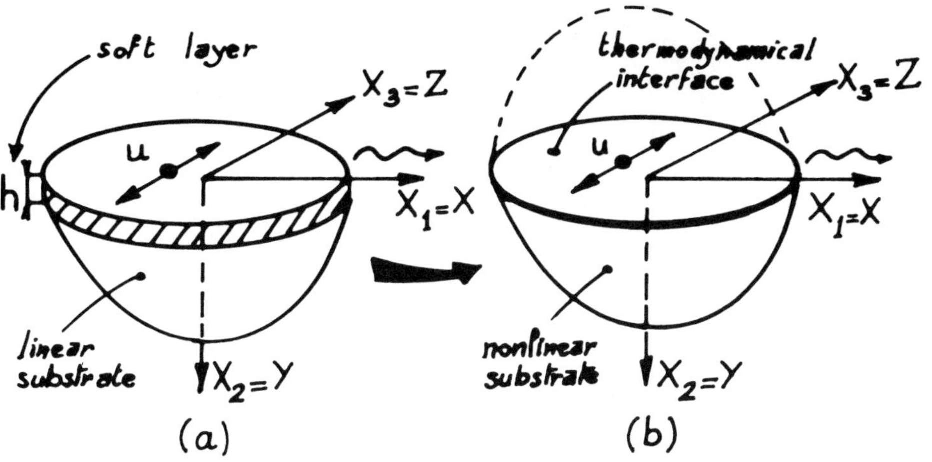

Figure 1. SH surface waves (Love and Murdoch waves)

2. Basic equations and linear solution

The wave propagates in the $X_1 = X$ direction, is polarized in the $X_3 = Z$ direction, and has an amplitude which decreases to zero as $X_2 = Y$ goes to infinity (Fig. 1). For isotropic substrate and layer it is possible to show [5] that the *nondimensional* equations of the problem written in Lagrangian coordinates are as follows (they are deduced from the theory of continua in which $X_2 = 0$ is a thermodynamical interface, [6], [7]):

$$\Box_B u - \Delta \beta^2 T_B^{\mathrm{NL}}(u) = 0, \quad Y > 0, \tag{1}$$

$$\Box_S \hat{u} - u_Y - \Delta \beta^2 T_S^{\mathrm{NL}}(u) = 0, \quad Y = 0, \tag{2}$$

$$u(X, Y = 0, T) = \hat{u}(X, T), \tag{3}$$

$$u(X, Y \to \infty, T) = 0, \tag{4}$$

where

$$\Box_B \equiv \beta^2 \frac{\partial^2}{\partial T^2} - \left(\frac{\partial^2}{\partial X^2} + \frac{\partial^2}{\partial Y^2} \right), \quad \Box_S \equiv \frac{\partial^2}{\partial T^2} - \frac{\partial^2}{\partial X^2}, \tag{5}$$

$$\left.\begin{aligned} &\beta = \frac{1}{c} = \frac{c_S}{c_T}, \quad c_S^2 = \frac{\mu_S}{\rho_S} = \frac{\hat{\mu}}{\hat{\rho}_0}, \quad c_T^2 = \frac{\mu}{\rho_0}, \quad \ell = \hat{\rho}_0/\rho_0, \\ &\Delta = c^2(\delta_{\mathrm{eff}}/\mu) = \frac{c^2}{\mu}\left[\delta + \left(\bar{\beta} + \frac{3}{2}\gamma\right) + \left(\mu + \frac{\lambda}{2}\right)\right], \end{aligned}\right\} \tag{6}$$

$$\left.\begin{aligned} &T_B^{\mathrm{NL}}(u) = \left[u_X\left(u_X^2 + u_Y^2\right)\right]_X + \left[u_Y\left(u_X^2 + u_Y^2\right)\right]_Y = O(u^3), \\ &T_S^{\mathrm{NL}}(u) = u_Y\left(u_X^2 + u_Y^2\right) = O(u^3). \end{aligned}\right\} \tag{7}$$

In these expressions left subscripts X and Y indicate partial spatial differentiation, a superimposed caret refers to surface properties, c_T and c_S are transverse elastic speeds, λ and μ are usual Lamé constants, $\bar{\beta}$ and γ are third-order elasticity coefficients, δ and δ_{eff} are the thermodynamical and effective fourth-order elasticity coefficients. Because of the peculiar SH nature of studied waves, Eqs. (1) and (2) in fact are *third-harmonic* generators. The two surviving parameters, β and Δ, account for dispersion and nonlinearity, respectively.

Linear problem

In the linear harmonic approximation of frequency ω and wave number k, the problem (1)-(4) provides Murdoch's solution [4] with *linear* dispersion relation:

$$\mathcal{D}_L(\omega, k) = \mathcal{D}_M(\omega, k) = \beta^2\,\omega^2 - k^2 + (\omega^2 - k^2)^2 = 0, \tag{8}$$

which is obtained by eliminating the depth coefficient χ between the linear 'bulk' and 'surface' conditions

$$\mathcal{D}_B(k, \omega; \chi) \equiv \beta^2\,\omega^2 - k^2 + \chi^2 = 0, \quad \mathcal{D}_S(k, \omega; \chi) \equiv \omega^2 - k^2 - \chi = 0, \tag{9}$$

deduced for solutions $u(X, Y, T) = A_0 \exp(-\chi Y) \exp i(kX - \omega T)$, $\chi > 0$. An alternate form of (8) is Murdoch's one:

$$\mathcal{D}_M(s, k) \equiv \ell^2\,k^2 - \frac{1 - s^2}{\left(s^2 - s_0^2\right)^2} = 0, \quad s = \frac{v}{C_T}, \quad s_0 = \beta, \quad v = \frac{\omega}{k}, \tag{10}$$

where ℓ is the previously introduced characteristic length. Typical dispersion relations (8) and (10) are represented in Fig. 2. The linear harmonic SH SAW solution exists only for $s_0 \leq s < 1$.

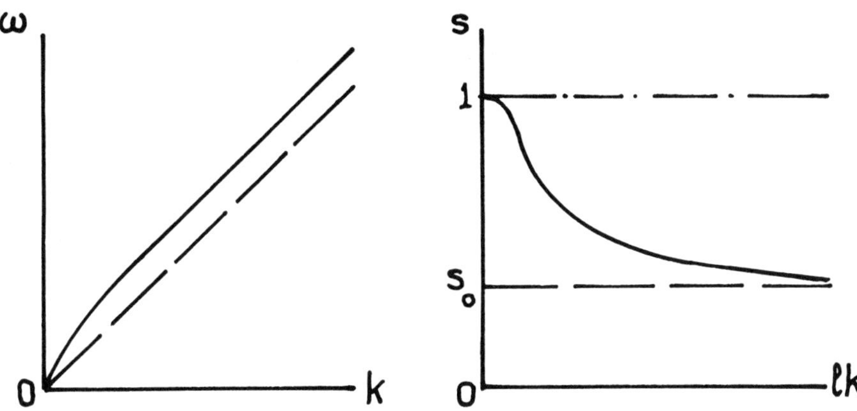

Figure 2. Murdoch linear dispersion relation for SH SAWs.

3. Nonlinear dispersion relation

In the nonlinear regime but for small amplitudes, trying solutions of the type $u = A \exp(-\chi Y) \cos\theta + ,\dots, \theta = kX - \omega T$, helps one to find out that T_B^{NL} and T_S^{NL} will produce terms of the form

$$T_B^{NL} \sim (9\chi^4 - 3k^4 + 2k^2\chi^2)\ \frac{A^3}{4} \exp(-3\chi Y)\cos\theta + (\text{term in } \cos 3\theta)_B$$

$$T_S^{NL} \sim -(\chi k^2 + 3\chi^3)\ \frac{A^3}{4} \cos\theta + (\text{term in } \cos 3\theta)_S \tag{11}$$

This allows one, for A *depending on* X, Y, T, to look for the *nonlinear* bulk and surface dispersion relations in the form

$$\mathcal{D}_B^{NL}(k, \omega, \chi, A) \equiv \mathcal{D}_B(k, \omega; \chi) + \frac{\Delta\beta^2}{4}(9\chi^4 - 3k^4 + 2k^2\chi^2)A^2\exp(-2\chi Y)$$

$$= \varepsilon\,\ell^{(1)} + \varepsilon^2\,\ell^{(2)} + \dots \qquad (12)_1$$

$$\mathcal{D}_S^{NL}(k, \omega, \chi, A) \equiv \mathcal{D}_S(k, \omega; \chi) - \frac{\Delta\beta^2}{4}(\chi\, k^2 + 3\chi^3)A^2$$
$$= \varepsilon\, m^{(1)} + \varepsilon^2\, m^{(2)} + \dots \qquad (12)_2$$

where the right-hand sides are perturbations to be determined. This is achieved through an asymptotic expansion by writing

$$x = \varepsilon X,\ y = \varepsilon Y,\ t = \varepsilon T,\ u(X, Y, T) = f(\theta, A) + \varepsilon u_1 + \varepsilon^2 u_2 + \dots, \qquad (13)$$

where θ is a generalized phase such that $\theta_X = k$, $\theta_T = -\omega$, $\theta_Y = i\chi$. On evaluating $\partial/\partial T$, $\partial/\partial X$, etc, we find out that, at order one,

$$u_0 = A \exp i(kX - \omega T + i\,\chi Y),$$

the linear SAW solution, and that, at order two, $u_1 = 0$, $\ell^{(1)} = 0$, $m^{(1)} = 0$, together with the generalized *wave action* conservation equations (compare [8], [9]),

$$\frac{\partial}{\partial t}(\beta^2\, \omega\, A^2) + \frac{\partial}{\partial x}(kA^2) + \frac{\partial}{\partial y}(i\chi A^2) = 0,\ y > 0,$$
$$(14)$$
$$\frac{\partial}{\partial t}(\omega A^2) + \frac{\partial}{\partial x}(kA^2) - \frac{i}{2}\frac{\partial}{\partial y}A^2 = 0,\ y = 0.$$

At the next order in ε we obtain the *secularity conditions*:

$$\left.\begin{aligned} &\mathcal{D}_S^{NL}(\omega, k, \chi, A) - \frac{\varepsilon^2}{A}\square_B A = 0,\ y > 0, \\ &\mathcal{D}_S^{NL}(\omega, k, \chi, A) - \frac{\varepsilon^2}{A}\square_S A = 0,\ y = 0. \end{aligned}\right\} \qquad (15)$$

These are *nonlinear* 'dispersive' dispersion relations which result from a *double* expansion in which ε and A are of the same order. Equations (14) and (15), together with the *kinematic-wave theory* condition,

$$\frac{\partial k}{\partial t} + \frac{\partial \omega}{\partial x} = 0, \qquad (16)$$

are the equations which govern ω, k and A once χ has been eliminated from Eqs.

(14) and (15). These equations could be deduced by means of the *averaged Lagrangian method* of Whitham [9] and Hayes [10] since the system (1)-(4) admits a Lagrangian [5].

4. Small amplitude, almost monochromatic limit

Assume that we are close to an harmonic regime (ω_0, k_0, χ_0) which satisfies Eqs. (9). That is, we write

$$k = k_0 + \varepsilon\varnothing_x, \quad \omega = \omega_0 - \varepsilon\varnothing_t, \quad \chi = \chi_0 - \varepsilon^i\varnothing_y. \tag{17}$$

We need to evaluate the slowly varying quantities $\varnothing$ and A. Introducing [11]

$$\xi = x - \omega_0' t + i\chi_0' y, \quad \tau = \varepsilon t, \quad A \to \varepsilon A, \tag{18}$$

where a prime denotes differentiation with respect to k_0, using Eqs. (14), (15) and the first two differentials of Eqs. (9) evaluated at (ω_0, k_0), and finally introducing the complex quantity $a = A\exp(i\varnothing)$, two nonlinear Schrödinger (NLS) equations follow between which χ_0'' is eliminated at $y = 0$ to reduce the whole problem to the following NLS equation:

$$i\, a_\tau + p(\omega_0, k_0) a_{\xi\xi} + q(\omega_0, k_0)\, |a|^2\, a = 0 \tag{19}$$

where

$$\tau = \varepsilon t, \quad \xi = x - \omega_0' t, \quad y = 0,$$

$$p(\omega_0, k_0) = \frac{1}{2}\omega_0''$$

$$q(\omega_0, k_0) = \frac{3}{8}\Delta\beta^2\, \frac{\beta^2\omega_0^2\left(\beta^2\omega_0^2 - 2k_0^2\right)}{\omega_0\left[\beta^2 + 2\left(\omega_0^2 - k_0^2\right)\right]} \tag{20}$$

and $(\omega_0, k_0) \in \mathcal{D}_L(\omega_0, k_0) = 0$. By the *inverse-scattering method a stable* solution of the *dark soliton* type is shown to exist for $pq < 0$ [12]. This condition here takes the form (for $\Delta > 0$)

$$\left(\omega_0^2 - k_0^2\right)\left[2\left(\omega_0^2\beta^4 - k_0^2\right) - \beta^2\left(\omega_0^2 - k_0^2\right)\right] > 0, \tag{21}$$

which finally yields the simple condition $\beta^2 < \frac{1}{2}$. Typically,

$$\hat{a}(X,T) = \left(|\omega_0''/q|\right)^{\frac{1}{2}} \varepsilon \tanh\left[\varepsilon\,(X - \omega_0'{}_T - X_0)\right] \exp i\,(k_0 X - \omega_0 T). \tag{22}$$

If lithium niobate $LiBnO_3$ is selected as the nonlinear substrate, then gold, silver, platinum and tantalum are materials which realize $\beta^2 < \frac{1}{2}$ for the superimposed layer while, e.g., aluminium, tungsten and nickel yield the stable bright soliton solution ($\beta^2 > \frac{1}{2}$).

References

[1] Maradudin, A.A. (1988). In: *Recent Developments in Surface Acoustic Waves*, (ed. D.F. Parker and G.A. Maugin), pp. 72–71, Springer-Verlag, Berlin.

[2] Maugin, G.A. (1983). In: *Advances in Applied Mechanics*, (ed. J.W. Hutchinson), vol. 23, pp. 373–434, Academic Press, New York.

[3] Maugin, G.A. (1988). In: *Recent Developments in Surface Acoustic Waves*, (ed. D.F. Parker and G.A. Maugin), pp. 158–172, Springer-Verlag, Berlin.

[4] Murdoch, A.I. (1976), *J. Mech. Phys. Solids*, **24**, 137–146.

[5] Hadouaj, H. and Maugin, G.A. (to be published).

[6] Daher, N. and Maugin, G.A. (1986), *Acta Mechanica*, **60**, 217–240.

[7] Daher, N. and Maugin, G.A. (1987), *Int. J. Engng. Sci.*, **25**, 1093–1129.

[8] Newell, A.C. (1985), *Solitons in Mathematics and Physics*, SIAM Philadelphia.

[9] Whitham, G.B. (1974), *Linear and Nonlinear Waves*, Wiley-Interscience, New York.

[10] Hayes, W.D., (1970), *Proc. Roy. Soc. Lond.*, **A320**, 187–208.

[11] Benney, D.J. and Newell, A.C. (1967). *J. Math. and Phys*. (now *Stud. Appl. Math.*) **46**, 133–135.

[12] Zakharov, V.E. and Shabat, A.B. (1972), *Soviet Physics-J.E.T.P.*, **34**, 62–69; (1973), *ibid*, **37**, 823–828.

G.A. Maugin and H. Hadouaj
Laboratoire de Modélisation en Mécanique,
Associé au CNRS, URA 229,
Université Pierre-et-Marie Curie,
Tour 66, 4 place Jussieu
75252 Paris, Cedex 05
FRANCE

J. POUGET

Nonlinear dynamics and instability of elastic domains patterns in martensitic-ferroelastic materials

Nonlinear dynamics and instability processes of a lattice model for martensitic-ferroelastic transformations are presented. The starting point of the study is the construction of a two-dimensional lattice model involving nonlinear and competing interactions. Because of the special kinds of interatomic potential thus introduced, the continuum model accounts for nonlinear and weakly nonlocal behaviours. The lattice distorsion proceeds by shearing the close-packed atomic layers stacked in a crystallographic direction leading them to strongly deformed regions modelled by a moving martensitic soliton. The complete two-dimensional model is examined next. We direct our attention to the instability process of a strain soliton with respect to the transverse direction, giving rise to a complex dynamics of domain patterns. The physical conjectures are also checked by means of numerical simulations.

1. Introduction

The goal of this work is to examine the *nonlinear dynamics of domain patterns* relating to *martensitic-ferroelastic transformations* on the basis of a lattice model. Our main purpose is to understand the underlying micro-physics which induces special kinds of strain transformations associated with elastic domain formation. The term *martensite* was originally used by metallurgists to designate the hard transformation formed in quenched steel. Martensitic is also applied to transformations taking place in crystalline solids by displacements of atoms over distances smaller than lattice spacing in the parent phase (or the high-temperature phase) [1]. For this reason the transformations are described as being diffusionless. Within the overall class of displacive diffusionless transformations, martensitic transformations are characterized by involving homogeneous lattice deformations and deviatoric (shear) displacements. The strain energy exerts then a dominant influence on the kinetic and on the martensite nucleation process, accordingly. As a consequence, the transition is of *first order* where spontaneous strain is the order parameter in the Landau theory of phase transition and the ferroelasticity is called proper [2]. Crystals such as In–Tl, Ni–Tl, Nb_3Sn, Fe–Pd etc. ... are good examples. On the other hand, crystals undergoing such transformations exhibit particularly interesting effects: elasticity; pseudo-elasticity; ferroelasticity with hysteresis; as well as *shape memory effects*, of which technological applications are especially promising. All these effects are in fact intimately connected with phase transition taking place in materials and they are only observed as the global behaviour of the crystal. Nevertheless, our task will be limited to small regions of perfect materials, that is low-dimensional materials, where only a few elastic domains coexist. The global behaviour is, in fact, the result of an averaging of all these regions with different orientations (product phases) involving statistical physics in order to introduce an adequate thermodynamics. The global behaviour is certainly monitored by the dynamics of twinned regions and the formation of coherent arrays of twins or ferroelastic domains is a consequence of the first-order transition. These domain patterns are usually observed experimentally by means of electron microscopy, and the morphology of martensitic twins seems to be very rich and complex [3]. This is a motivation for examining the problem of the elastic domain patterns at the microscopic level.

In the present study the emphasis is placed on martensitic-ferroelastic transformations in order to examine an intrinsic mechanism of the elastic twin formation as well as their dynamics. The nucleation process can be seen as pre-transitional modulated structures developing in the high-temperature phase which forms thus a periodic array of parallel twin bands. It follows therefore that the coherent motion of the twin boundaries is of great importance to understand martensitic transformations. In the mechanical point of view, we are concerned with the competition between the *nonlocal elasticity* (inhomogeneous state of defomration) and *local nonlinear* elastic energy (homogeneous state of deformation with stable, unstable or metastable regions), which plays a crucial role in motion of coherent elastic twin boundaries. A special attention is devoted to a two-dimensional lattice model which allows us to describe a cubic-tetragonal transformation. The latter transformations is mostly characterized by the shearing motion of the close-packed atomic planes $(1\bar{1}0)$ stacked in the [110] direction. Because of the anisotropy and, of course, of the special kinds of interactions the material exhibits anomalously low anisotropy *shear modulus softening* (the transverse acoustic branch in the [110] direction) which is strongly temperature dependent. The phonon mode softening usually occurs at non-zero wave vector in the phonon dispersion spectrum [4]. Structural modulations can then develop within the parent phase with periodic deformation patterns of which the wavelength corresponds to that of phonon anomalies [3, 5]. The origin of such patterns emerges from competing and nonlinear interactions. Accordingly, a rather fine scale of the lattice description seems to be necessary.

2. Construction of the microscopic model

Let us consider an atomic plane extracted from a cubic lattice (austenistic or high-temperature phase), for instance the f.c.c. symmetry of In–Tl [6]. However, we assume homogeneous deformation of the lattice along a perpendicular direction to the atomic plane, say the [001] direction, which means we deal with plane deformation in the (001) plane. Then, our lattice, in its undeformed state, consists of squares parallel to the [100] and [010] crystallographic directions (see Fig. 1). A particle of the plane is located by (i, j) in the co-ordinate system {[100], [010]} or (i, j) or by (I, J) in {[110], [1$\bar{1}$0]} or (I, J) deduced from the former by a rotation 45° clockwise. After deformation of the lattice the particles perform displacements in the lattice plane defined by $u(i, j)$ and $v(i, j)$, which are the longitudinal and transverse displacements, respectively.

We define next the interatomic forces acting on the particles of the lattice. We assume that the particle at a site (i, j) interacts with the first nearest neighbours surrounding it. Two types of interactions are considered: (i) interactions by *pairs* between the first nearest particles in the four directions i, j, I, J and (ii) interactions of the *three-body* type in the same directions [7]. The latter interactions acts on particles as *non-central forces*, which is equivalent to bond-bending or torsional forces [7, 8]. The lattice potential must then account for the change in angle between bond segments joining particle pairs, which represents, at the microscopic level, twisting and bending of the unit crystalline cell.

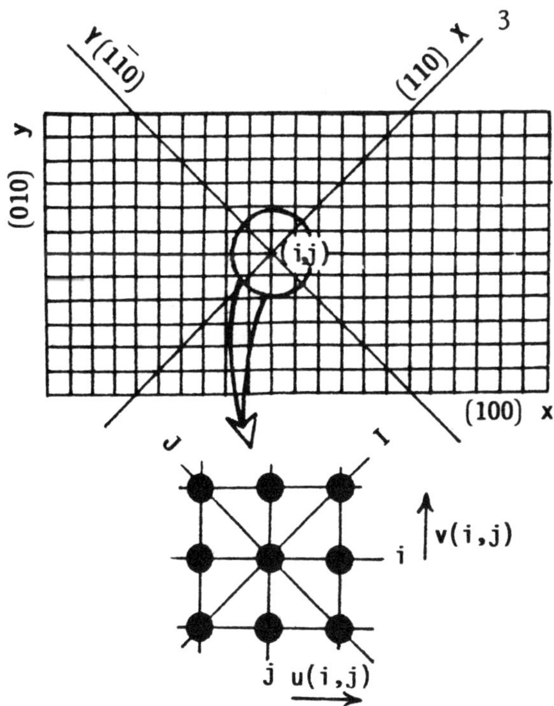

Figure 1. Geometry of the lattice model, two-dimension system made of squares. Interactions of the particle at (i, j) with the first neighbours.

Moreover, the lattice anisotropy is characterized by potentials of different strengths in the four directions. Since the lattice energy must be translationally and rotationally invariant, it depends only on the modulus of the relative particle positions. In fact, after some algebraic manipulations we are able to show that the lattice energy is a function of discrete Lagrangian deformation tensor including geometric nonlinearities and also of the first order finite differences of these discrete deformations. The lattice energy can be written as [7]

$$\mathcal{V} = \sum_{i,j} \{\varphi_1(\in_{11}(i,j)) + \varphi_2(\in_{22}(i,j)) + \varphi_3(\in_{11}(i,j),\ \in_{22}(i,j),\ \in_{12}(i,j))$$

$$+ \varphi_4(\in_{11}(i,j),\ \in_{22}(i,j),\ \in_{12}(i,j)) + \psi(\Delta_L \in_{11}(i,j),\ \Delta_T \in_{11}(i,j),$$

$$\Delta_L \in_{22}(i,j),\ \Delta_T \in_{22}(i,j),\ \Delta_L \in_{12}(i,j),\ \Delta_T \in_{12}(i,j))\} \tag{1}$$

where the discrete components of the Lagrangian strain tensor $\in_{pq}$ are defined exactly as usually in continuum mechanics and are functions of the first-order finite differences of the discrete displacements Δ_L and Δ_T in the i and j directions, respectively. The potentials φ_α (α = 1, 2, 3, 4) are functions of the particle pairs between the nearest neighbours in [100], [010], [110] and [1$\bar{1}$0] respectively, and the potential ψ derives from the three-body interactions in the same directions. On introducing appropriate symmetrical strain components for the cubic symmetry

$$e_1 = (\in_{11} + \in_{22}) / \sqrt{2},\quad e_2 = (\in_{11} - \in_{22}) / \sqrt{2},\quad e_3 = \in_{12} \tag{2}$$

where the index (i, j) has been omitted, we can write the lattice energy as a function of the deviatoric and dilatational parts of the (discrete) strain tensor and their first-order finite differences.

3. One-dimensional version

(a) Hamiltonian of the system

We assume next that the lattice displacements depend only on the I index ($I = i + j$). Important reductions of the model are then obtained and significant physical situations can be emphasized. After some algebra the Hamiltonian of the discrete system can be written (in dimensionless notation) as

$$\mathcal{H} = \sum_I \frac{1}{2} \dot{V}^2(I) + \sum_I \Big[\frac{1}{2} \alpha S^2(I) - \frac{1}{3} S^3(I) + \frac{1}{4} S^4(I) + \frac{1}{2} \beta (S(I+1) + S(I))^2$$

$$+\frac{1}{2}\delta(S(I)-S(I-1))^2+\frac{1}{2}\eta(S(I+1)+S(I)-S(I-1)-S(I-2))^2\Big] \quad (3)$$

where $V(I)$ represents the transverse displacement in the J direction (or $[1\bar{1}0]$) and $S(I)$ is the single strain component, connected with $V(I)$ through

$$S(I)=V(I+1)-V(I). \quad (4)$$

The lattice parameters α, β, δ and η are linear combinations of the second-order derivatives of the potentials φ_α and ψ with respect to the particle positions taken at equilibrium. The first part of the Hamiltonian (3) is the kinetic energy, the next three terms derive from the expansion of the lattice energy up to the fourth order with respect to the strain. The fifth term of (3) describes the interactions between second nearest neighbouring atomic planes and the last two terms emerge from the non-central interactions between first and second nearest neighbouring atomic planes. The combination $A=\alpha+4\beta$ corresponds to the elastic modulus $(C_{11}-C_{12})/2$ [9]. Along with the Landau theory of phase transitions for ferroelastic crystal the elastic modulus A depends on the temperature according to the usual Curie-Weiss law. We remark that the lattice deformation is now well described by $S(I)$, which represents the *relative shear displacement* of the close-packed atomic planes $(1\bar{1}0)$ in the stacking direction [110].

(b) Equations of motion

The equations of motion for the transverse displacement $V(n)$ are easily obtained by using the Hamiltonian principle and the definition of the discrete strain (4). This yields the following difference equations for the discrete shear deformation $S(n)$:

$$\ddot{S}(n)=\Delta^2(\sigma(n)-\Delta\chi(n)), \quad (5a)$$

where we have set

$$\sigma(n)=\alpha S(n)-S^2(n)+S^3(n)+\beta(S(n+1)+2S(n)+S(n-1)) \quad (5b)$$

$$\chi(n)=-+\delta\Delta S(n)+\eta\Delta(S(n+2)+4S(n+1)+6S(n)+4S(n-1)+S(n-2)) \quad (5c)$$

Eq. (5b) defines the discrete macro-stress, and the micro-stress due to the noncentral interaction is given by Eq. (5c). This is a set of coupled nonlinear ordinary differential equations which governs the shear deformation travelling perpendicularly to the close-packed atomic planes $(1\bar{1}0)$. In the discrete case, these equations are not usually tractable except for the linearized situation. The strongly nonlinear problem can, however, be investigated by means of numerical simulations, but an alternative situation happens in the case of the continuum approximation.

(c) Dispersion curves

Now, we consider linearized equations (5) about a uniform deformation S_0, where S_0 must satisfy one of the minima of the lattice potential. Solutions of the linearized discrete equations yield the dispersion curves depicted in Fig. 2 for different values of the lattice parameters α, β, δ and η, which places the importance of the competing interactions in evidence. Curve (a) differs slightly from the classical case (one-dimensional model of the monoatomic chain). The dispersion branch (b) has a small bend and curve (c) with a deeper bend undergoes a softening at *nonzero wave-number*, which can be interpreted as a *precursor of martensitic transformation* or premartenistic transition [5]. This softening is commonly observed by means of neutron inelastic scattering techniques [5], [6]. Finally, curve (d) exhibits an upward convexity, this situation is important for the existence of strain solitons.

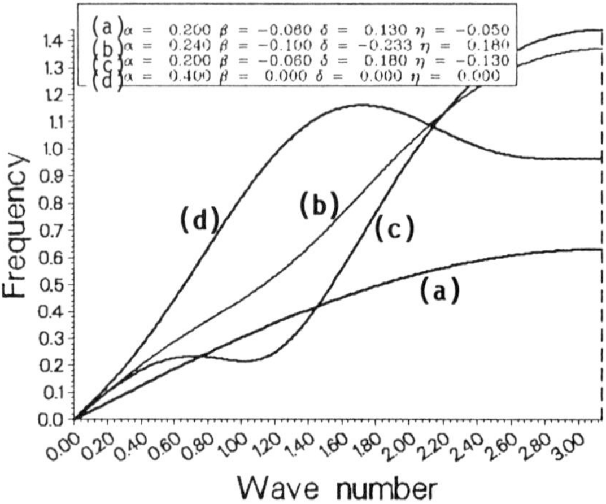

Figure 2. Dispersion curves for the transverse phonons in the [110] direction.

4. Continuum approximation

We assume now that the discrete functions (displacements, strains) are slowly varying over a lattice spacing. This means that we can find a continuous function $S(x, t)$, having derivatives with respect to x up to the fourth order, which is a *good interpolation* of the sampling $(nb, S_n(t))$ such that $S(x = nb, t) = S_n(t)$. After some classical algebra we arrive at

$$S_{,tt} = \Sigma_{,xx} \tag{6a}$$

where

$$\Sigma = \sigma - \chi_{,x} \tag{6b}$$

is the Cauchy stress, σ the macro-stress and χ the micro-stress. These stresses derive from an elastic potential as follows:

$$\sigma = \frac{\partial \psi}{\partial S} \quad \text{and} \quad \chi = \frac{\partial \psi}{\partial (S_{,x})}. \tag{7}$$

The elastic potential is given by

$$\psi(S, S_{,x}, \theta) = \frac{1}{2}AS^2 - \frac{1}{3}S^3 + \frac{1}{4}S^4 + \frac{1}{2}\gamma(S_{,x})^2 \tag{8}$$

where we have set

$$\gamma = \delta + 16\eta - \beta - A/12. \tag{9}$$

Eq. (6a) is a nonlinear dispersive partial derivative equation governing the continuous strain. The form of the elastic potential corresponds exactly to the Ginzburg-Landau free energy including a cubic term in strain which induces a first-order phase transition [10]. Note that the micro-stress is due to the noncentral interatomic interaction of the microscopic model. We can therefore compare the continuum approximation to the nonlinear elasticity exhibiting weakly nonlocal behaviour (strain gradient elasticity) [8].

Since we have introduced stresses we must define the corresponding boundary conditions. We will consider for numerical applications periodic boundary conditions which are compatible with the mechanical conditions on stresses and with the equation of motion. Moreover, because of the special type of motion $U = U_0$ and $V = V(x)$, we have an incompressible motion. On the other hand the strain compatibility conditions are also satisfied.

5. Nonlinear excitation solutions and strain solitons

Since we have in mind to examine the possible existence of nonlinear excitations, we will solve the set of equations (6a,b, 7-9) with appropriate boundary conditions. Subsonic strain solitons can travel for $c^2 < A$, but such nonlinear excitations exist if we have $\gamma > 0$, this condition is just that of the upward curvature of the dispersion curve. In addition, we have obviously other conditions on the soliton amplitude.

We consider now the fully dynamic processes. At this extreme the numerical scheme is provided by the set of discrete equations (5) with periodic boundary conditions at each end of the lattice. We begin with a martensitic soliton as initial conditions. The evolution of the wave is given in Fig. 3. This illustration represents a small layer of martensite (deformed lattice) moving in the parent phase (undeformed lattice). The corresponding deformed two-dimensional lattice is given in Fig. 4; the martensitic band is sweeping across the crystal in the stacking direction [110] [11]. The results obtained above concerning domain structures are similar to those of Falk, but the latter considers a symmetric free energy (the Landau condition is fulfilled) expanded up to the sixth order with respect to the strain [12].

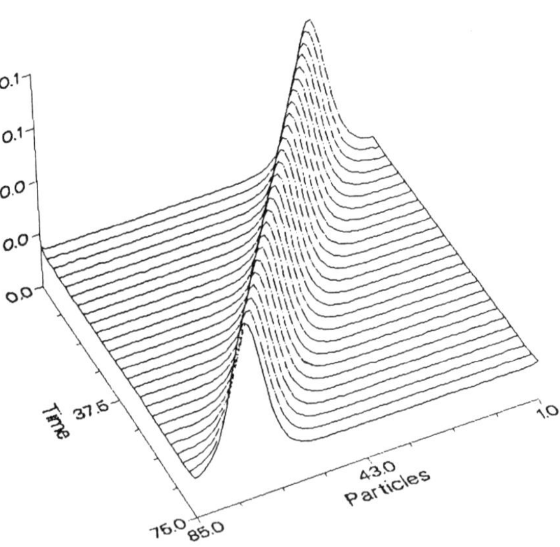

Figure 3. Martensitic soliton moving in an austenitic matrix.

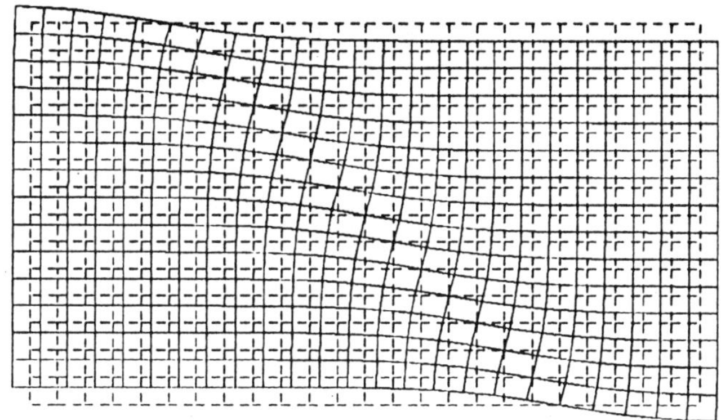

Figure 4. The corresponding lattice to the martensitic soliton sweeping across the crystal.

6. Two-dimensional problem

Now, we suppose that all the discrete quantities, i.e. strains and displacements, depend on i and j (or I and J if we use the (I, J) coordinate system). The lattice energy (1) is still valid, nevertheless we consider a simple case where the energy is a function of the discrete spherical and deviatoric parts of the strain tensor. In addition, we expand the lattice energy up to the fourth order in strain components and up to harmonic order for the first-order finite differences of the deformation. The resulting expression must satisfy the symmetry of the cubic-tetragonal transformation. A further step ito the simplification consists of considering a special case of transformation for which the lattice potential is a function only of the deviatoric part, say $S(i, j)$, and keeping the spherical part constant. As for the one-dimensional problem we construct the Lagrangian of the discrete system and the equations of motion are deduced from the Lagrangian. This yields

$$\ddot{S}(i,j) = \Delta_L^2 \Sigma(i,j) + \Delta_T^2 \Sigma(i,j) \tag{10}$$

where we have introduced

$$\Sigma(i,j) = \sigma(i,j) - \Delta_L \chi_L(i,j) - \Delta_T \chi_T(i,j) \tag{11a}$$

and

$$\sigma(i,j) = \alpha S(i,j) - S^2(i,j) + S^3(i,j) \tag{11b}$$

$$\chi_L(i,j) = \delta(S(i,j) - S(i-1,j)) \tag{11c}$$

$$\chi_T(i,j) = \delta(S(i,j) - S(i,j-1)) \tag{11d}$$

where Δ_L and Δ_T are the first-order finite differences in the i and j directions respectively, and Δ_L^2 and Δ_T^2 are the second-order finite differences in the same directions. Eq. (11a) represents the total discrete stress, where $\sigma(i,j)$, defined by Eq. (11b), is due to the linear and nonlinear parts of the lattice energy, and $\chi_L(i,j)$ and $\chi_T(i,j)$ are the discrete micro-stresses deduced from the noncentral interactions. Moreover, the moment of momentum balance is also satisfied when the continuum approximation is considered.

The problem is now to examine the existence of localized nonlinear coherent structures, for instance the formation of elastic domains. The idea is in fact, to see into the stability of a strain soliton given by the one-dimensional model (see the preceding sections) with respect to the transverse direction, i.e. [010] direction. We investigate the problem by means of numerical simulations. The numerical scheme is provided, once more, by the set of discrete equations (10, 11a–d) with periodic boundary conditions in both directions. Since the problem deals with a dynamical process we must consider initial conditions, which are given by the solution of the one-dimensional system extended to the transverse direction. Fig. 5 collects together the deformation state of the lattice for different times. The initial condition is presented in Fig. 5(a), which is a shaded-contour map of the strain in the lattice plane where each pattern corresponds to different gaps of strain strengths. We note that the initial condition which is a strain or martensitic soliton does not depend on the transverse variable. A little later, small perturbations are developing in both directions while the soliton is moving (see Fig. 5(b)). In particular, we observe that the strain amplitude is modulated with a period which seems to be the transverse length of the lattice. It turns out that the strain soliton is *no longer stable with respect to the transverse direction.* The instabilities are growing; this effect leads to a bifurcation of the initial solution and the strain soliton becomes a *localized elastic domain* in both directions (Fig. 5(c)).

However, the instability is not the result of discreteness effects but certainly the result of nonlinear instabilities monitored by the strain amplitude and lattice parameters. The analytical problem of such an instability process, which is a very complex study, will be presented elsewhere.

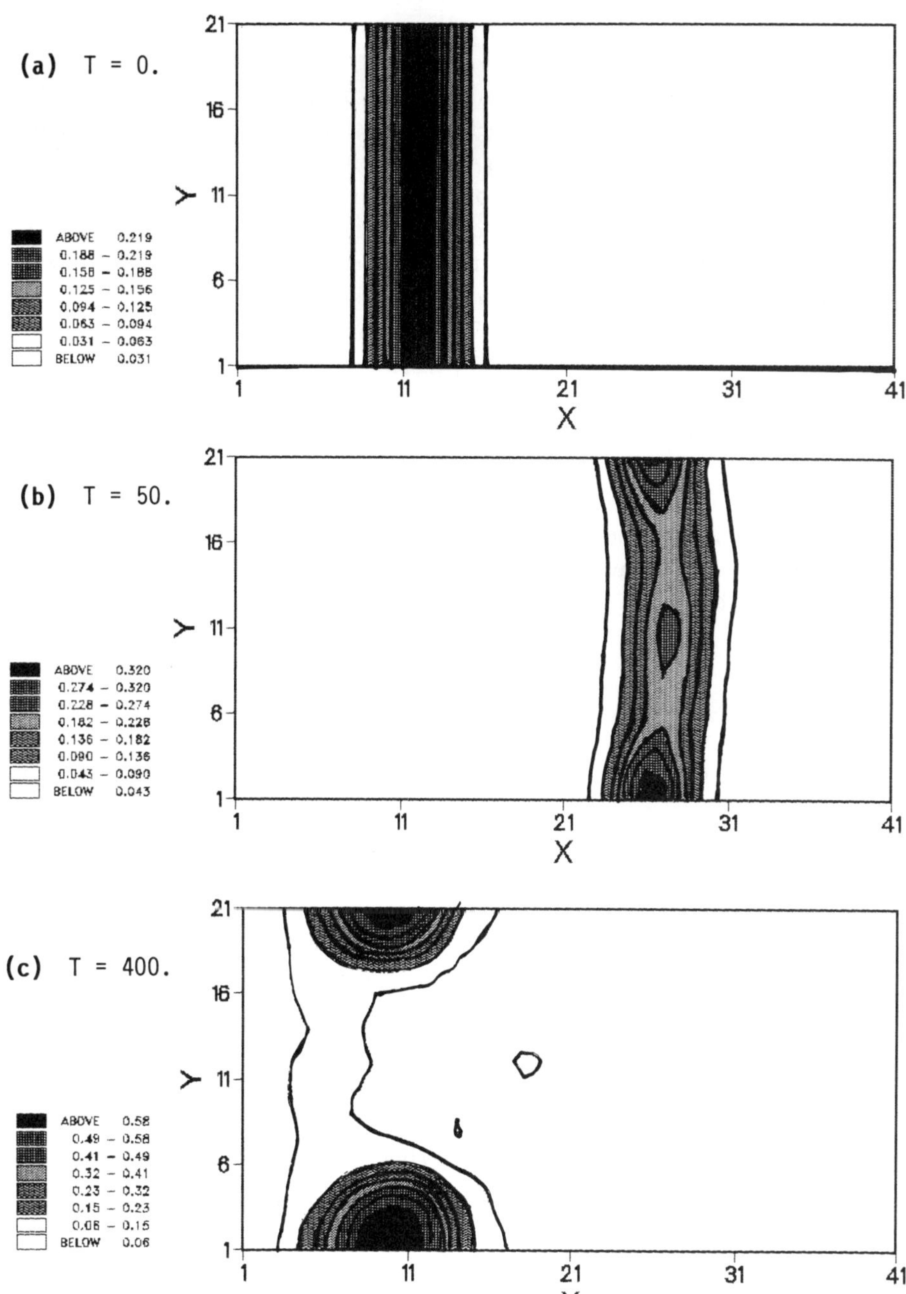

Figure 5. Instability process for the two-dimensional system: (a) initial condition, strain soliton, (b) small perturbations are developing in the transverse direction, (c) formation of a localized coherent structure (ferroelastic domain).

7. Conclusions

In the present work we have aimed at studying the pattern formation related to elastic domains as met in martensitic-ferroelastic transformations exhibiting first-order phase transition (the order parameter is a strain). With this in view, we have considered a lattice model on the basis of a two-dimensional lattice model possessing all the ingredients which allow for nonlinear excitation propagation. Important results have been pointed out which characterize some features of the transformation (softening of the transverse acoustic phonons, martensitic soliton as shearing motion of atomic planes, instability process). The patterns thus obtained correspond to the domain structures commonly observed by means of electron microscopy in various alloys such as Ni-Ti, Ti-Mn, In-TI, just to quote a few examples [3, 13, 14]. Nevertheless, further work can be investigated. We can study, on the basis of the lattice model, the discreteness effects or the ground states of deformation which minimize the lattice energy in the case of rather tiny martensitic domains. It is also worthwhile examining the influence of an external force and damping on the dynamics of such structures, leading to a particularly complex domain patterns ('zig-zag' shape twin bands).

References

[1] Wayman, C.M., 1981. *Proceedings of an Int. Conf. on Solid-Solid Phase Transformations* (The Metall. Soc. of AIME), p. 1119.

[2] Landau, L.D. and Lifshitz, E., 1980. *Statistical Physics*, Pergamon Press, Oxford.

[3] Christian, J.W. and Knowles, K.M., 1981. *Proceedings of an Int. Conf. on Solid-Solid Phase Transformations* (The Metall. Soc. of AIME), p. 1185.

[4] Mori, M., Yamanda, Y. and Shirane, G., 1975. *Solid State Comm.* **17**, 127.

[5] Shapiro, S.M., Lerese, J.Z., Noda, Y., Moss, S.C. and Tanner, L.E., 1986. *Phys. Rev. Letters* **57**, 3199.

[6] Finlayson, T.R., Mostoller, M., Reichardt, W., Smith, H.G., 1985, *Solid State Comm.* **53**, 461.

[7] Krumhansl, J.A., 1963, in *Lattice Dynamics* (ed. R.F. Wallis) Pergamon, Oxford.

[8] Mindlin, R.D. and Eshel, N.N., 1968, *Int. J. Solid Struct.* **4** 109.

[9] Gunton, D.J., Saunders, G.A., 1974. *Solid State Comm.* **14**, 865.

[10] Anderson, R.W., Blount, E.I., 1965. *Phys. Rev. Lett.* **14**, 217.

[11] Pouget, J., 1989. *Proceedings of the Symposium on Continuum Models and Discrete Systems* 6 (Longman, U.K., in press).

[12] Falk, F., 1983. *Z. Physik B-Condensed Matter* **51**, 177.

[13] Knowles, K.M., Christian, J.M. and Smith, D.A., 1982, *J. de Physique*, **43**, C4. 185.

[14] Tanner, L.E., Pelton, A.R. and Gronsky, R., 1982. *J. de Physique* **43**, C4. 169.

J. Pouget
Laboratoire de Modélisation en Mécanique,
Associé au C.N.R.S.
Université Pierre et Marie Curie,
Tour 66,
4, Place Jussieu
75252 Paris
FRANCE

M. SEISL, A. STEINDL AND H. TROGER

A numerical study of the transition to chaos for perturbed sine-Gordon equations

1. Introduction

Experimental findings ([1]) concerning the asymptotic time behaviour of infinite-dimensional dynamical systems, even if chaotic, show that often it can be described by the flow on a low-dimensional attractor. Hence, it is a question of great physical importance to specify those few active modes which allow representation of the attractor, and hence determine the dynamics.

To be able to approach the problem of determining the active modes in a more specific way we restrict our attention to the investigation of problems in which, under variation of one parameter through a series of successive bifurcations, a transition from a simple regular state to a complicated irregular one can occur. The simplest case where such transition phenomena can be studied are slightly perturbed bifurcation problems at multiple eigenvalues. For such problems already small parameter variations in the neighbourhood of a multiple bifurcation point can lead to a sequence of bifurcations. Centre manifold theory ([2]) can be used to investigate the interaction of competition of the various critical modes. This approach, however, fails if the parameter variations from the critical value are no longer small or if a system with large aspect ratio is given ([3]), even if the variations are kept small. For systems with large aspect ratio mode-selection principles must be introduced ([4, 5]) to be able to find the active modes.

There are two possibilities indicated in [3] how to study a transition to chaos for the case of large parameter variations. First by the use of a Galerkin approximation. However, it is well known from examples (Benard problem-Lorenz equations) ([3]) that often the *a priori* selected number of modes is the parameter that influences most seriously the behaviour of the truncated system. Probably the recently introduced concept of inertial and integral manifolds ([6, 7]) will lead to a substantial progress in the application of Galerkin approximations. The second possibility concerns a restricted class of systems, namely those which are close to integrable ones. The class of completely integrable systems is continuously increasing. This is due to the fact that the recently discovered soliton revolution produces a steadily increasing list ([8-12]). Now it is conjectured that these soliton solutions can be taken as active modes and, hence, the solutions of perturbed non-integrable cases can be obtained as superposition of the soliton solutions.

We want to discuss two different cases of perturbation of the sine-Gordon equation. The sine-Gordon equation is an integrable Hamiltonian system ([9-10])

and by means of a mechanical model of a chain of pendula an illustrative interpretation of the results can be given. The two types of perturbation are realized as external and parametric excitations, respectively. In both cases we study the attractors by a numerical analysis in order to show whether the conjecture made above is correct.

2. Mechanical model and governing equations

We consider the mechanical model of Fig. 1 which consists of a circular chain of pendula coupled to each other by elastic torsional springs. It is well known ([8]) that such a chain can be interpreted as a discrete model for the sine-Gordon equation

$$\varphi_{tt} - \varphi_{xx} + \sin \varphi = 0, \tag{1}$$

with periodic boundary conditions. The solutions of (1) can be obtained from a superposition of elementary soliton solutions. There are four basic solutions for (1)

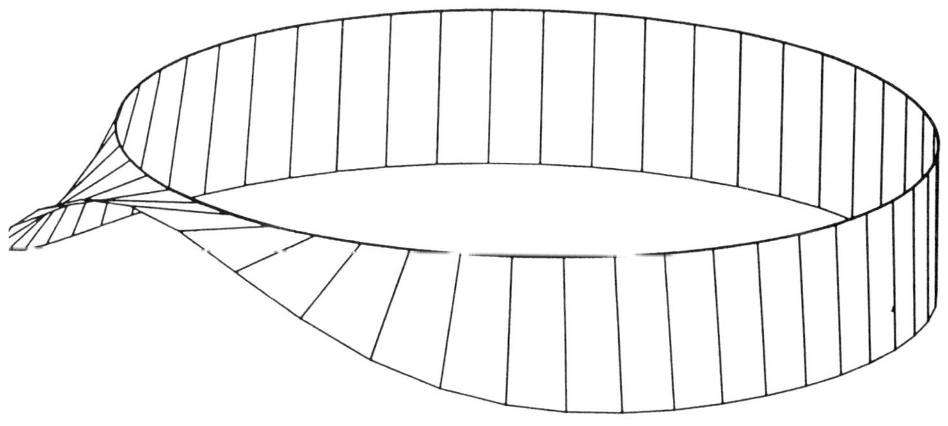

Figure 1. Circular chain of torsionally coupled pendula serving as a discrete mechanical model for (1). Shown is a breather oscillation.

on an infinite domain ([10]): (i) A radiation wave train, consisting of oscillations of the pendula with amplitudes smaller than π. (ii) Localized rotations by 2π, called kinks, and by -2π, called anti-kinks, propagating with constant speed. (iii) A breather, which consists of a kink-antikink pair bound together in an oscillatory state.

For the chain of pendula this means that almost all pendula are at rest and only a few perform a localized breathing oscillation (Fig. 1). (iv) A wave train, which consists of a periodic array of kinks and results in a helical type of rotation of the pendula.

As perturbations of (1) we consider two different excitation processes. First, a periodic external excitation, which gives rise to an equation

$$\varphi_{tt} - \varphi_{xx} + \sin\varphi = a\cos\omega t - \varepsilon b\varphi_t. \tag{2}$$

Again we have periodic boundary conditions. In addition also linear external damping is included.

Second, a periodic parametric excitation, which gives rise to an equation of the form

$$\varphi_{tt} - \varphi_{xx} + \varepsilon(b\varphi_t - c\varphi_{xxt}) + (1 + a\cos 2t)\sin\varphi = 0. \tag{3}$$

In (3) also internal damping due to the relative motion of two adjacent pendula is included. Again periodic boundary conditions hold.

The physical interpretation of (2) is given by moments acting at the suspension points of the pendula, whereas for (3) the ring is shaken periodically in the vertical direction.

3. Results

3.1. External excitation

Equation (2) has been extensively studied in [11, 12]. We use for the numerical analysis a Runge-Kutta procedure. The continuous structure is discretized in 160 points. The amplitude of the external excitation is increased quasistatically and the initial conditions are one stationary breather (Fig. 1). From Fig. 2 follows that after a few bifurcations a chaotic state is reached. However, it is remarkable that the behaviour not only changes in time but also in its spatial structure (Fig. 3). The growing complexity of the spatial structure still can be explained by a superposition of the elementary soliton solutions mentioned above. To make the behaviour in Fig. 3(f) in the chaotic regime better comprehensible the motion of two points is shown for a larger period of time in Fig. 4. Obviously it is of intermittent chaotic type.

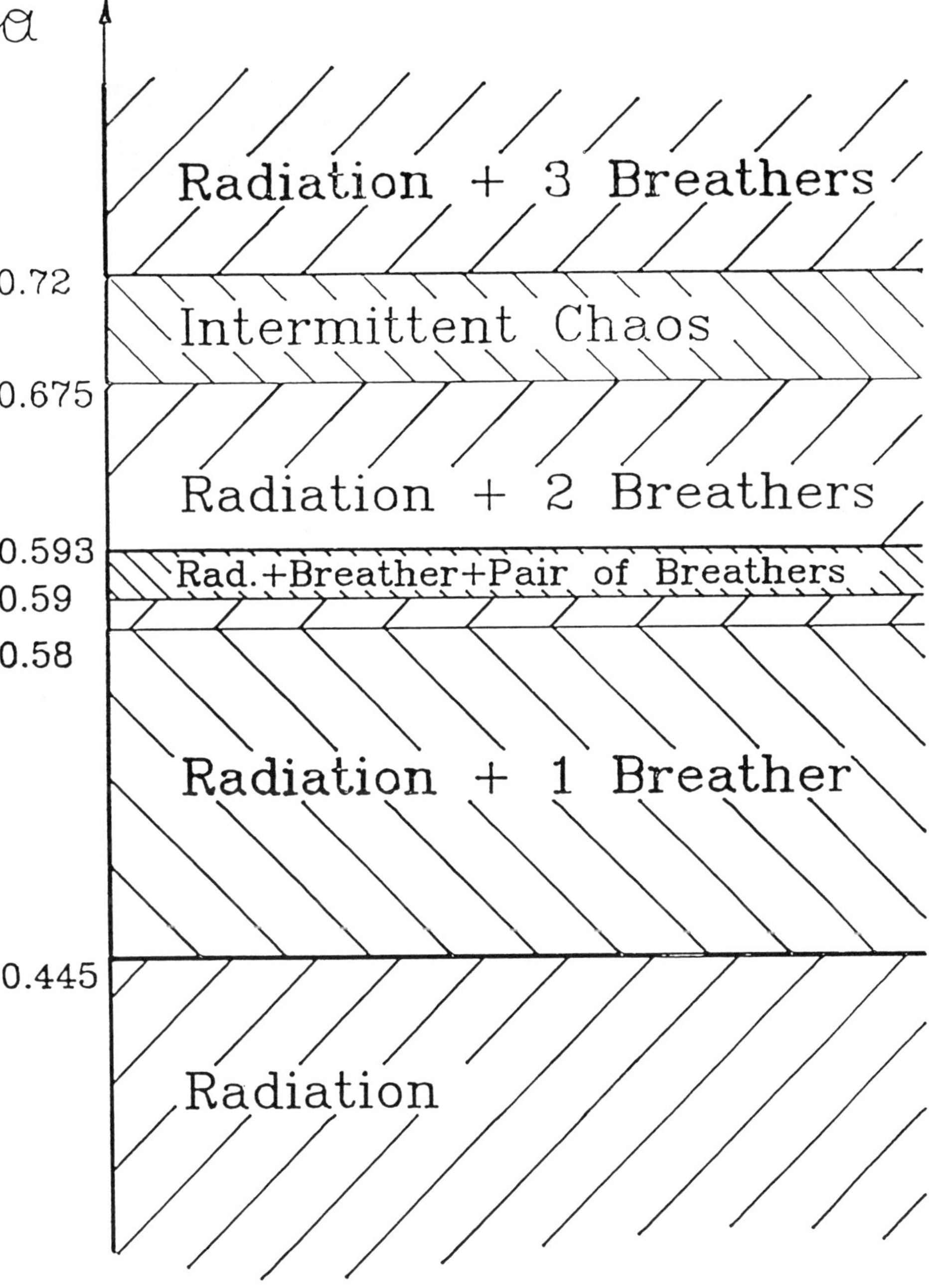

Figure 2. Bifurcation diagram describing the transition from a periodic to a chaotic behaviour for a circule chain under external excitation.

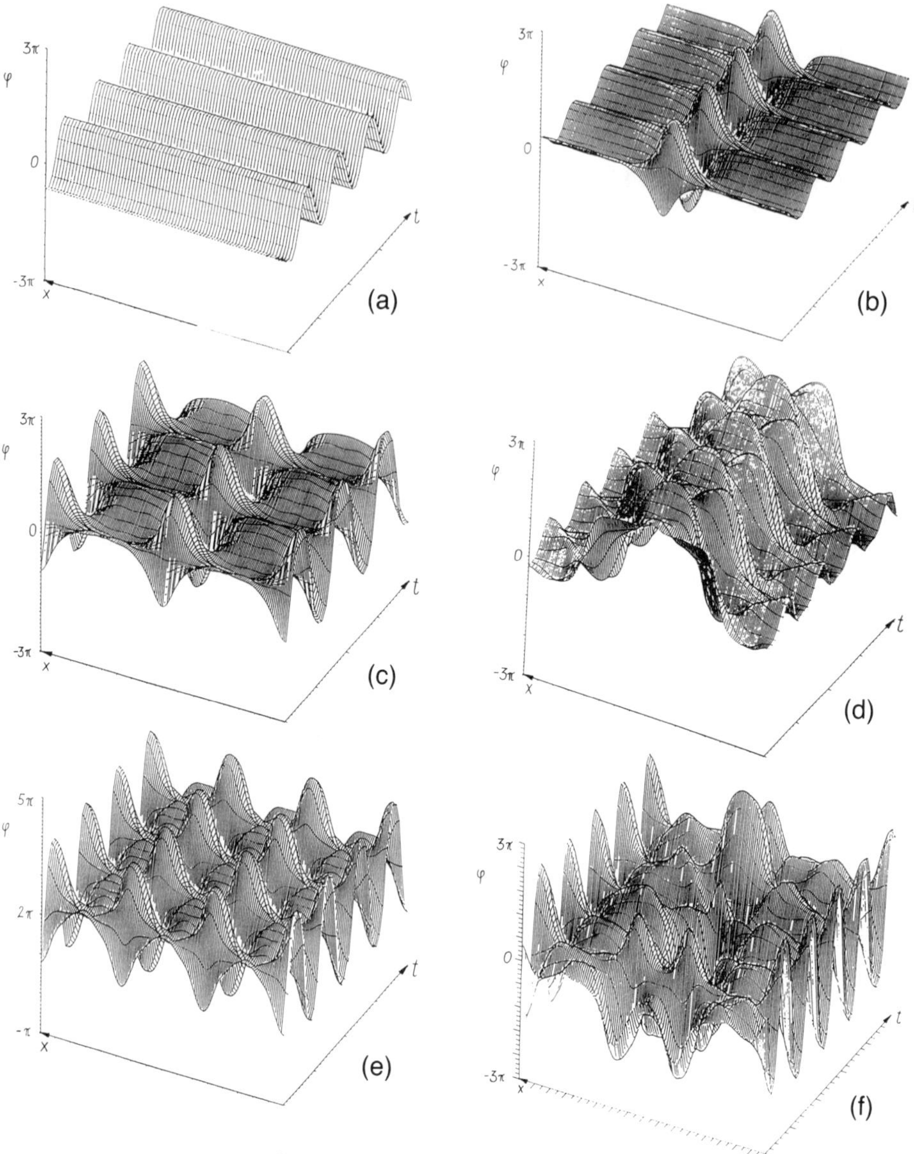

Figure 3. Motion of the whole chain.

(a) Flat state (radiation) ; all pendula oscillate in phase (a = 0.4 in Fig. 2) for $950 \leq t \leq 990$;

(b) one breather superposed to the oscillating flat state (a = 0.5) for $820 \leq t \leq 860$;

(c) two breathers superposed to (a) (a = 0.6) for $820 \leq t \leq 860$;

(d) one breather stationary in space and a pair of breathers in a periodic oscillating state (a = 0.591) for $940 \leq t \leq 980$;

(e) three breathers superposed to (a) (a = 0.8) for $820 \leq t \leq 860$;

(f) chaotic state (a = 0.7) for $450 \leq t \leq 500$.

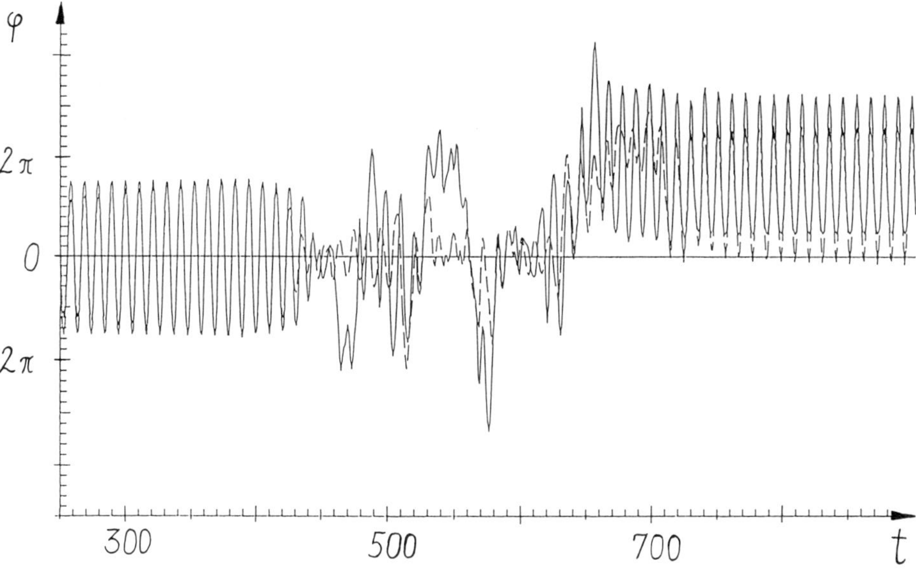

Figure 4. Intermittent chaotic oscillations of points 40 and 80 corresponding to Fig. 3(f) but for $250 \le t \le 900$.

3.2. Parametric excitation

The frequency of the vertical oscillations of the ring is twice the eigenfrequency. Thus, if the excitation amplitude a is big enough a parametrically excited oscillation of the ring which initially is flat will occur. That is, the spatial structure is flat and all pendula are vibrating in phase (comparable to Fig. 3(a)). This synchronous oscillation becomes unstable if a reaches the tongue in Fig. 5.

Now an oscillation which has period three in space (Fig. 6) is superimposed to the flat oscillation. We have studied this bifurcation by a local analysis. It corresponds to an eigenvalue $+1$ and is of the isola type (Fig. 7). This means that for increasing a further again the flat oscillation is found. The same behaviour then holds for bifurcations to solutions of period four and period five in space until a reaches the next boundary in the $a - \delta$ plane (Fig. 5) (δ is the wave number). Now a sudden onset of a chaotic solution is found. This can be seen from the Poincaré map of Fig. 8. Furthermore a positive Liapunov exponent was obtained.

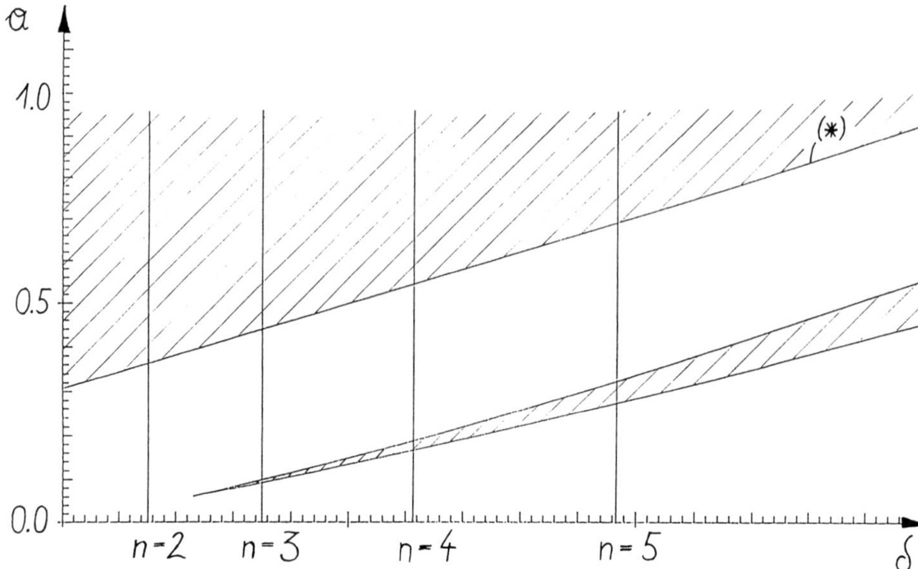

Figure 5. Stability chart for the flat oscillations of the parametrically excited pendulum chain, showing that the first instability of the flat vibration is of period three. If the second boundary (*) is reached suddenly chaotic behaviour is found (unstable domains are hatched).

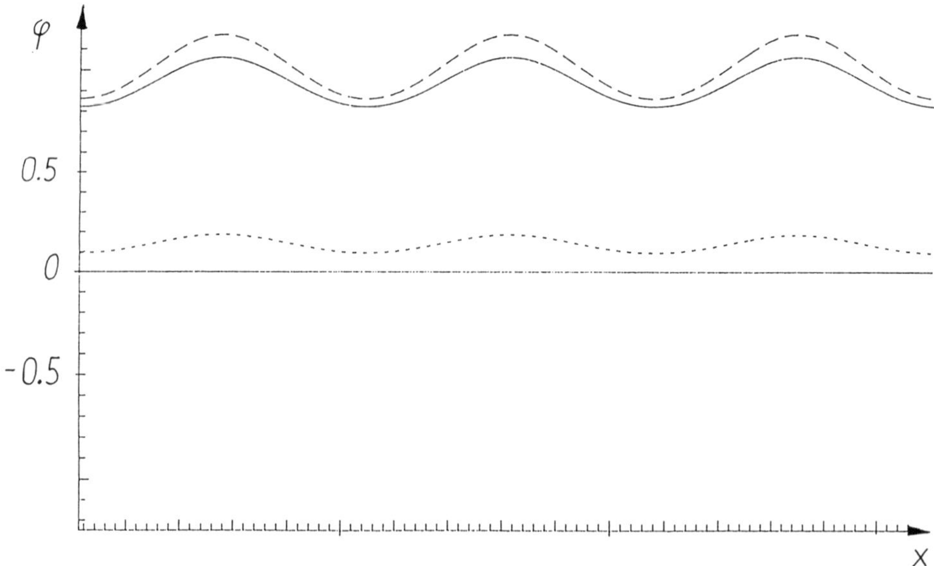

Figure 6. Period three oscillations corresponding to Fig. 5 (a = 0.094).

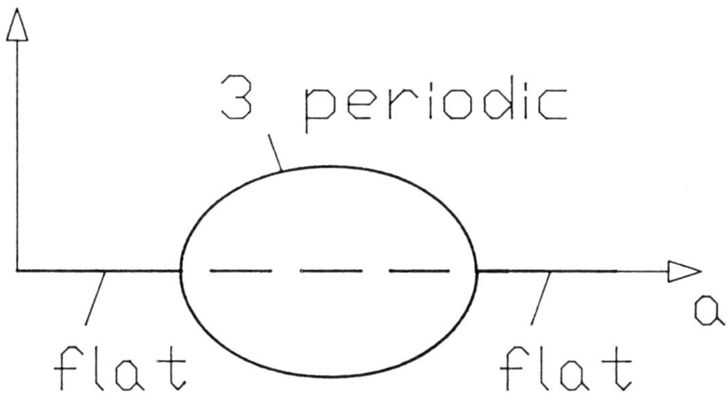

Figure 7. Isola bifurcation to the 3-periodic state corresponding to Fig. 6.

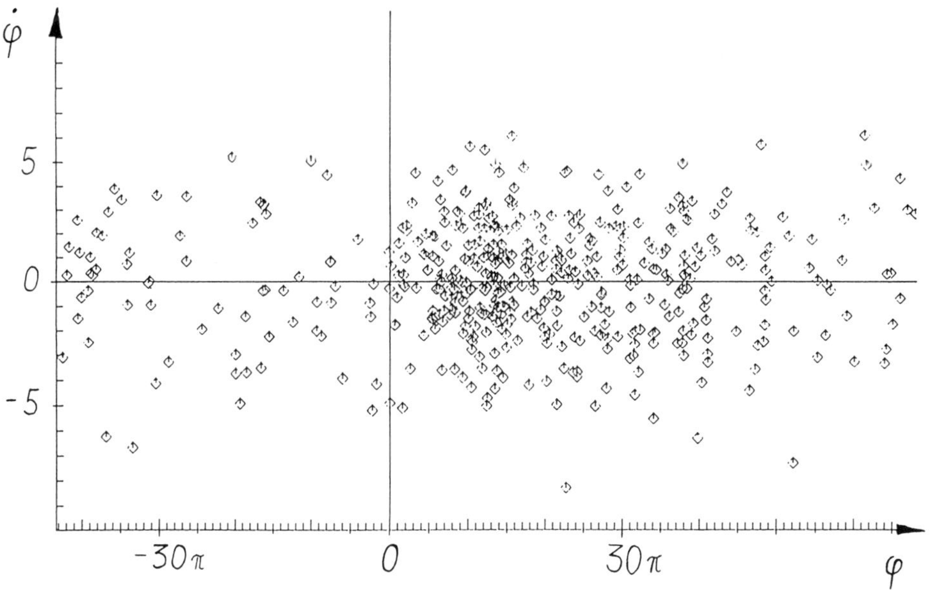

Figure 8. Poincaré section for the motion of one pendulum in the chaotic regime (a = 0.342).

4. Conclusions

Whereas in the case of external excitation the spatial structure in the transition to chaos can easily be understood from the basic soliton solutions; this does not seem to be the case for the parametric excitation. In the former case a nice scenario for the transition to chaos is given, because solutions with one, two and three breathers appear and only the occurrence of the third breather introduces enough spatial complexity to obtain time chaotic motions. It is further interesting to note that this clear picture is spoiled by a domain in the parameter space (Fig. 2) where a more complicated spatial structure (Fig. 3d) appears but still with a time periodic behaviour.

However, in the case of parametric excitation it is not possible to identify the soliton structure in the solutions - A fact which still deserves further investigations.

5. Acknowledgement

This research has been partly supported by the 'Fonds zur Förderung der wissenschaftlichen Forschung' in Austria under project P07003 PH.

References

[1] Swinney, H.L. and Golub, J.P. (eds.), *Hydrodynamic Instabilities and the Transition to Turbulence*. Topics in Applied Physics, vol. 45, Springer-Verlag, Berlin, Heidelberg, New York 1981.

[2] Carr, J. *Applications of Centre Manifold Theory*, Appl. Math. Sciences 35, Springer-Verlag, Berlin, Heidelberg, New York 1981.

[3] Aceves, A., Adachihara, H., Jones, Ch., Lerman, J.C., McLaughlin, D.W., Moloney, J.V. and Newell A.C., Chaos and Coherent Structures in Partial Differential Equations, *Physica* **18D** (1986), 85-112.

[4] Potier-Ferry, M., *Foundation of Elastic Postbuckling Theory*, Lecture Notes in Physics 288, Springer-Verlag, Berlin, Heidelberg, New York 1987, 1-82.

[5] Brusse, F.H., Patterns of convection in spherical shells, *J. Fluid Mechanics* **72** (1975), 67-85.

[6] Temam, R., *Infinite Dimensional Dynamical Systems in Mechanics and Physics*, Appl. Math. Sciences 68, Springer-Verlag, Berlin, Heidelberg, New York 1988.

[7] Constantin, P., Foias, C., Nicolaenko, B. and Temam, R., *Integral Manifolds and Inertial Manifolds for dissipative Partial Differential Equations*, Appl. Math. Sciences 70, Springer-Verlag, Berlin, Heidelberg, New York 1989.

[8] Scott, A.C., *Active and Nonliner Wave Propagation in Electronics*, Wiley Interscience, New York 1970.

[9] Drazin, P.G., *Solitons*, London Math. Soc. Lec. Notes Series 85, Camb. Univ. Press 1983.

[10] McLaughlin, D.W. and Scott, A.C., Perturbation analysis of fluxon dynamics, *Physical Review* A, **18**, (1978), 1652-1680.

[11] Bishop, A.R., Feser, K., Lomdahl, P.S. and Trullinger, S.E. Influence of solitons in the initial state on chaos in the driven, damped sine-Gordon system, *Physics* **7D** (1983), 259-279.

[12] Forest, M.G. and McLaughlin, D.W., Spectral theory for the periodic sine-Gordon equation: A concrete viewpoint, *J. Math. Phys.* **23** (7), 1982, 1248-1277.

M. Seisl, A. Steindl and H. Troger
Institute of Mechanics (E325)
Technical University of Vienna
Wiedner Hauptstraße 8-10
A-1040 Vienna
AUSTRIA

A.K. BELYAEV* * AND F. ZIEGLER

Traffic-noise-excited uniaxial waves in complex structures*

Abstract

By means of an integral method and an expansion technique in local modes of substructures, the mean vibrational response of a heterogeneous one-dimensional body is derived. A spectral decomposition and the WKB-approximation render the amplitude distribution of waves in a semi-infinite rod in closed form. Application to a railway-noise-excited tall building serves as a practical illustration of the powerful method. A majorant of the actual deformation which is independent fof the source intensity is derived as well.

1. Introduction

A slender heterogeneous structure is considered with respect to uni-axial wave propagation. The illustrating example is a high-rise building subjected to traffic-noise from railways. The spectral representation in conjunction with the WKB-method applies equally well to similar problems arising in space structures as well as in aircraft, rockets and ships. The structure is quite naturally subdivided into a finite number of substructures which are the elementary components with known sets of 'local' normal modes. Material damping is included by a model of dry friction. An extremely smooth function of gross displacement is singled out to produce an integral description of the vibrational field and the results for waves in a semi-infinite structure (no reflections are taken into account) are evaluated in closed form. Most importantly, a majorant of the deformation is determined which solely depends on the properties of the structure and the distance from the source of excitation. It is independent of the intensity of the noise input.

2. Boundary-value problem of an extended complex structure

An actual mechanical system rather complicated in its composition is considered. Fixing of the structural components to one another or to carrier surfaces results in the formation of a structure considerably heterogeneous in its rigidity and mass characteristics. Accurate determination of the vibration field of such structures is

---

* This paper is dedicated to Professor E. Kröner on the occasion of his 70th birthday.

** A.K. Belyaev: On leave 1988/89 from Leningrad Polytechnic Institute, USSR.

absolutely hopeless. In this paper we suggest an integral method based on the idea of finding some mean vibration field.

Let the structure under consideration be a body extending along the axis x. In the case of propagation of longitudinal waves it can be represented as a one-dimensional heterogeneous rod. Let us divide the structure of length L at the cross-sections x_n into substructures L_n $(n = 1,2,...,N)$. Substructures are elementary structural components. Thus the system of surfaces and substructures will be built up in a natural way. For instance, a storey of a tall building is such a separate substructure.

Let us consider a typical substructure L_n. The genuine displacement $a_n(x, t)$ is sought in the form of the expansion in terms of normal modes. A function $u(x, t)$ is extremely smooth with respect to x within the whole system L, and will be singled out in the following:

$$a_n(x, t) = u(x, t) + \sum_{k=1}^{\infty} u_{nk}(x) q_{nk}(t), \quad x \in L_n. \tag{2.1}$$

Here $q_{nk}(t)$ are generalized coordinates, and normal modes $u_{nk}(x)$ are assumed to satisfy the zero kinematic conditions on the substructures' boundaries. Hence, the function u coincides on the substructures' boundaries with the genuine displacement, but it has this property of extreme smoothness and may be called the displacement of the carrier structure.

The kinetic energy of the structure with consideration of [1] allows the representation

$$T = \frac{1}{2} \sum_{n=1}^{N} \int_{L_n} m\dot{a}_n^2 \, \mathrm{d}x - \frac{1}{2} \sum_{n=1}^{N} \sum_{k=1}^{\infty} \dot{q}_{nk}^2 + \sum_{n=1}^{N} \sum_{k=1}^{\infty} \dot{q}_{nk} \int_{L_n} m\dot{u} u_{nk} \, \mathrm{d}x + \frac{1}{2} \int_L m\dot{u}^2 \, \mathrm{d}x, \tag{2.2}$$

where m is the mass per unit of length, and normal modes are assumed to be orthonormal within each substructure. In view of the extreme spatial smoothness of u and the essential heterogeneity of the structure the next estimation proved to be correct:

$$\int_L m\dot{u}^2 \, \mathrm{d}x = \sum_{n=1}^{N} (\dot{u})_n^2 \int_{L_n} m \, \mathrm{d}x = \sum_{n=1}^{N} \langle m \rangle (\dot{u})_n^2 L_n = \int_L \langle m \rangle \dot{u}^2 \, \mathrm{d}x, \tag{2.3}$$

where $\langle m \rangle = \frac{1}{L} \int_L m \, \mathrm{d}x$ is the average density of the whole structure. The right-hand equality in (2.3) expresses the standard transition from the Riemann-Stieltjes sum to the corresponding integral. This transition is quite admissible for large N. In the second summand in (2.2) $m(x)$ and $u_{nk}(x)$ are highly oscillating functions of x, while u is an extremely smooth one. If we introduce the average displacement of the centre of mass of the substructure of length L_n, when it moves according to mode k:

$$\langle u_{nk}\rangle = \frac{1}{\langle m\rangle_n L_n} \int_{L_n} m u_{nk}\, dx,$$

the kinetic energy (2.2) can be rewritten as

$$T = \frac{1}{2} \sum_{n=1}^{N} \sum_{k=1}^{\infty} (\dot{q}_{nk}^2 + 2\langle m\rangle_n L_n \langle u_{nk}\rangle \dot{q}_{nk}\dot{u}) + \frac{1}{2} \int_L \langle m\rangle \dot{u}^2\, dx. \tag{2.4}$$

The representations of the potential energy and of the work of the external loads can be obtained analogously [1]. After that, using the Hamiltonian variational principle one obtains the following boundary-value problem

$$\langle EF\rangle u'' - \langle m\rangle(\ddot{u} + \sum_{k=1}^{\infty} \langle u_{nk}\rangle \ddot{q}_{nk}) + h = 0, \quad x \in L_n, \tag{2.5}$$

$$\ddot{q}_{nk} + \omega_{nk}^2 q_{nk} = p_{nk} - \langle m\rangle_n \langle u_{nk}\rangle \ddot{u}, \tag{2.6}$$

$$x = 0 - \langle EF\rangle u' = f_0(t); \quad x = L\colon \langle EF\rangle u' = f_L(t). \tag{2.7}$$

Here E is Young's modulus, F is the cross-sectional area, p_{nk} are generalized forces, h is a load per unit of length. The ends $x = 0$ and $x = L$ are assumed to be under the corresponding external forces f_0 and f_L.

The boundary-value problem (2.5)-(2.7) has a considerable shortcoming, i.e., it does not take into account the internal friction in the structural material and the dry friction between the components of the structure. Both types have a definite nonlinear character. To describe the friction we shall use the theory of the Ishlinsky-elastoplastic-material [2]. Its rheological model is composed of an infinite number of elements each having an elastic spring and a dry damper. The behaviour of the Ishlinsky-material in case of uniaxial deformation is described in the following way:

$$\sigma = E\varepsilon - E \int_0^{\infty} \varepsilon_y R(y)\, dy; \quad y \,\mathrm{sign}\, \dot{\varepsilon}_y + \varepsilon_y = \varepsilon. \tag{2.8}$$

Here σ and ε are the stress and deformation of the material, ε_y is the plastic deformation in the element, $R(y)$ is the density of distribution of the dimensionless yield strength y.

Keeping in mind the desire to create the mathematical apparatus that would be equally suitable for investigating stationary and nonstationary, sure and random dynamic processes in complex structures [3], let us make use of the following spectral decompositions:

$$u(x,t) = \int_{-\infty}^{+\infty} u(x,\omega)e^{i\omega t}\,d\omega;\; f_0(t) = \int_{-\infty}^{+\infty} f_0(\omega)e^{i\omega t}\,d\omega;\; q_{nk}(t) = \int_{-\infty}^{+\infty} q_{nk}e^{i\omega t}\,d\omega, \text{ etc.,} \tag{2.9}$$

where from now on the same designations for the spectra will be retained. Linearization by the describing function method of the essential nonlinear dependence (2.8) results in the complex Young's modulus [2]:

$$E_c = E\left[1 - \int_0^1 \left(1 - \eta^2 - i\eta\sqrt{1-\eta^2}\right) R\left(\frac{\pi a\eta}{4}\right)\frac{\pi a}{4}\,d\eta\right] = E(1 + i\chi)^2. \tag{2.10}$$

Here a is the deformation amplitude and, since $\chi(a)$ is small, the asymptotic estimation applies:

$$\chi(a) = \frac{1}{2}\int_0^1 \eta\sqrt{1-\eta^2}\, R\left(\frac{\pi a\eta}{4}\right)\frac{\pi a}{4}\,d\eta. \tag{2.11}$$

Obtaining q_{nk} from (2.6) and substituting it into (2.5) we get the following differential equation:

$$\langle E_c F\rangle u'' + \omega^2 M(\omega)u + h_e = 0, \tag{2.12}$$

where

$$h_e = h + \langle m\rangle\omega^2 \sum_{k=1}^{\infty} \frac{p_{nk}\langle u_{nk}\rangle}{-\omega^2 + \omega_{nk}^2(1+i\chi)^2},$$

$$M(\omega) = \langle m\rangle\left[1 + \langle m\rangle\omega^2 \sum_{k=1}^{\infty} \frac{\langle u_{nk}\rangle^2}{-\omega^2 + \omega_{nk}^2(1+i\chi)^2}\right]. \tag{2.13}$$

Here h_e is the spectrum of the effective load per unit of length. $M(\omega)$ is called the mass inertia per unit of length [1]. If the condition

$$\Delta\omega_{nk} = |\,\omega_{nk+1} - \omega_{nk}\,| \le \chi\,|\,\omega_{nk+1} + \omega_{nk}\,| \text{ or } \frac{\Delta\omega_{nk}}{\omega_{nk}} \le 2\chi \tag{2.14}$$

is satisfied, i.e., if the eigenfrequencies of the substructure are located so densely that the sum in (2.13) can be replaced by the integral with a locally smooth distribution function of the eigenfrequencies $\Phi(\alpha)$, then

$$M(\omega) = \langle m \rangle \left[1 + \omega^2 \int_0^\infty \frac{\Phi(\alpha)d\alpha}{-\omega^2 + \alpha^2(1+i\chi)^2} \right]. \tag{2.15}$$

We assume that the spectral properties are identical in all cross-sections of the structure, since more information can be obtained only as a result of some experiments of harmonic wave propagation in the whole structure.

The condition (2.14) tells us that for each structure, characterized by a certain relative density of the natural frequency spectrum, there is a certain critical value of damping. If it is exceeded, then the structure acts as a mechanical system with a continuous spectrum of natural frequencies.

3. Wave propagation in a semi-infinite rod

An extended complex structure may be considered as a semi-infinite rod, if

(a) it has a continuous spectrum of eigenfrequencies, and if
(b) it is possible to neglect the waves, reflected from the other end of the structure.

These conditions are satisfied in the high-frequency domain. The absence of body forces is assumed $(h = 0)$ and the boundary-value problem (2.12), (2.7) is transformed into the following form

$$x > 0: \quad \langle E_c F \rangle u'' + \omega^2 M(\omega) u = 0 \tag{3.1}$$

$$x = 0: \quad -\langle E_c F \rangle u' = f_0. \tag{3.2}$$

The differential equation (3.1) is actually nonlinear, since χ and $M(\omega)$ depend on the amplitude deformation a, which is a function of x itself. Let a be a slowly varying function of x: Then $\chi(a)$ and $M(\omega)$ have the same property, because in (2.11), (2.15) $a(x)$ is replaced in the integrand. It means the further smoothing of the dependence on x. For smooth functions of x the effective solution may be found using the WKB method [4]. A new variable y and a new unknown function $U(y)$ are introduced in the following manner

$$y = \int_0^x \sqrt{\frac{M(\omega)}{\langle E_c F \rangle}}\, dx, \quad \mathrm{Im}\, y < 0; \quad u(x) = \langle E_c F \rangle^{-1/4}\, U(y). \tag{3.3}$$

Instead of (3.1) we will have

$$\frac{d^2 U}{dy^2} + \left[\omega^2 - \langle E_c F \rangle^{-1/4} \frac{d^2}{dy^2} \langle E_c F \rangle^{1/4} \right] U = 0. \tag{3.4}$$

In the high-frequency domain the addend may be neglected in comparison with augend. Hence, the displacement $u(x)$, decreasing for $x \to \infty$, has the following form:

$$u(x) = D \langle E_c F \rangle^{-1/4}\, e^{-i\omega y}, \tag{3.5}$$

where D is still to be determined. Expression (3.5) is not the final solution, because E_c and y depend on a, but the latter is to be found through the displacement u. So the result (3.5) at the most may be considered as the equation which determines a. Obtaining a^2 we get

$$a^2 = \omega^2 D^2 \langle E_c F \rangle^{-3/2} \,|\, M(\omega) \,|\, e^{2\omega\, \mathrm{Im}\, y}. \tag{3.6}$$

The resulting equation is an integral one, because of the dependences (2.10), (2.15). Taking the logarithmic derivative from a^2 gives

$$\frac{a'}{a} = \omega\, \mathrm{Im} \sqrt{\frac{M(\omega)}{\langle E_c F \rangle}}. \tag{3.7}$$

While obtaining (3.7), terms of χ^2, $\chi\chi'$ etc. have been neglected since they are asymptotically small. If we use the following representation [1]

$$M(\omega) = \langle m \rangle \, [\, 1 - i\, \kappa(\omega) \,]^2 \tag{3.8}$$

then, after comparing (2.15) and (3.8) the imaginary part becomes the infinite integral

$$\kappa(\omega) = \chi\omega^2 \int_0^\infty \frac{\alpha^2 \Phi(\alpha)\, d\alpha}{\left[-\omega^2 + \alpha^2 (1 - \chi^2) \right]^2 + 4\alpha^4 \chi^2}. \tag{3.9}$$

The small value of $\chi(\chi << 1)$ and local smoothness of $\Phi(\alpha)$ are assumed, hence, the integral (3.9) can be determined by the methods of the theory of random vibrations [4], i.e.

$$\kappa(\omega) = \frac{\pi\omega\Phi(\omega)}{2}. \tag{3.10}$$

It obviously follows that $\kappa(\omega)$ does not depend on a. Taking into account the explicit expression for $\chi(a)$ (2.11), we obtain a nonlinear differential equation

$$\frac{a'}{a} = -\frac{\omega}{c}\left[\kappa(\omega) + \chi(a)\right] = -\frac{\omega}{c}\left[\kappa(\omega) + \frac{1}{2}\int_0^1 \eta\sqrt{1-\eta^2}\, R\left(\frac{\pi a\eta}{4}\right)\frac{\pi a}{4}\mathrm{d}\eta\right], \tag{3.11}$$

where $c = \sqrt{\dfrac{<EF>}{<m>}}$, is the velocity of the energetic centre of the propagating disturbance.

We borrow the distribution function $R(z)$ from the theory of internal friction, i.e., $R(z) = \beta H z^{\beta-1}$ $(H > 0,\ \beta > 0)$ [2], [6]. Evaluation of the integral in (3.11) renders $\chi(a) = g a^{\beta}$, where $g = \dfrac{\beta H}{2}\left(\dfrac{\pi}{4}\right)^{\beta} B\left(\dfrac{\beta+1}{2}, \dfrac{3}{2}\right)$, $B(\,;\,)$ being the Eulerian beta-function. The equation (3.11) will look like

$$\frac{a'}{a} = -\frac{\omega\kappa}{c} - \frac{\omega g}{c} a^{\beta}. \tag{3.12}$$

The explicit solution of the equation (3.12) is the following:

$$a(x) = \left[\left(a(0)^{-\beta} + \frac{g}{\kappa}\right)\exp\left(\frac{\beta\omega\kappa}{c}x\right) - \frac{g}{\kappa}\right]^{-1/\beta}. \tag{3.13}$$

Using the boundary condition (3.2) we immediately get $a(0) = |f_0| <EF>^{-1}$ $(\chi << 1)$. Hence, the final expression for $a(x)$ may be rewritten in the following way:

$$a(x) = \left\{\left[\left(\frac{|f_0|}{<EF>}\right)^{-\beta} + \frac{g}{\kappa}\right]\exp\left(\frac{\beta\omega\kappa}{c}x\right) - \frac{g}{\kappa}\right\}^{-1/\beta}. \tag{3.14}$$

4. The principal regularities of propagation of the uniaxial waves

The analysis of the dependence $a(x)$ makes it possible to determine the basic tendencies of the dynamic process in complex structures. For large values of x the addend in the figure bracket can be neglected in comparison with the augend and then

$$a(x) = \left[\left(\frac{|f_0|}{<EF>} \right)^{-\beta} + \frac{g}{\kappa} \right]^{-1/\beta} \exp\left(-\frac{\omega\kappa}{c} x\right) = \frac{|f_0|}{<EF>} \left\{ 1 + \frac{\chi[a(0)]}{\kappa} \right\}^{-1/\beta} \exp\left(-\frac{\omega\kappa}{c} x\right). \tag{4.1}$$

If the losses of the mechanical energy of the propagating wave on the plastic deformations and on the dry friction of the structural elements are negligible in comparison with the resonance absorption $(\chi(a) \leq \chi[a(0)] << \kappa)$, then

$$a(x) = \frac{f_0}{<EF>} \exp\left(-\frac{\omega\kappa}{c} x\right). \tag{4.2}$$

It is seen from the formula obtained that the absorption of the energy is determined by the value of $\kappa(\omega)$. The latter is given through the function of the eigenfrequency distribution $\Phi(\omega)$ (see (3.10)). Hence, the internal degrees of freedom of each substructure correspond to a set of dynamic absorbers with respect to the carrier structure, thus providing considerable spatial absorption of the energy of propagating waves in the whole high-frequency domain.

If the condition $\chi << \kappa$ is violated, the structure exhibits the evident nonlinear properties. For example, for any amplitude of the external force the following inequality holds:

$$a(x) < \left\{ \frac{g}{\kappa} \left[\exp\left(\frac{\beta\omega\kappa}{c} x \right) - 1 \right] \right\}^{-1/\beta} = a_m(x). \tag{4.3}$$

The last formula indicates the limit of the level of amplitude of deformation in any point of the structure even if the power of the external source of excitation grows beyond any bounds. Hence, this level does not depend on the source power, but is a function only of the spatial distance from the source and of the mechanical characteristics of the structure. It means that for any complex structure we can plot some universal curve which is a majorant of the curve of the amplitude of the actual deformation.

5. An example: vibration of a tall building under railway-noise excitation

As was shown in [7], [8], railway vehicles generate mainly the frequencies 30–125 Hz. Let us make up the mind for the frequency 50 Hz ($\omega = 2\pi \times 50$ rad/s). A tall building having the following parameters: $c = 10^3$ m/s, $\kappa = 0.2$ [2] will be under consideration. Yield strength y is assumed to be uniformly distributed, i.e. $\beta = 1$. Hence, $\chi(a) = ga$. Let us put $g = 5$, which corresponds to internal friction $\chi = 5 \times 10^{-2}$, while the deformation amplitude $a = 10^{-2}$. The result of the numerical calculation of $a(x)$ (3.14) for three values of $a(0)$ is depicted in Fig. 1 as well as the majorant $a_m(x)$ (4.3).

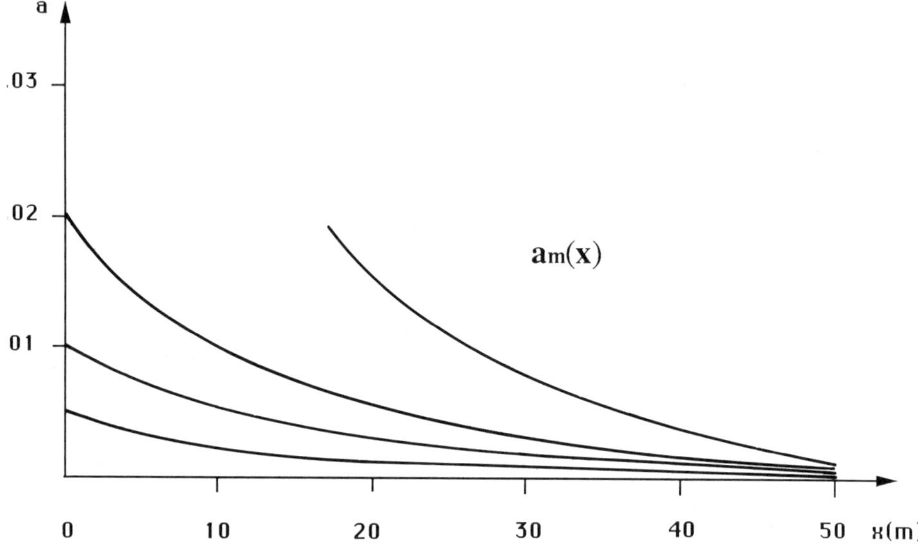

Figure 1. The curves of actual deformation and the majorant.

6. Concluding remarks

Uniaxial waves in complex structures are considered. A very specific region of the high-frequency vibration is shown to exist for each actual structure. In this domain the resonance curves of substructures merge, thus, the structure acts as a mechanical system with a continuous spectrum of eigenfrequencies. The internal degrees of freedom of secondary structures correspond to a set of dynamic absorbers with respect to the primary structure. It means that the secondary structures may be damaged. The results obtained may prove to be useful for predicting the vibration of secondary structures. It is worth also to mention the existence of a certain universal curve, which is a majorant of the amplitude of the actual deformation. It may turn out to be very convenient, since no information about the intensity of the external loads is needed.

References

[1] Belyaev, A.K. and Palmov, V.A. Integral theories of random vibration of complex structures. In: I. Elishakoff and R.H. Lyon (Eds.). *Random Vibration - Status and Recent Developments,* Elsevier, Amsterdam, 1986, pp. 19–38.

[2] Palmov, V.A. *Vibrations of Elastoplastic Bodies* (in Russian), Nauka, Moscow 1976.

[3] Ziegler, F. Random vibrations: A spectral method for linear and nonlinear structures. *Journal of Probabilistic Engineering Mechanics*, 1987, **2** (2), 92–99.

[4] Heading, J. *An Introduction to Phase-Integral Methods*, Wiley, New York, 1962.

[5] Bolotin, V.V. *Statistical Methods in Structural Mechancis,* Holden-Day, San Francisco, 1969.

[6] Kröner, E. *Statistical Continuum Mechanics.* International Centre for Mechanical Science, Courses and Lectures. Springer-Verlag, Vienna and New York, 1972.

[7] Flesch, R. Schwingungen im Wohn- und Industriebau. In: M. Steinwachs (Hrsg.). *Ausbreitung von Erschütterungen im Boden und Bauwerk.* 3. Jtg. DGEB, Trans. Tech. Publications Clausthal, 1988.

[8] Flesch, R., Schoitsch, G. and Fischbach, A. *Erschütterungs- und Körperschallisolation im Eisenbahnbau - Ergebnisse der Versuchsreihe Kledering*. Manuscript, BVFA-Vienna, 1986.

A.K. Belyaev
Department of Dynamics & Strength of Machines
Leningrad Politechnic Institute
Politeckhnicheskaya, Dom, 2
Leningrad 195251
USSR

F. Ziegler
Technical University of Vienna
Institut für Allgemeine Mechanik (E201)
Wiedner Hauptstraβe 8-10
A-1040 Wien
AUSTRIA

Part III
Asymptotic Theory of Viscous Flow

M.A. GOLDSHTIK AND V.N. SHTERN

Asymptotic analysis of collapse phenomenon in viscous flows

Such a fluid property as viscosity usually causes a smoothing of a velocity field even if initially it has been irregular. But in some special problems of steady motion of viscous fluid, the velocity field becomes singular at a finite value of the Reynolds number. A convergent motion yields such a strong cumulation of momentum and vorticity that the velocity becomes infinite. This phenomenon ('collapse') is found in a few conically similar flows, heat convection problems and some MHD flows. Here these problems are studied numerically and by asymptotic methods. For the near collapse situation the inner and outer expansions are given and analytic estimates of solutions and critical parameters are performed. Ways of overcoming the paradox are discussed and some speculations are outlined to model jet-like flows in the ocean, atmosphere and space by the conical solutions.

1. Introduction

Non-smooth solutions of the Euler equations are well known. But solutions of the Navier-Stokes equations are regular inside the flow region as a rule. A general theorem on the existence of a smooth unsteady three-dimensional solution of the Navier-Stokes equations has not been proved yet and hence it is not excluded that an initially regular velocity field may become singular at a finite time [1]. But a singularity appearance in a steady axisymmetric problem seems to be paradoxical at the first sight. Such a paradox has been found and mathematically studied in the problem of interaction of a rigid plane with a vortex line [2]. It has been proved that at a Reynolds number (which is a circulation/vorticity ratio) being less than $\mathrm{Re}_* = 5.53$, a solution exists having a bounded axial velocity and at $\mathrm{Re} > \mathrm{Re}_*$ there is a singularity inside the flow region. Serrin [3] has generalized the problem allowing a logarithmic singularity of longitudinal velocity at the symmetry axis. Introduction of a new parameter P characterizing the singularity makes it possible to model a rather complex kinematics of natural tornadoes. But in the Serrin problem also the collapse occurs at some values of Re and P.

The simplest example of the collapse may be exposed by the Squire analytic solution [4]. Squire has interpreted his solution as a jet emerging from a hole in a plane wall. But Schneider [5] and Zauner [6] have shown theoretically and experimentally that such an interpretation is wrong. The adherence condition is not satisfied by the Squire solution, which significantly alters the flow pattern. There is another interpretation [7, 8] of the Squire solution: the jet-like flow is driven by convergent motion of the plane matter. Such situations happen in natural circumstances in the ocean, atmosphere and space. In this interpretation the motion driver is measured by the Reynolds number, which is a surface discharge/viscosity ratio. Then it follows form the analytic solution that at $Re > Re_* = 7.67$ the singularity apears inside the flow region.

Sozou [9] has found a similar phenomenon in a problem of a conducting fluid flow induced by an electric current discharge placed at the fluid surface. Fluid flows toward the electrode near the surface and then outflows to infinity forming a near-axis jet. At a finite value of the current, the axial velocity becomes infinite and the regular solution loses its existence. The collapse also has been found in a few problems of heat convection in frames of the Boussinesq equations [10]. All mentioned examples belong to a conical self-similar class of viscous fluid flows in which velocity is inversely proportional to the distance from the origin.

The physical mechanism of momentum and vorticity cumulation near the symmetry axis is rather clear. The convergent motion induces a convective transport to the axis but viscous diffusion induces a transfer from the axis. If the motion is intense enough, then this causes a concentration of momentum and vorticity near the axis. But the fact that diffusion may not balance convection and the collapse occurs at finite values of the Reynolds number seems to be paradoxical. Possible ways of overcoming the paradox are discussed here.

2. Conical class

Steady flows of a viscous incompressible fluid are considered, for which velocity and pressure fields have the representation in spherical coordinates (r, θ, φ)

$$v_r = -\frac{\nu}{r}\,y(x),\ v_\theta = -\frac{\nu}{r}\frac{y(x)}{r\sin\theta},\ v_\varphi = \frac{\nu\Gamma(x)}{r\sin\theta},\ p = p + \frac{\rho\nu^2}{r^2}q(x),\ x = \cos\theta. \quad (1)$$

Here ρ, ν are density and kinematic viscosity, prime denotes differentiation with respect to x. After substituting (1) in the Navier-Stokes equations and some simple transformations the ODE system is obtained

$$(1 - x^2)y' + 2xy - y^2/2 = F\,; \quad (2)$$

$$(1 - x^2)\Gamma'' = y\Gamma'\,; \quad (3)$$

$$F''' = 2\Gamma\Gamma'/(1 - x^2) + f_r' - 2f_\theta/\sin\theta. \quad (4)$$

Here f_r, f_θ correspond to an outer force which for the sake of self-similarity must have the representation $\rho\nu^2 r^{-3}\,\mathbf{f}(x)$ and in our case $f_\varphi = 0$. With the help of the relation

$$y = -2(1-x^2)\,U'/U \tag{5}$$

mechanism of the collapse has been commented in Section 1. At $\mathrm{Re} = \mathrm{Re}_x$ ac Eq. (2) is transformed to

$$U'' + F(x)\,[2(1-x^2)^2]^{-1}\ U = 0. \tag{6}$$

When $f_r = f_\theta = \Gamma = 0$ and the symmetry axis is free, then there is an analytical solution [4]

$$y = \mathrm{Re}(1-x)/z,\ z = \lambda \cot[\lambda \ln(1+x)] - 1/2,\ \ \lambda = 1/2\,(2\,\mathrm{Re} - 1)^{1/2}, \tag{7}$$

satisfying the conditions $y(0) = 0$, $y'(0) = \mathrm{Re}$. The Reynolds number Re characterizes a radial velocity value at the plane $x = 0$. Contrary to Squire here a plane matter motion is thought to be a driver of the flow described by (7). For instance, water inflowing into the central sink at the plane entrains an air above the plane that causes a jet-like air flow of type (7).

3. Collapse

In such an interpretation solution (7) has a paradoxical feature. When $\mathrm{Re} < \mathrm{Re}_* = 7.67$ zeroes of $z(x)$ are placed out of $0 \le x \le 1$. But if Re is increased formally above Re_*, then the first zero passes $x = 1$ and penetrates into the interval. It means the velocity field becomes singular and the solution (7) loses a physical meaning. When Re approaches Re_* from below, then a strong upward jet is formed and at $\mathrm{Re} = \mathrm{Re}_*$ the axial velocity turns to infinity. The physical cording to (7) $y(1) = 4$, that means there is a sink of fluid at the axis with a discharge per unit length $8\pi\nu$ as in the Schlichting solution for a round submerged jet [11]. Though the jet momentum becomes infinite, the jet entrainment ability remains bounded. Thus the collapse phenomenon is realized not only in solution (7) but it seems to be typical for the conical viscous flows.

4. Inner expansion

Let us introduce a small parameter $\varepsilon = \mathrm{Re}_* - \mathrm{Re}$ and an inner coordinate $\eta = (1-x)/\varepsilon$ into (2) and let ε tend to zero. The function F is supposed to have a zero at $x = 1$ and a bounded derivative. This takes place not only in the case of the free axis but also when the axis is a vortex line and moreover when the axial velocity has a logarithmic singularity as in the Serrin problem [3]. Then $F \to 0$ with $\varepsilon \to 0$ and equation (2) is separated from the system (2) - (4). It has the solution

$y = 4\eta/(4+\eta)$ which corresponds to the Schlichting jet. A boundary layer type solution for circulation is $\Gamma = 4\,\Gamma_1/(4+\eta)$ [12] in the vortex-line case and $\Gamma = \Gamma_0\,\eta/(4+\eta)$ in the free axis case where Γ_1 and Γ_0 are circulation values at the axis and in the outer region respectively. One sees that in the vortex-line case the collapse correlates to the circulation locking near the axis. If a thin cylinder being normal to a wall rotates, then it swirls fluid far from the axis at small angular velocity only. If the velocity becomes large, then fluid is swirled in a near-cylinder region but in the ambient region the rotation vanishes. In the free axis case the collapse correlates to circulation penetration up to the axis where a vortex line is formed in the limit.

5. Outer expansion in the Serrin problem

Naturally, outer expansions are not so universal as the inner one and a special analysis is needed for each problem. Let a flow be driven by a vortex line of circulation Γ_1 and axial force $4\pi A\rho\nu^2/r$. It corresponds to the boundary conditions $y(1) = 0$, $\Gamma(1) = \Gamma_1$, $(1-x)y''(x) \to -A$ with $x \to 1$ [3]. The flow region is bounded by a conical wall $x = x_w$ where the adherence conditions are to be satisfied: $y(x_w) = y'(x_w) = \Gamma(x_w) = 0$. Integration of (4) using the boundary conditions and $f_r = f_\theta = 0$ yields

$$F(x) = C(x-x_w)(1-x) - (1-x)^2 \int_{x_w}^{x} \frac{t\Gamma^2\,dt}{(1-t^2)^2} + x_w \frac{1-x}{1-x_w} \int_{x_w}^{x} \frac{\Gamma^2\,dt}{(1+t)^2} - \frac{x-x_w}{1-x_w} \int_{x}^{1} \frac{\Gamma^2\,dt}{(1+t)^2};$$

$$C = \frac{\Gamma_1^2 + 4A}{2(1-x_w)} - \frac{x_w}{(1-x_w)^2} \int_{x_w}^{1} \frac{\Gamma^2\,dt}{(1+t)^2}.$$

At the collapse $\Gamma(x) \to 0$ in $x_w \le x < 1$ and all the integrals turn to zero. As a result the leading term of the outer expansion for $y(x)$ is ruled by the equation

$$(1-x^2)\,y'_0 + 2xy_0 - y_0^2/2 = C(x-x_w)(1-x); \tag{8}$$

$$y_0(1) = 4,\; y'_0(1) = 2 - (1-x_w)C/4,\; y_0(x_w) = 0,\; C = (\Gamma_1^2 + 4A)/[2(1-x_w)].$$

Integrating (8) as an initial-value problem form $x = 1$ to $x = x_w$ we find $C(x_w)$

by shooting to fulfil $y(x_w) = 0$. Upper ad lower bounds for $C(x_w)$ are obtained analytically. Transforming (6) to the integral form, using the normalization $U_0(x_w) = 1$ and the collapse condition $U_0(1) = 0$ we obtain

$$C = 2\left\{ \int_{x_w}^{1} \frac{t - x_w}{(1+t)^2} U_0(t)\,dt \right\}^{-1}.$$

It may be easily shown that $(1 - x)/(1 - x_w) \leq U_0(x) \leq 1$ and hence we have $C_\ell \leq C \leq C_u$, where

$$C_\ell = 4\left\{2\,\ell n \frac{2}{1+x_w} - 1 + x_w\right\}^{-1}, \quad C_u = \left\{\frac{3+x_w}{2(1-x_w)}\,\ell n \frac{2}{1+x_w} - 1\right\}^{-1}.$$

These estimates together with numerical results of integration (8) are shown in Fig. 1. Particularly in the case $x_w = 0$ the collapse curve in the plane (Γ_1, A) corresponds to relation $\Gamma_1^2 + 4A = 30.6$.

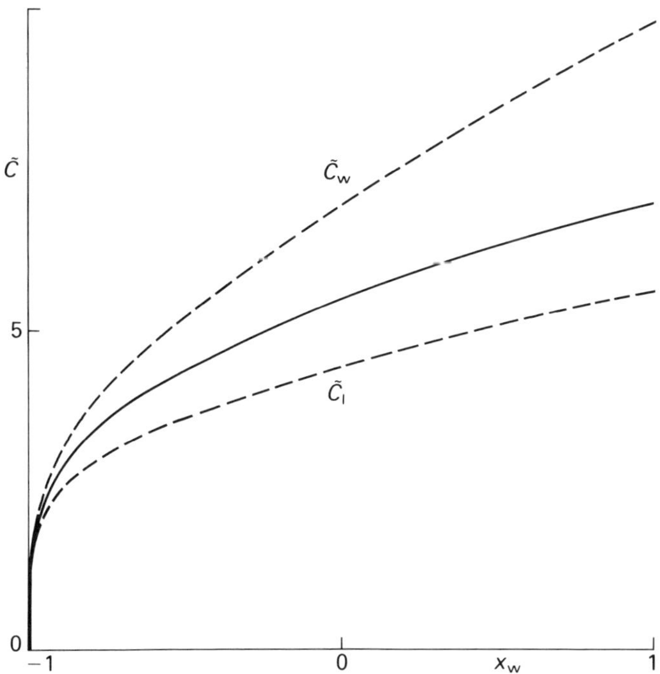

Figure 1. Collapse condition for the generalized Serrin problem.
Calculation results (solid curve) and analytic estimates.
$\tilde{C} = (1 - x_w)\,[2C(x_w)]^{1/2}$.

6. Free vortex

Now the flow region is $-1 \le x \le x_c$ and motion drivers are placed at the cone $x = x_c$: $\Gamma(x_c) = \Gamma_1$ (circulation) and $(1 - x_c)y''(x_c) = -A$ (radial friction). The axis $x = -1$ is free. This problem has been studied in [13, 14]. It follows from (4) $(f_r = f_\theta = 0)$ that

$$F(x) = \frac{2xx_c - 1 - x^2}{2(1+x_c)} \int_{-1}^{x} \frac{\Gamma^2 - \Gamma_1^2}{(1-t)^2}\, dt - \left(\frac{1+x}{1+x_c}\right)^2 \int_{x}^{x_c} \frac{x_c - t}{1 - t^2}\, \Gamma\Gamma'\, dt -$$

$$- \frac{(x_c - x)(1+x)(\Gamma^2 - \Gamma_1^2)}{2(1+x_c)(1-x)} - \frac{\Gamma_1^2}{4}\,\frac{1+x}{1+x_c}\,(1 - x + 2x_c) - \frac{A}{2}(1+x)^2.$$

For the collapse case $\Gamma(x) \equiv \Gamma_1$, and all terms except the last ones become zero. Thus the leading outer expansion equation has the form

$$(1 - x^2)y'_0 + 2xy_0 - y_0^2/2 = 2b(x - x_0)(1 + x); \qquad (9)$$

$$y_0(-1) = -4,\; y'_0(-1) = 2 - b(1 + x_0)/2,\; y_0(x_c) = 0,$$

where

$$b = \Gamma_1^2(1 + s)/[8(1 + x_c)],\; s = -2A\,\Gamma_1^2(1 + x_c),\; x_0 = -1 + 2(1 + x_c)/(1 + s).$$

Integrating (9) as an initial-value problem from $x = -1$ to $x = x_c$ we find $A_*(\Gamma_1, x_c)$ to satisfy $y_0(x_c) = 0$. Using similar methods as in Section 5 we obtain estimates $b_\ell \le b \le b_u$. The expressions for b_ℓ, b_u, being rather bulky, are not written here. The estimates together with calculation results for the case $x_c = 0$ are shown in Fig. 2. The value $q = q_* = -A/2 - \Gamma_1^2/4$ characterizes the radial velocity at the plane $(q = -rv_r(0)/\nu)$. Bounds Y_ℓ, Y_u from [14] are also shown. Curve 2 separates regions of one- and two-cell regimes respectively. In the limit $x_c \to 1$ the bounds b_ℓ and b_u coincide and provide $\Gamma_1^2 + 4A = 0$.

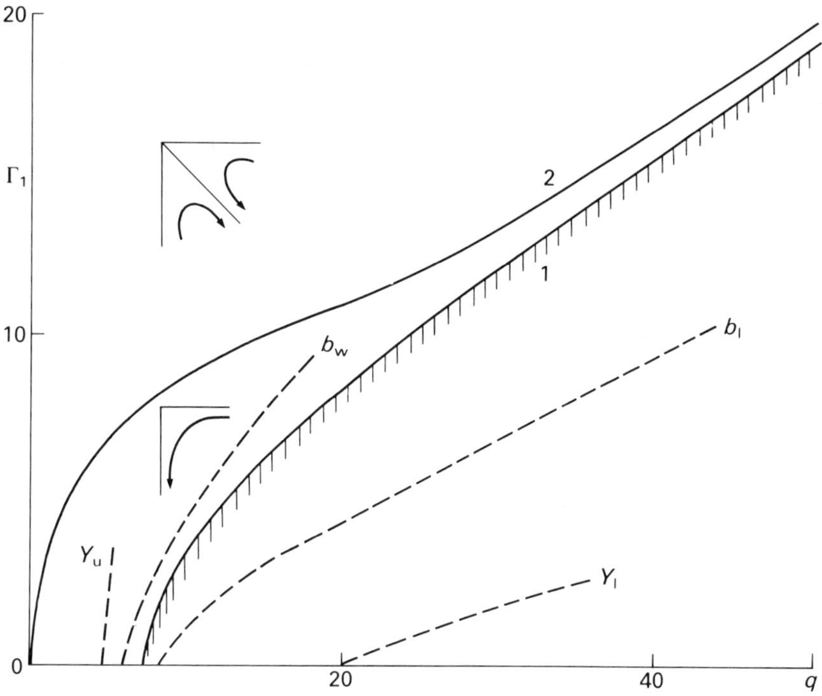

Figure 2. Regime map for the generalized Squire problem.

Calculated collapse curve (ШШ), its bounds according to [14] (Y) and the present authors (b). Curve 2 separates the regions of one- and two-cell regimes.

7. Heat convection

In this problem a motion driver is not placed at boundaries but the motion driver is the buoyancy force. We suppose that a thermal quadrupole is placed at the origin: $T = T_\infty + \gamma\theta(x)r^{-3}$ and put the Prandtl number Pr equal to zero. Then the heat equation is separated and $\theta = 3x^2 - 1$. Equation (4) at $\Gamma = 0$ has the form $F''' = \mathrm{Gr}(15x^2 - 3)$, where the Grashof number $\mathrm{Gr} = \beta\gamma\, g/\nu^2$, β is the thermal expansion coefficient, g is the gravity acceleration. The flow region $x_w \le x \le 1$ is bounded by the conical wall $x = x_w$, the axis $x = 1$ is free. Then equation (2) is reduced to

$$(1 - x^2)y' + 2xy - y^2/2 = \frac{1}{4}\mathrm{Gr}(1 - x)^2\,[x(1 + x)^2 - x_w(1 + x_w)^2] \qquad (10)$$

with the collapse conditions $y_0(1) = 4$, $y'_0(1) = 2$, $y_0(x_w) = 0$.

The latter may be used to determine Gr_* after integrating (10) from $x = 1$ to $x = x_w$. Numerical results (solid lines) and analytic estimates (dashed curves being produced as described above) are shown in Fig. 3. The regular solutions exist in the region bounded by the solid lines, the sketches illustrate regime patterns. The collapse does not happen with $|\mathrm{Gr}|$ growth only if $\mathrm{Gr} < 0$ and $-1/3 \leq x \leq 1$.

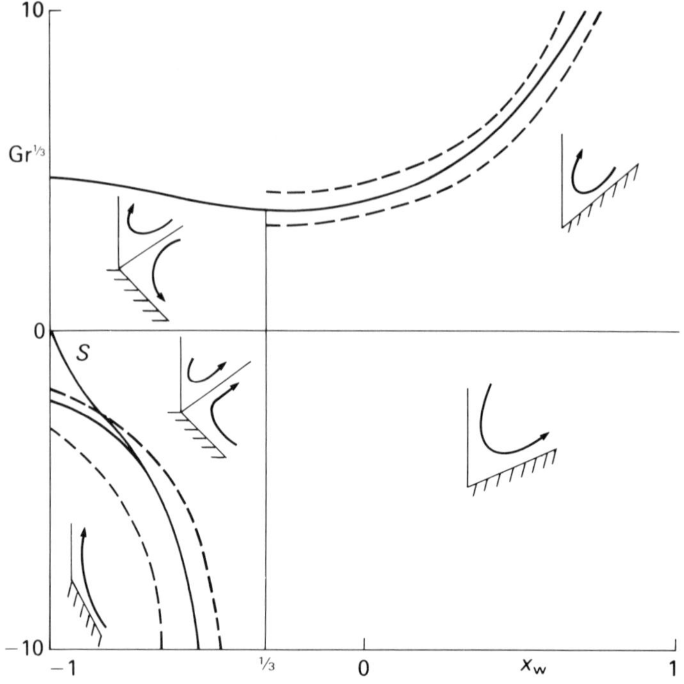

Figure 3. Regime map for the problem of quadrupolar convector. Dashed curves correspond to the analytic bounds of the collapse (solid) lines.

One more example of a self-similar solution of the Boussinesq equations may be constructed if the density depends on temperature by the anomalous way: $\rho/\rho_\infty = 1 - \beta_3(T - T_\infty)^3$. This relation is a rather relevant approximation for water near the maximum point of $\rho(T)$ at $T = 4°\mathrm{C}$. Let there be a heat source in the origin, $T = T_\infty + \gamma\theta(x)/r$. If $\mathrm{Pr} = 0$, then $\theta \equiv 1$ and equation (2) has the form

$$(1 - x^2)y' + 2xy - y^2/2 = \mathrm{Gr}_0(x - x_\mathrm{w})\,(1 - x^2)/2, \; \mathrm{Gr}_0 = \beta_3\,\gamma^3\,g/\nu^2.$$

The collapse conditions are

$$y_0(1) = 4, \;\; y'_0(1) = 2 + \mathrm{Gr}_{0*}(1 - x_\mathrm{w})/4, \;\; y_0(x_\mathrm{w}) = 0.$$

Analytical estimates provide $\mathrm{Gr}_{0\ell} \leq Gr_{0*} \leq Gr_{0\mathrm{u}}$, where

$$Gr_{0\ell} = 8\,\{8(2 + x_\mathrm{w})/(1 - x_\mathrm{w})\,\ln[2/(1 + x_\mathrm{w})] - x_\mathrm{w} - 11\}^{-1};$$

$$\mathrm{Gr}_{0u} = 4\,\{(3 + x_w)\ \ln[2/(1 + x_w)] - 2(1 - x_w)\}^{-1}.$$

The results are shown in Fig. 4 where $\mathrm{Gr} = (Q/\lambda)^3\, \beta_3\, g/\nu^2$, Q is the heat flux, λ is the thermal conductivity. This solution may model free convection near thermal sources at the ocean bottom (so called 'black smokers'). At the critical value of the heat flux the upward jet becomes infinitely thin and strong and the laminar solution ceases to exist. If $x_w \to -1$, then $\mathrm{Gr}_* \to 0$, hence for a free heat source the solution does not exist for an arbitrarily small heat flux.

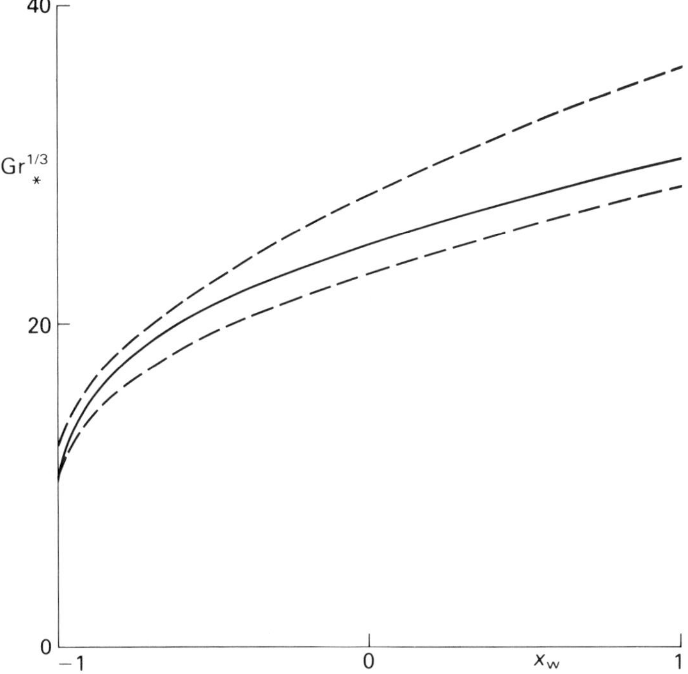

Figure 4. Collapse condition for the convection problem for a medium with thermal anomaly. Calculation results (solid curve) and analytic estimates

In the Sozou problem [9] an external force is of electrodynamic nature but the collapse phenomenon is the same. The collapse does not disappear even when the self-induction effect is taken into account.

Recently the authors have found that there is a hydromagnetic dynamo in the Sozou problem and so the paradox may be overcome.

8. Regularization

The fact that the velocity tends to infinity indicates that the flow model (or medium model) becomes irrelevant. What may be a small modification of the model to recover a regular solution? In free convection problems it is sufficient to put $Pr \neq 0$. In the problem on the vortex line, if one replaces the vortex line by a vortex cone of arbitrarily small angle, then regular solutions exist for all values of Γ_1 and A. Such a modification is rather natural as the vortex line is an idealization of objects having a finite width, for instance a thin rotating cylinder or a tornado core.

The situation is more sophisticated in the free vortex case. Here it seems to be a rather adequate way to take into account instability and transition to turbulence which may hapen before the initial solution ceases to exist. Indeed, jet-like flows are known to be unstable at small Reynolds numbers. If an eddy viscosity is introduced to model turbulence, then the collapse may be avoided. Such an approach has been used in [7] and an intriguing effect has been found - a possibility of spontaneous onset of rotational motion as the result of a bifurcation. The self-swirling is realized for both free [7] and rigid [15] boundaries.

9. Possible applications

The problem of the vortex line with the plane interaction has been considered by Serrin to model tornadoes [3]. The present authors have found that the Serrin problem solution is not unique and the nonuniqueness is conserved after replacing the vortex line by a thin cone. A hysteresis phenomenon takes place and this suggests a possible mechanism for the sudden generation or breakdown of tornadoes [16].

The free vortex problem may serve as the simplest model of astrophysical jets [7, 8]. Buoyancy-driven flow problems simulate qualitatively a free convection near volcanoes, glaciers and thermal outflows at the ocean bottom.

References

[1] Ladyzhenskaya, O.A. (1970). *Mathematical Aspects of Viscous Incompressible Fluid Dynamics*, Nauka, Moscow, 288 p.

[2] Goldshtik, M.A. (1960). One paradoxical solution of the Navier-Stokes equations, *Appl. Math. Mech.* (Sov.). No. 4, 610-621.

[3] Serrin, J. (1972). The swirling vortex, *Phil. Trans. R. Soc. Lond.* A **271**, 325-360.

[4] Squire, B. (1952). Some viscous flow problems, 1. Jet emerging from a hole in a plane wall. *Phil. Mag*. **43**, 942-945.

[5] Schneider, W. (1981). Flow induced by jets and plumes, *J. Fluid Mech*. **108**, 55-65.

[6] Zauner, E. (1985). Vizualization of the viscous flow induced by a round jet. *J. Fluid Mech*. **154**, 111-120.

[7] Goldshtik, M.A. and Shtern V.N. (1988). Conical flows of fluid with variable viscosity. *Proc. R. Soc. Lond.* A **419**, 91-106.

[8] Goldshtik, M.A. and Shtern, V.N. (1989). On a mechanism of astrophysic

jets. *Sov. Phys. Doklady*. **304**, 1069-1072.

[9] Sozou, C. (1971). On fluid motion induced by electric current source. *J. Fluid Mech*. **46**, 25-32.

[10] Goldshtik, M.A., Shtern, V.N. and Yavorsky, N.I. (1989). *Viscous Flows with Paradoxical Features*, Nauka, Novosibirsk, 340 p. (In Russian).

[11] Schlichting, H. (1965). *Grenzschicht Theorie*, G. Braun, Karlsruhe, 742 pp.

[12] Zubtsov, A.V. (1984). On a self-similar solution for a weak-swirling jet. *Mech. Fluid and Gas*. (Sov.). No. 4, 45-50.

[13] Goldshtik, M.A. (1979). On swirling jets. *Mech. Fluid and Gas*. (Sov.). No. 1, 26-36.

[14] Yih, C.-S., Wu, F., Garg, A.K., Leibovich, S. (1982). Conical vortices: A class of exact solutions of the Navier-Stokes equations, Phys. Fluids, **25**, 2147-2157.

[15] Goldshtik, M.A. and Shtern, V.N. (1989). Turbulent vortex dynamo, *Appl. Math. Mech*. (Sov.), No. 4, 613-624.

[16] Goldshtik, M.A. and Shtern V.N. (1990). A model of sudden generation and break-down of tornadoes, *Phys. Atm. and Ocean*. (Sov.), **26**, 27-34.

M.A. Goldshtik and V.N. Shtern
Institute of Thermophysics
Novosibirsk 630090
USSR

H. HERWIG

Regular perturbation problems in viscous flows

Abstract

Regular perturbation solutions, their general background as well as practical applications are reviewed. Special attention is given to a detailed description of its application in momentum and heat transfer problems. As an example fully developed, laminar pipe flow with (slightly) variable viscosity is considered.

1. Introduction

Among those methods for solving the basic equations of momentum and heat transfer asymptotic methods will probably provide the most general results.

Asymptotic solutions hold in the limt $\varepsilon \to 0$ of some pertinent perturbation quantity ε of a problem under consideration. This quantity is either linked to a parameter of the problem (parameter expansion) or to a coordinate (coordinate expansion). In both cases the limit solutions for $\varepsilon \to 0$ can be used as an approximation for solutions in which ε is small (but not zero). Furthermore the asymptotic theory can provide higher-order approximations by which these approximate solutions can be improved systematically.

A famous example for perturbation methods is Prandtl's boundary layer theory (see Schlichting (1982)) which turns out to be of the singular perturbation type. The mathematical foundation of perturbation theory nowadays is undoubtedly sound with major contributions in the 1960s and 1970s, see for example Kaplun (1967), Nayfeh (1973), Van Dyke (1975a), Schneider (1978), Kevorkian and Cole (1981), and Aziz and Na (1984).

2. Regular/singular perturbation problems

Perturbation problems are either regular or singular. Often they are combined in more complex problems. For example boundary layer theory is of the singular type. Accounting for the influence of variable viscosity within the boundary layer may be accomplished by a regular perturbation procedure - as will be demonstrated hereafter.

The fundamental difference between regular and singular perturbation problems can be stated like this: in regular perturbation problems you can find an asymptotic expansion of the solution that is *uniformly valid* within the whole domain of solution, including all boundaries. Here 'uniformly valid' means that the error is of the same order of magnitude throughout the whole solution domain.

In singular perturbation problems - in contrast to this - there is no uniformly valid expansion that covers the whole domain of solution. Often near physical boundaries regions of nonuniformity appear, which can be treated in a number of different ways. The most important of these methods is the *method of matched asymptotic expansions*, to which Prandtl's boundary layer theory is closely related, see for example M. Van Dyke (1975a).

There are two criteria which can be used to decide whether a perturbation problem is singular or regular. It will be singular if at least one of the two following criteria is fulfilled:

(I) The basic equations *degenerate* in the limit $\varepsilon \to 0$, i.e. in this limit the equations can no longer fulfill all physical boundary conditions. Often this happens because the degree of a partial differential equation is decreased by the loss of its highest derivative in the $\varepsilon = 0$ limit.

(II) The perturbation parameter ε is the *ratio of two characteristic (length or time) scales* s_1 and s_2 of the problem under consideration. Each scale is associated with a certain physical effect. Singular behaviour occurs when $\varepsilon := s_1/s_2 = 0$ or $s_2/s_1 = 0$.

If neither of these criteria holds, the problem should be of the regular perturbation type. This situation is assumed for the rest of this paper.

3. Regular perturbation problems: examples

Perturbation methods are applicable to a wide variety of physical and mathematical problems, though they are particularly popular in fluid mechanics.

A spring-mass-damping system with small damping is just one example for a regular perturbation problem from general mechanics. (The same problem for small mass would be a singular perturbation problem, see Kevorkian and Cole (1981).)

Ten examples from fluid mechanics may illustrate typical regular perturbation problems (p.p. denotes perturbation parameter).

(1) Flow around a slightly distorted body, p.p.: body distortion; Van Dyke (1975a, p. 13).
(2) Flow through channels of slowly varying cross section, p.p.: change in channel width; Van Dyke (1987).
(3) Laminar flow through a loosely coiled pipe, p.p.: Dean number; Van Dyke (1978).
(4) Slight shear flow around a body, p.p.: vorticity of the oncoming stream; Van Dyke (1975a, p. 11).
(5) Laminar mixed convection in a vertical pipe, p.p.: gradient of wall

temperature; Axiz and Na (1984, p. 29).

(6) Compressible flow around a thin airfoil, p.p.: thickness ratio; Schneider (1978, p. 59).

(7) Slightly compressible boundary layer flow, p.p.: Mach number; Herwig (1987).

(8) Natural convection flow in a highly porous matrix, p.p.: porosity; Herwig and Koch (1989).

(9) Pipe flow of a slightly non-Newtonian fluid, p.p.: shear rate; Herwig and Körber (1989).

(10) Flow with temperature-dependent properties, p.p.: wall heat flux, Herwig (1985).

Several special examples from the last category will be given later in this paper.

4. Regular perturbation solutions: general aspects

The basic idea with perturbation solutions is a reduction in the number of solution parameters for a certain problem under consideration. Asymptotic solutions for $\varepsilon \to 0$ of a certain problem no longer explicitly depend on the parameter which was chosen to define ε with. Once the asymptotic solution is found, it holds (as an asymptotic approximation) for *all* (small) values of ε. Here and in what follows it is assumed that the solution smoothly depends on ε (i.e. no singularities, no bifurcations).

The perturbation equations often are much simpler from a mathematical point of view than the equations of the full problem.

The prize for the more general as well as simpler solutions is the fact that these solutions are only approximations with respect to the perturbation parameter. But, they are *rational approximations*, which means that their error is known (in an asymptotic sense) to be of the order of the first neglected term. From this immediately follows that these solutions can be improved if necessary. Thus the degree of accuracy can be adapted to the requirements of the specific situation.

As will be shown in the following examples, it often is not only the perturbation parameter itself that can be separated from the general solution. In the course of expanding dependent parameters of a problem one should always try to find certain combinations of the perturbation parameter with other (parameter like) constants of the problem (and sometimes also combinations with coordinates) which then altogether can be extracted as explicit factors in the particular expansion formulae of the dependent variables. An example for this is given in Eq. (E-10) below.

5. Regular perturbation approach: general procedure

Basically the perturbation procedure is determined by assuming a solution in form of a series expansion (often: power series) with respect to the perturbation parameter ε.

To explain the procedure in detail it will be subdivided into four steps Step 1 - Step 4. To give a more vivid description it will be applied to an illustrative example step by step. Fully developed laminar pipe flow (see Fig. 1), has been chosen for illustration of the method. Under the influence of a constant wall heat flux $q_w{}^*$ the influence of temperature-depenent viscosity will be investigated. This is a typical example for a regular perturbation problem. Momentum and heat transfer at constant viscosity are slightly changed by variations in the viscosity due to its temperature dependence. The zero-order solution is that for constant viscosity with the well-known results of a parabolic velocity profile. Accounting for variable viscosity effects will be accomplished by the following 4-step procedure.

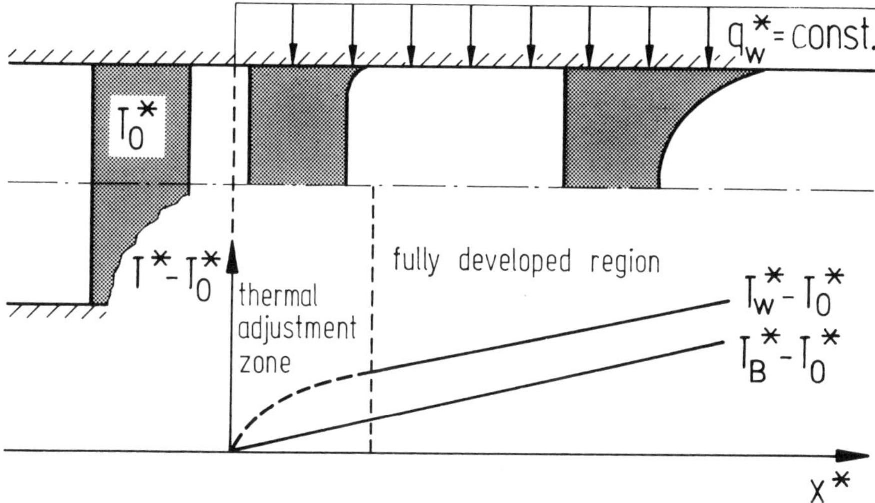

Figure 1. Development of the temperature profile

Step 1. Nondimensionalize the basic equations

Perturbation procedures always have to start from the basic equations of the problem under consideration. Nondimensionalizing them with characteristic scales of the problem will reveal the nondimensional parameters of the problem. For momentum and heat transfer problems that will be parameters like: Reynolds number Re, Prandtl number Pr, Eckert number Ec, Mach number Ma, and so on.

Parameter perturbations are then based on one (or sometimes more than one) of these parameters which are chosen as perturbation parameter(s).

Pipe flow example; step 1

The Navier-Stokes equations for fully developed pipe flow (no viscous heating, no buoyancy forces) nondimensionalized and transformed according to Table 1 reads (see Herwig, 1985):

continuity: $$\frac{\partial u}{\partial x} + \frac{1}{r}\frac{\partial}{\partial r}(rv) = 0 \qquad \text{(E-1)}$$

momentum: $$u\frac{\partial u}{\partial x} + v\frac{\partial u}{\partial r} = -\frac{\mathrm{d}p}{\mathrm{d}x} + \frac{1}{r}\frac{\partial}{\partial r}\left[r\frac{\partial u}{\partial r}\right] \qquad \text{(E-2)}$$

energy (enthalpy): $$u\frac{\partial h}{\partial x} + v\frac{\partial h}{\partial r} = \frac{\mathrm{Pr}_0^{-1}}{r}\frac{\partial}{\partial r}\left(r\frac{\partial h}{\partial r}\right) \qquad \text{(E-3)}$$

with associated boundary conditions:

$$r = 0 \quad : \quad v = \frac{\partial u}{\partial r} = \frac{\partial h}{\partial r} = 0 \qquad \text{(E-4)}$$

$$r = 1 \quad : \quad u = v = \frac{\partial h}{\partial r} - q_{\mathrm{w}} = 0 \qquad \text{(E-5)}$$

All physical properties are constant except η, i.e. $\rho = \lambda = c_{\mathrm{p}} = 1$.

Step 2. Identify the perturbation parameter and expand all dependent variables

In perturbation problems one is interested in solutions for certain parameters (or coordinates) of the problem which are close to certain limits (usually zero or infinity). Let the paramter be P, then the perturbation parameter is defined by

$$\varepsilon := P^n \qquad (1)$$

with $n > 0$ in the limit $P \to 0$ and $n < 0$ in the limit $P \to \infty$. When two or more parameters are considered simultaneously, they may or may not be linked in their limiting behaviour. If they are linked, they are called similarlity parameters, see for example Van Dyke (1975, p. 21). The exponent n often emerges quite naturally from the problem (sometimes by trial and error).

All dependent variables of the problem are expanded in asymptotic series, the form of which again is obvious (or can be found by trial and error). For a dependent variable $D(x_j, \varepsilon)$, with independent variables x_j, it reads:

$$D(x_j, \varepsilon) = \sum_{i=0}^{\infty} g_i(\varepsilon) D_i(x_j); \quad \lim_{\varepsilon \to 0} \frac{g_{i+1}}{g_i} = 0. \qquad (2)$$

The functions $g_i(\varepsilon)$ are called *gauge functions.*

Pipe-flow example; step 2

We are interested in the influence of variable viscosity, i.e. in situations in which η is different from $\eta = 1$. A Taylor series expansion with respect to temperature around $\eta = 1$ reads:

$$\eta = 1 + K_\eta T + \ldots; \quad K_\eta = \left(\frac{T^* \partial \eta^*}{\eta^* \partial T^*} \right)_0 . \tag{E-6}$$

Since heat transfer across the wall with $q_w = \text{const}$ is the reason for non-isothermal flow, q_w should be used to re-nondimensionalize T and h. With

$$\theta := \frac{T}{q_w} \text{ and } \bar{h} := \frac{h}{q_w} \tag{E-7}$$

equation (E-6) now reads:

$$\eta = 1 + q_w K_\eta \theta + \ldots . \tag{E-8}$$

The isothermal case ($\eta = 1$ throughout the whole flow field) holds for $q_w \to 0$, so that q_w is the parameter to define the perturbation parameter with. According to equation (1) we set ($n = 1$)

$$\varepsilon := q_w \tag{E-9}$$

Next all dependent variables D (i.e.: u, v, p, ..) are expanded according to equation (2):

$$D(x, r, \varepsilon) = \sum_{i=1}^{\infty} g_i D_i = D_0(x, r) + \varepsilon K_\eta D_1(x, r) + O(\varepsilon^2) \tag{E-10}$$

Here $g_0(\varepsilon) = 1$, $g_1(\varepsilon) = \varepsilon K_\eta$. The constant K_η is absorbed in g_1 as it is suggested by equation (E-8).

Step 3. Derive the 0,1,... -order equations and solve them

All asymptotic expansion series from Step 2 are inserted into the basic equations. Collecting terms of equal asymptotic order the basic equations can be written as sums of terms which are of descending order of magnitude asymptotically. For a power series expansion, for example, the basic equations will be of the general form

$$\{\,.....\}_0 + \varepsilon\{\,...\,\}_1 + \varepsilon^2\{\,.....\}_2 + ... = 0 \tag{3}$$

Equation (3) is fulfilled under the condition

$$\{.....\}_i = 0, \quad i = 0, 1, ... \tag{4}$$

Equations (4) are the 0, 1, ... -order equations which by definition are free of ε and can be solved after all boundary conditions are treated likewise. If certain constants are absorbed in $g_i(\varepsilon)$, they should also be extracted from $\{\,...\,\}_i$ in order to allow for more general results.

Pipe-flow example; step 3

Inserting (E-8) and (E-10) into the basic equations (E-1)-(E-3) and collecting terms of equal magnitude for the continuity equation (E-3), for example, results in

$$\left\{\frac{\partial u_0}{\partial x} + \frac{1}{r}\frac{\partial}{\partial r}(rv_0)\right\} + \varepsilon K_\eta \left\{\frac{\partial u_1}{\partial x} + \frac{1}{r}\frac{\partial}{\partial r}(rv_1)\right\} + O(\varepsilon^2) = 0 \tag{E-11}$$

so that the zero and first order continuity equations are

$$\frac{\partial u_0}{\partial x} + \frac{1}{r}\frac{\partial}{\partial r}(rv_0) = 0; \quad \frac{\partial u_1}{\partial x} + \frac{1}{r}\frac{\partial}{\partial r}(rv_1) = 0. \tag{E-12}$$

Together with the corresponding equations for momentum and enthalpy (here: $\bar{h}_i = \theta_i$) zero and first order systems of equations arise which have to be solved. For this simple example analytical solutions exist (see Herwig (1985)):

$$u_0 = 2(1 - r^2); \quad u_1 = \frac{1}{12}(-2r^6 + 12r^4 - 13r^2 + 3) \tag{E-13}$$

$$v_0 = 0; \quad v_1 = 0 \tag{E-14}$$

$$\theta_0 = \frac{2x}{\mathrm{Pr}_0} + \frac{1}{4}(-r^4 + 4r^2 - \frac{7}{6}); \quad \theta_1 = \frac{1}{576}(-3r^8 + 32r^6 - 78r^4 + 72r^2 - 5) \tag{E-15}$$

The boundary condition (E-5), $\partial h/\partial r - q_w = 0$, first was rewritten as $\partial\bar{h}/\partial r = 1$ or $\partial\theta/\partial r = 1$, respectively, and then expanded according to (E-10):

$$\frac{\partial \theta_0}{\partial r} + \varepsilon K_\eta \frac{\partial \theta_1}{\partial r} + O(\varepsilon^2) = 1 \tag{E-16}$$

From (E-16) the zero and first order b.c. are

$$\frac{\partial \theta_0}{\partial r} = 1; \quad \frac{\partial \theta_1}{\partial r} = 0 \tag{E-17}$$

Step 4. Set up the final asymptotic results

As far as momentum and heat transfer are concerned, the final results will be adequately set up in terms of skin friction factor f and Nuβelt number Nu. According to the expansion procedure they will also be of the general form of equation (2). Especially when higher-order equations $(i \geq 2)$ are considered certain combinations of the original gauge functions will appear in the final results. Again their specific form is obvious and arises quite naturally from the overall procedure. Let $\hat{c}$ be a dimensionless coefficient (like f or Nu), the final results are of the general form

$$\hat{c}(x_j, \varepsilon) = \sum_{i=0}^{\infty} G_i(\varepsilon)\, \hat{c}_i(x_j). \tag{5}$$

Here G_i are the original gauge functions or combinations of them (including constants absorbed in g_i).

Pipe-flow example; step 4

With the definition of f, $f := (-2\,\tau_w{}^*)/(\rho_0{}^*\, U_0^{*2})$, together with Newton's law $\tau^* = \eta^*\, \partial u^*/\partial r^*$ the asymptotic result for the friction factor f reads

$$f\ \mathrm{Re_B} = 8 + \varepsilon\, K_\eta 2 + O(\varepsilon^2). \tag{E-18}$$

The index B indicates that all properties now have to be evaluated at the local bulk temperature; for details see Herwig (1985).

The Nuβelt number $\mathrm{Nu} := (-\,q_w^*\, 2R^*)/(\lambda_0^*(T_w^* - T_B^*))$ together with Fourier's law $q^* = -\,\lambda^*\, \partial T^*/\partial r^*$ in its asymptotic form reads:

$$\mathrm{Nu} = \frac{48}{11} - \varepsilon K_\eta \frac{26}{121} + O(\varepsilon^2). \tag{E-19}$$

Deviations from the constant property case are better revealed by rewriting (E-18) and (E-19) with f_{cp}, Nu_{cp} defined as the coefficients for constant properties. The final results then are:

$$\frac{f\,Re_B}{(f\,Re_B)_{cp}} = 1 + \varepsilon K_\eta \frac{1}{4} + O(\varepsilon^2); \quad \frac{Nu}{Nu_{cp}} = 1 - \varepsilon K_\eta \frac{13}{264} + O(\varepsilon^2). \qquad \text{(E-20)}$$

It should be pointed out that (E-18)-(E-20) hold for all (Newtonian) fluids and all (small) heat transfer rates. Only in the final results ε and K_η will be specified for the particular situation under consideration.

Extension to the temperature dependence of all physical properties again is straightforward (Herwig (1985b)) with some peculiarities with the thermal boundary condition of constant wall temperature (Herwig *et al.* (1989)).

6. Higher-order effects

The pipe flow example in Section 5 only accounted for variable property effects that were linear in ε. Therefore the theory should be called *linear theory*. Carrying the expansions to higher orders is straightforward. Actually, due to this fact the asymptotic approach is a *rational approximation*. On the other hand two problems arise when the expansion is continued to nonlinear terms:

(1) Arithmetic labour rapidly grows because there is a rapid increase in the number of terms for each increase in the asymptotic order. This is no real problem since the arithmetic labour can be delegated to the computer, see for example Van Dyke (1975b), and three examples in Section 7.

(2) The range of applicability of the resulting series ordinarily is very limited. The reason is that asymptotic series are not necessarily convergent for fixed values of ε. This means that adding more terms may increase the error when ε is fixed. Only for $\varepsilon \to 0$ the error will always tend to zero. Sometimes one can improve the rate and radius of convergence or even render a divergent series convergent (Van Dyke (1975a)), but these methods will not always work.

For practical applications, however, there often is no real need for higher-order results because the crucial effects are covered by the linear theory already. The pipe flow of Section 5 may serve as an example again.

Pipe-flow example: property ratio method

There is an empirical method (well established in the heat-transfer literature) to account for variable property effects called the *property ratio method* . In this method empirical correction formulae are assumed. For skin friction, for example,

$$\frac{f\,\mathrm{Re_B}}{(f\,\mathrm{Re_B})_{cp}} = \left[\frac{\eta^*(T^*_w)}{\eta^*(T^*_B)}\right]^n \;;\quad n = \text{const.} \tag{E-21}$$

All variable property effects are absorbed in the (empirical) exponent n. For fully developed pipe flow under the thermal boundary condition q^*_w = constant empirical values for n are (Kays and Crawford (1980)):

$$n = 0.50 \text{ (cooling)};\qquad n = 0.58 \text{ (heating)}. \tag{E-22}$$

By means of the expansion (E-8) the empirical result (E-21) can easily be compared to the asymptotic result (E-20):

empirical:
$$\frac{f\,\mathrm{Re_B}}{(f\,\mathrm{Re_B})_{cp}} = 1 + \varepsilon K_\eta\,(\theta_w - \theta_B)n + \ldots\;;\quad \theta_w - \theta_B = \frac{11}{24} \tag{E-23}$$

asymptotic:
$$\frac{f\,\mathrm{Re_B}}{(f\,\mathrm{Re_B})_{cp}} = 1 + \varepsilon K_\eta\,\frac{1}{4} + O(\varepsilon^2) \tag{E-24}$$

Comparing (E-23) and (E-24), two conclusions can be drawn:

(1) The exponent n obviously is $n = 6/11 = 0.545$. There is no need to have recourse to empirical data. All information is extracted from the basic equations. In this sense the property ratio method no longer is an empirical method if it is based on the asymptotic expansions.

(2) If n is assumed to be a constant the same value holds for cooling and heating. Asymptotically n would be $n(\varepsilon)$ if it would be correlated to nonlinear terms in (E-24). Only then cooling could be told from heating through the exponent n. From these considerations it seems obvious that assigning different values to n for cooling and heating is unsuitable to account for nonlinear effects of that kind.

Altogether we can conclude that the linear theory accounts for all effects that are claimed to be covered by the property ratio method.

7. Computer extended series

Though for practical applications often a few terms are calculated only (see Section 6), it may be very illustrative to extend the series to far more terms. For particular problems this has been done with the help of the computer. Survey papers have been written by Van Dyke (1975b), (1980). Three examples may give an impression of that approach.

(1) For natural convection flow in a long horizontal pipe with adiabatic walls and differentially heated ends Tsan–Hsing Shih (1981) extended the Nußelt number solution to 24 terms based on an expansion of all dependent variables into power series of the Rayleigh number.

(2) Starting from Stokes series for small Reynolds numbers Van Dyke (1977) extended the series to about 20 terms (in several different examples) with the aim of an analytical continuation to large Reynolds numbers. Thus the laminar boundary layer limit (which is singular in a direct approach) may be reached via a regular perturbation starting from the small Reynolds number end.

(3) For a loosely coiled pipe Van Dyke (1978) extended the series to 24 terms. He concluded that the flux ratio (flow rate through the coiled pipe divided by that in a straight pipe under the same pressure gradient) decays for large Dean number as $\mathrm{De}^{-1/4}$. However, numerical solutions suggest $\mathrm{De}^{-1/2}$, a disagreement which is still unresolved.

8. Conclusions

It was demonstrated that regular perturbation technique is quite simple to adapt to problems in fluid mechanics. That may be different for singular perturbation problems because then two or more series have to be matched asymptotically. For regular perturbation problems a straightforward four–step procedure was recommended. As demonstrated for fully developed laminar pipe flow asymptotic results provide quite general results which hold for a whole class of flow situations.

Table 1 : Dimensionless variables and parameters (* ≅ dimensional quantities)

x	r	u	v	p	T	h	η
$\dfrac{x^*}{R^*\mathrm{Re}_0}$	$\dfrac{r^*}{R^*}$	$\dfrac{u^*}{U_0^*}$	$\dfrac{v^*\mathrm{Re}_0}{U_0^*}$	$\dfrac{p^* - p_0^*}{\rho_0^* U_0^{*2}}$	$\dfrac{T^* - T_0^*}{T_0^*}$	$\dfrac{h^* - h_0^*}{c_{po}^* T_0^*}$	$\dfrac{\eta^*}{\eta_0^*}$

Re_0	Pr_0	0 denotes reference condition
$\dfrac{\rho_0^* U_0^* R^*}{\eta_0^*}$	$\dfrac{\eta_0^* c_{p0}^*}{\lambda_0^*}$	U_0^* denotes mean velocity at ref. cond. R^* denotes pipe radius

References

Aziz, A. and Na, T.Y. (1984). *Perturbation Methods in Heat Transfer*, Hemisphere Publishing Corp., Washington.

Herwig, H. (1985a). *Asymptotische Theorie zur Erfassung des Einflusses variabler Stoffwerte auf Impuls- und Wärmeübertragung*, VDI Fortschritt-Berichte, Reihe 7, No. 93, VDI-Verlag, Düsseldorf (FRG).

Herwig, H. (1985b). The Effect of Variable Properties on Momentum and Heat Transfer in a Tube with Constant Heat Flux across the Wall, *Int. J. Heat Mass Transf.*, **28**, 423-431.

Herwig, H. (1987). An asymptotic approach to compressible boundary-layer flow, *Int. J. Heat Mass Transfer,* **30**, 59-68.

Herwig, H. and Koch, M. (1989). An asymptotic approach to natural convection momentum and heat transfer in saturated highly porous media to be published in *J. Heat Transfer*, 1991.

Herwig, H. and Körber, J. (1989). An asymptotic theory for laminar pipe flow of a non-Newtonian fluid, *Wärme- und Stoffübertragung*, **24**, 87-95.

Herwig, H., Voigt, M. and Bauhaus, F.-J. (1989). The Effect of Variable Properties on Momentum and Heat Transfer in a Tube with Constant Wall Temperature, Int. J. Heat Mass Transfer **32**, 1907-1915.

Kaplun, S. (1967). In: *Fluid Mechanics and Singular Perturbations* (Eds.: P.A. Lagerstrom, L.N. Howarth and C.S. Liu), Academic Press, New York.

Kevorkian, J. and Cole, J.D. (1981). *Perturbation Methods in Applied Mathematics*, Springer-Verlag, New York.

Nayfeh, A.H. (1973). *Perturbation Methods*, Wiley-Interscience, New York.

Schlichting, H. (1982). Grenzschicht-Theorie, G. Braun Verlag, Karlsruhe.

Schneider, W. (1978). Mathematische Methoden der Strömungsmechanik, Vieweg-Verlag, Braunschweig, FRG.

Tsang-Hsing Shih (1981). Computer-Extended Series: Natural Convection in a Long Horizontal Pipe with Different End Temperatures, *Int. J. Heat Mass Transfer*, **24**, 1295-1303.

Van Dyke, M. (1975a). *Perturbation Methods in Fluid Mechanics*, Parabolic Press, Palo Alto, California.

Van Dyke, M. (1975b). Computer extension of perturbation series in fluid mechanics, *SIAM J. Appl. Math.*, **28**, 720-734.

Van Dyke, M. (1977). From zero to infinite Reynolds number by computer extension of Stokes series. In: *Lecture Notes in Math.*, No. 594, 506-517, Springer-Verlag, Berlin.

Van Dyke, M. (1978). Extended Stokes series: laminar flow through a loosely coiled pipe, *J. Fluid Mech.* **86**, 129-145.

Van Dyke, M. (1980). Successes and surprises with computer-extended series. In: *Lecture Notes in Physics*, No. 141, 405-410, Springer-Verlag, Berlin.

Van Dyke, M. (1987). Slow variations in continuum mechanics, *Adv. Appl. Mech.*, **25**, 1-45.

H. Herwig
Ruhr Universität Bochum
Institut für Thermo und Fluiddynamik
Theoretische Strömungsmechanik
Gebäude IB-6/42
Postfach 10 21 48
D-4630 Bochum
GERMANY

G. HACKMÜLLER AND A. KLUWICK

Marginal separation in quasi-two-dimensional flow

Abstract

The theory of marginally separated flows is generalized to cover the effects of surface–mounted obstacles on the interaction process. It is assumed that the direction of the free stream velocity and the normal to the leading edge of the wing include a finite sweep angle and, furthermore, that the shape of the surface–mounted obstacle varies in the spanwise direction. The resulting interaction equation is solved for a one–parameter family of hills/dents the height/depth of which varies slowly over distances comparable in magnitude with the nose radius.

1. Problem formulation

Ruban [1] and independently Stewartson *et al.* [2] showed that the equations for an incompressible boundary layer admit a solution which can be extended continuously through the point of vanishing skin friction although it is singular at this point (marginal separation). In contrast to the case of a strong Goldstein singularity such a singularity can be removed by taking into account the interaction between the boundary layer and the outer inviscid flow. The appropriate length of the interaction region is of order $\mathrm{Re}^{-1/5}$ (Fig. 1). Inside the interaction region three layers (decks) exhibiting different physical properties have to be distinguished. Viscous effects are found to be of importance inside the lower deck only. The role of the main deck is to transmit displacement effects caused by the lower deck to the external flow region (upper deck) where they lead to an inviscid pressure response.

An example of such a situation is provided by the flow past a slender airfoil at incidence. As the angle of incidence k is increased up to a critical angle k_0 the singularity is assumed to occur at $x = 0$, where x measures the distance along the wall.

The aim of the present study is to analyse this type of interaction in the case of a swept wing configuration with a surface mounted obstacle in the interaction region, the shape of which varies in the spanwise direction. To ensure that the additional pressure disturbances caused by the obstacle and those needed to remove the singularity, entering the classical boundary layer solution, are comparable in magnitude the height H of the obstacle must be of order $\mathrm{Re}^{-7/10}$ (Fig. 1).

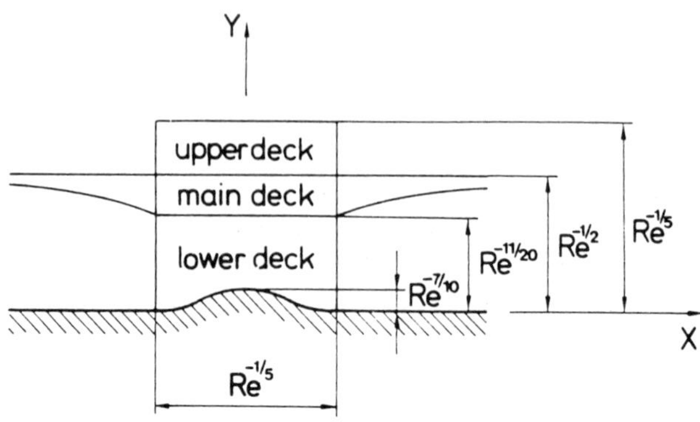

Figure 1. Asymptotic structure of the interaction region

Coordinates and velocity components are nondimensionalized with the noseradius $\tilde{l}$ of the profile and the component of the freestream velocity normal to the leading edge of the airfoil $\tilde{u}_\infty$ respectively. To formulate the asymptotic expansions for the various field quantities in the lower deck region holding in the limit of large Reynolds number

$$\mathrm{Re} = \frac{\tilde{u}_\infty \tilde{l}}{\tilde{\nu}_\infty} \gg 1, \tag{1}$$

where $\tilde{\nu}_\infty$ is the (constant) kinematic viscosity, it is convenient to introduce the stretched coordinates

$$x_* = \mathrm{Re}^{1/5} x \quad y_* = \mathrm{Re}^{11/20}\ y - \mathrm{Re}^{-3/20}\, h_0(x_*, z). \tag{2}$$

Here y_* and $h_0(x_*, z)$ characterize the distance from the surface and the shape of the surface mounted obstacle, respectively. Matching to the solution of the classical boundary layer equations outside the interaction region and applying the Prandtl transposition theorem then gives the expansion for the velocity components

$$u \sim \mathrm{Re}^{-1/10}\, \tfrac{1}{2}\, p_{00}\, y_*^2 + \mathrm{Re}^{-1/4}\, A_1(x_*, z) y_* + \mathrm{Re}^{-2/5}\, u_2^*(x_*, y_*, z) + \dots$$

$$v \sim \mathrm{Re}^{-3/5}\, \tfrac{1}{2}\left[-A_{1_{x_*}}(x_*, z) + p_{00}\, h_{0_{x_*}}(x_*, z) \right] y_*^2$$

$$+ \mathrm{Re}^{-3/4}\left[v_2^*(x_*, y_*, z) + A_1(x_*, z) h_{0_{x_*}}(x_*, z)\, y_* + C_1\, p_{00}\, h_{0_z}(x_*, z)\, y_* \right] + \dots \tag{3}$$

$$w \sim \mathrm{Re}^{-1/20}\, C_1 p_{00}\, y_* + \mathrm{Re}^{-1/5}\, w_1^*(x_*, y_*, z) + \dots ,$$

where p_{00} denotes the imposed pressure gradient for $k = k_0$ at $x_* = 0$ and the constant C_1 has to be determined by matching this local solution to the global flow field.

The function $A_1(x_*, z)$ which has the meaning of the wall shear stress distribution remains arbitrary if terms up to second order are considered only. In order to determine $A_1(x_*, z)$, it is necessary to investigate the properties of the next terms in these expansions. The requirement that the results can be matched to the maindeck solution together with the relation for the induced pressure gradient then yields the nonlinear integro-differential equation

$$A^2(X, Z) - X^2 + \Gamma + 2 \int_{-\infty}^{X} \bar{A}_{\bar{Z}}(\xi, \bar{Z})\, d\xi = \int_X^{\infty} \frac{\bar{A}_{\xi\xi}(\xi, \bar{Z}) - h_{\xi\xi}(\xi, \bar{Z})}{\sqrt{\xi - X}}\, d\xi \tag{4}$$

if x_*, z, h_0 and A_1 are replaced by suitable scaled quantities $X, \bar{Z}, h$ and $\bar{A}$. The parameter $\Gamma \propto \mathrm{Re}^{2/5}(k - k_0)$ measures the angle of incidence.

As a first application of Eq. (4) we investigate the flow past obstacles, the shape of which changes slowly over distances of the order of the nose radius $\tilde{l} : \bar{A} = \bar{A}(X, Z; \varepsilon)$, $Z = \varepsilon \bar{Z}$, $\varepsilon \ll 1$. Expansion of the wall shear stress distribution in terms of the small perturbation parameter ε then yields the equation

$$A^2(X, Z) - X^2 + \Gamma = \int_X^{\infty} \frac{A_{\xi\xi}(\xi, Z) - h_{\xi\xi}(\xi, Z)}{\sqrt{\xi - X}}\, d\xi \tag{5}$$

for the first term $A(X, Z)$ of this expansion.

2. Results

The lateral coordinate Z is seen to enter Eq. (5) in the form of a parameter only. As a consequence, the flow quantities in different planes $Z = \text{const}$ can be calculated independently. This leads to a significant reduction of the computational efforts needed to solve the quasi-two-dimensional interaction problem as compared to the more general case governed by Eq. (4).

Numerical results have been obtained for obstacles of the form

$$h(X, Z) = \begin{cases} \alpha(Z)\,(X^2 - 1)^3 & |X| < 1 \\ 0 & |X| \geq 1 \end{cases} \tag{6}$$

describing protrusions ($\alpha < 0$) or indentations ($\alpha > 0$). The variation of the wall shear stress $A(0)$ at the origin $X = 0$ with α, for $\Gamma = 1$ is depicted in Fig. 2. It is found that solutions to the interaction Eq. (5) cease to exist if the height/depth of the protrusion/indentation exceeds the critical value -5.19/2.45, respectively; the flow properties for $\alpha < -5.19$ and $\alpha > 2.45$ are not known at present. Furthermore, the numerical computations indicate that Eq. (4) admits two different solutions for values $-5.19 < \alpha < -1.77$, $-1.24 < \alpha < -0.93$ and $0.76 < \alpha < 2.45$ while four different solutions exist for $-1.77 < \alpha < -1.24$. So far, numerical difficulties have prevented the calculation of separated flow solutions for $-0.93 < \alpha < 0.76$.

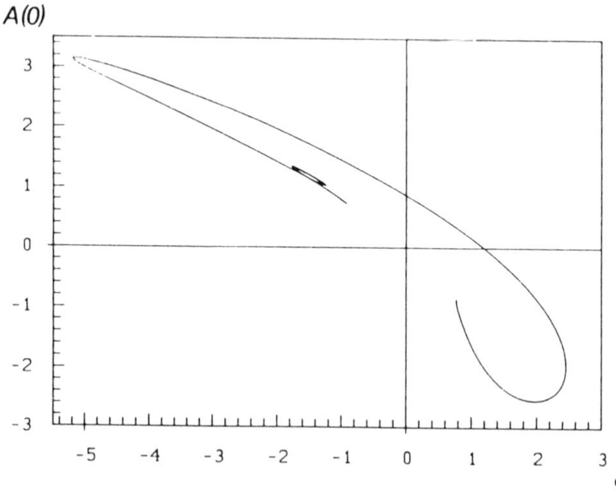

Figure 2. X-component of the wall shear stress at $X = 0$.

Wall shear stress distributions for $\Gamma = 1$ and various values of $\alpha \le 0$ are depicted in Fig. 3. For $\alpha = 0$ the flow (described by the upper branch of Fig. 2) is attached and incipient separation was found to occur at $\alpha \approx -2.87$. As α increases further, a separation bubble forms on the leeward side of the protrusion and the limiting solution $\alpha \approx -5.19$ is finally reached. Moving on along the lower branch of Fig. 2 leads to protrusions of decreasing height while the streamwise extent of the recirculation region still increases and, presumably, assumes its maximum length at $\alpha = 0$.

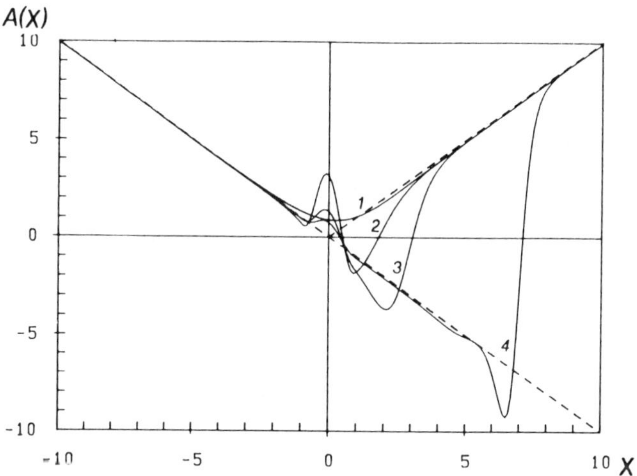

Figure 3. Distribution of the X-component of the wall shear stress for various values of α 1: $\alpha = 0$, 2: $\alpha = -5.19$, 3: $\alpha = -1.8$ 4: $\alpha = 0.93$

In Fig. 4 the wall stream line pattern is plotted for an obstacle the height of which is periodic with respect to Z

$$\alpha = \alpha_a \cos(2\pi Z) + \alpha_m \tag{7}$$

$$\alpha_a = -2.45 \quad \alpha_m = -2.73 \quad \varepsilon = 0.02.$$

The dotted lines are contour lines of the hill and the dot-dash line corresponds to a height of the hill for which separation would occur in the limiting case of strictly two-dimensional flow.

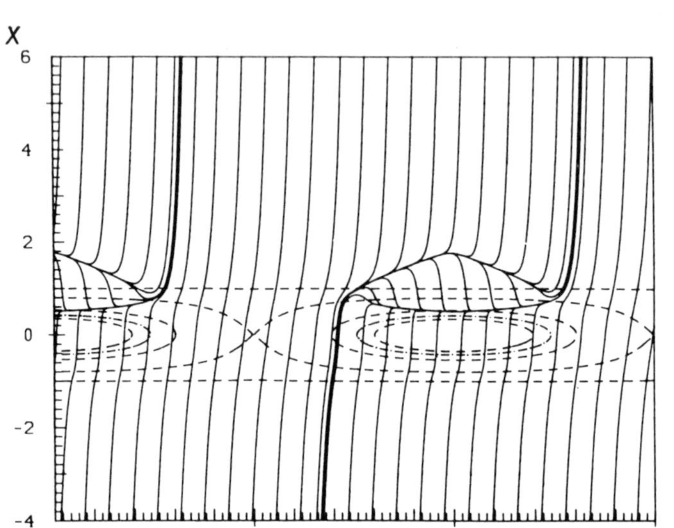

Figure 4. Wall streamline pattern due to the periodic protrusion (6)

The hill is high enough to cause a region of recirculation for a limited range of values of Z only. In this region a small open separation bubble is formed behind the obstacle, with a mass flux into the bubble from the left.

It should be noted that no lines which have the meaning of separation/ reattachment lines can be identified in Fig. 5, which shows a blow-up of the open end of the separation bubble. There are regions of high density of wall streamlines as a consequence of displacement of the fluid normal to the surface, but no envelope exists.

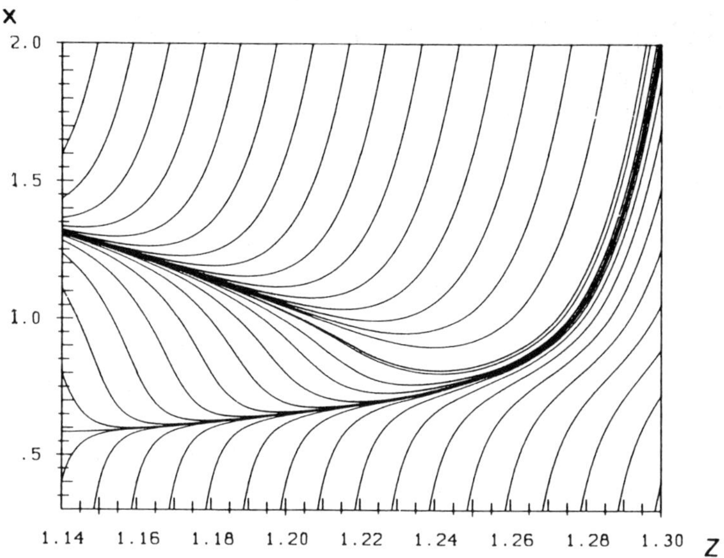

Figure 5. Wall streamline pattern near the open end of the separated flow region

References

[1] Ruban, A.I., (1981). Asymptotic theory of short separation regions on the leading edge of a slender airfoil. *Izv. Akad. Nauk. SSSR,* Mekh. Zhidk. Gaza, No. 1, 42–51.

[2] Stewartson, K., Smith, F.T. and Kaups, K., (1982). Marginal Separation. *Studies in Applied Mathematics* **67**, 45–61.

G. Hackmüller and A. Kluwick,
Institut für Strömungslehre und
Wärmeübertragung (Inst. 322)
Technical University of Vienna
Wiedner Hauptstraβe 7
A-1040 Wien
AUSTRIA

U. SCHAFLINGER

Asymptotic theory of particle settling in compartment centrifuges

Abstract

The separation of a suspension in a centrifuge is strongly influenced by the Coriolis force which has a negative influence on the efficiency of the settling process as long as the cylindrical container has no instalments. Recently, it has been found that inserted walls yield an additional convective flow caused by the Coriolis force, and separation becomes significantly enhanced. Due to continuity, a layer of clear liquid with a relatively high speed forms adjacent to each radial front wall similar to gravity settling in tilted vessels. Furthermore, compartment walls block the circumferential flow and separation becomes more sensitive to the shape of the end-caps. By means of a generalized theory of kinematic waves we shall obtain results for the asymptotic limit of $EkRo \Rightarrow 0$ and $Ro^2 \Rightarrow 0$.

1. Bulk flow

Centrifugal separation of a mixture that consists of a continuous liquid and a heavier, dispersed phase has been investigated in several papers *cf.* Amberg *et al.* (1986), Greenspan (1983) and Schaflinger and Stibi (1987). It is well known that, when the Coriolis force is dominant, the flow within the core is unaffected by the geometry. Therefore, stacks of conical-shaped end-caps, often used in industrial designs, do not improve automatically the separation process as one would expect from gravity settling, where inclined walls cause a convective flow and the separation is enhanced. By installing radial walls, the complete azimuthal motion of the mixture is blocked and the process becomes sensitive to the shape of the end-caps and a convective flow in the meridional plane emerges (Greenspan and Ungarish, 1985). Such radial walls, however, also produce an additional convective flow in the cross-sectional plane (Schaflinger *et al.* 1986 and Schaflinger 1987) and settling is even more augmented.

We investigate separation in cylindrical centrifuges having conical end-caps and radial compartment walls. The theoretical results are restricted to small density differences between the continuous and the dispersed phase. This assumption is commonly fulfilled in practical applications and guarantees that spin-up effects can be neglected. The analysis utilizes the general kinematic-wave concept introduced by Schneider (1982) and is based upon the continuity equations for the dispersed phase and the mixture, respectively, and on a drift flux relation describing the relative motion of the two phases under consideration of the Coriolis force. In the case of batch settling, discontinuities in the concentration propagate as kinematic shocks that separate the mixture from the bulk of the purified clear liquid and from the maximally concentrated layer of dispersed phase, deposed at the outer wall. In order to derive simple results of the bulk flow, the possibilities of further kinematic shocks originating within the mixture and centred waves emerging from the outer cylinder are not taken into account (Anestis and Schneider, 1983). Hence, the initial volume fraction of the dispersed phase is assumed to be small. The Coriolis force causes a boundary-layer adjacent to the radial front wall of each compartment and, if the conical-shaped end-caps are obtuse-angled, the centrifugal force creates a further thin layer of clear liquid at the end-cap. The kinematic wave theory requires the bulk flow to be independent of the boundary-layer flows. Thus, viscous effects have to be small and are bound to very thin sublayers. The entrainment of the dispersed phase into the boundary layers of clear liquid is negligible and a normal components of the flow of dispersed phase have to vanish near the front wall or end-cap.

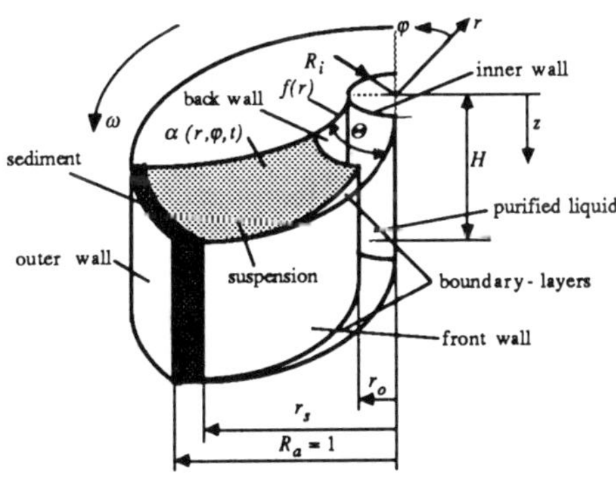

Figure 1. Definition of variables

Consider a mixture of an incompressible, viscous, Newtonian fluid that is denoted by subscript 1 and a dispersed phase consisting of droplets or solid particles of uniform shape and size indicated by subscript 2. Dimensionless variables are introduced by referring all lengths to the outer radius R_a, and the velocities and fluxes in radial (u, j_r), azimuthal (v, j_φ) and axial direction (w, j_z) to $\omega \mathrm{Ro} R_a$. A dimensionless time t is introduced by referring to $(\omega\ \mathrm{Ro})^{-1}$ and the dimensionless pressure p is defined according to

$$p = \left(2\bar{p} - \omega^2 R_a^2 \rho_1 \right) / \omega^2 R_a^2 \rho_1 ,$$

where $\bar{p}$ indicates the dimensional pressure within the mixture. Here, ω denotes the constant angular speed of the container and ρ the density (Fig. 1). From an order of magnitude analysis the Rossby number

$$\mathrm{Ro} = \mathrm{Ta}\ \varepsilon\ (1 + \mathrm{Ta}) / \left(1 + 4\mathrm{Ta}^2 \right), \tag{1}$$

is found to be a definite function of the Taylor number

$$\mathrm{Ta} = \omega\, d^2 \left(\nu_1 \rho_1 + \nu_2 \rho_2 \right) / \, 6\nu_1 \left(2\nu_1 \rho_1 + 3\nu_2 \rho_2 \right), \tag{2}$$

which compares the Coriolis force and the drag force acting on a single particle, and the relative density difference

$$\varepsilon = \left(\rho_2 - \rho_1 \right) / \rho_1 . \tag{3}$$

In Eq. (2) we made use of the Rybczinski–Hadamard formula. The symbol ν denotes the kinematic viscosity and d is the diameter of a single droplet or particle. The Ekman number

$$\mathrm{Ek} = \nu_1 / \omega\, \mathrm{Ra}^2 \tag{4}$$

compares the influence of viscosity and the Coriolis force on the mixture. Both Ro and Ek are usually very small in applications of centrifugal separation. One can ignore the viscosity of the suspension and inertial effects if

$$\mathrm{Ek\ Ro} \Rightarrow 0 \ \text{ and } \ \mathrm{Ro}^2 \Rightarrow 0.$$

Assuming that $\mathrm{Ro}^2 \Rightarrow 0$ the Coriolis terms in the momentum equations of the mixture can be neglected. The Coriolis influence on the relative motion, however, remains. The bulk flow is independent of the flows within the boundary layers if

$$\mathrm{Ek}\ \sqrt{\alpha_0}\ /\ \mathrm{Ro}^4 \Rightarrow 0 \quad \text{and} \quad \frac{\mathrm{d}f(r)}{\mathrm{d}r} = f'(r) \Rightarrow O(\mathrm{Ro}),$$

where α_0 is the initial volume fraction. Using cylindrical coordinates r, φ, z in a frame of reference rotating with constant angular speed ω, the simplified set of governing equations in terms of volume flux densities $\mathbf{j}_1 = \mathbf{v}_1(1 - \alpha)$, $\mathbf{j}_2 = \mathbf{v}_2\,\alpha$ and drift fluxes $\mathbf{j}_{21} = \mathbf{j}_2 - \alpha\mathbf{j}$ are

$$\operatorname{div}\mathbf{j} = 0, \tag{5}$$

$$j_{21_r} = G\left(\alpha, \mathrm{Ta}\right)\mathrm{Ta}\ \varepsilon\left(1 - \alpha\right)\alpha r\ /\ \mathrm{Ro}, \tag{6a}$$

$$j_{21_\varphi} = -\ 2\mathrm{Ta}^2\,\varepsilon\, G\left(\alpha, \mathrm{Ta}\right) F\left(\alpha\right) r\ /\ \mathrm{Ro}, \tag{6b}$$

$$j_{21_z} = 0. \tag{6c}$$

In addition we have

$$\operatorname{grad}\alpha = 0, \tag{7a}$$

and

$$\partial\alpha/\partial t = -\ 2G\left(\alpha, \mathrm{Ta}\right)\mathrm{Ta}\ \varepsilon\left(1 - \alpha\right)\alpha/\mathrm{Ro}, \tag{7b}$$

with the abbreviation

$$G\left(\alpha, \mathrm{Ta}\right) = \alpha\left(1 - \alpha\right) F\left(\alpha\right) / \left(\alpha^2\left(1 - \alpha\right)^2 + 4\mathrm{Ta}^2 F\left(\alpha\right)^2\right), \tag{8}$$

where $F(\alpha)$ describes the small-scale interactions between the continuous fluid and the dispersed phase

$$\lim_{\alpha \Rightarrow 0} F\left(\alpha\right) = \alpha, \tag{9}$$

however, a particular choice is not essential in what follows.

Eq. (7a) is a consequence of the momentum equations for the mixture. For smaller particle concentrations no centred waves appear (Anestis and Schneider, 1983) and the concentration α is only a function of time t and the generalized kinematic-wave equation reduces to (7b). In order to solve Eqs. (5) and (7b) we have to specify initial and boundary conditions. At time $t = 0$ we assume to have a homogeneous mixture and full body rotation. Further, the total flux has to be zero at the sediment, the top end-cap and the radial back wall and in addition at the bottom end-cap and the radial front wall the flux of dispersed phase has to vanish. In detail the initial and boundary conditions are:

$$t = 0 \quad : \quad \alpha = \alpha_0, \tag{10a}$$

$$r = r_s \quad : \quad j_r = 0, \tag{10b}$$

$$\varphi = 0 \quad : \quad j_\varphi = 0, \tag{10c}$$

$$z = f(r) \quad : \quad j_z = j_r f'(r), \tag{10d}$$

$$\varphi = \Theta \quad : \quad j_\varphi = -j_{21_\varphi}/\alpha, \tag{10e}$$

$$z = H + f(r) \quad : \quad j_z = (j_r + j_{21_r}/\alpha) f'(r), \tag{10f}$$

with Θ being the sectorial angle of one compartment and r_s the location of the outer shock.

As carried out elsewhere (Schneider, 1982 and Schaflinger, 1987) the radial component of the total flux is

$$j_r = j_r(r, t) \tag{11}$$

and therefore integration of Eq. (5) with respect to (10c)-(10f) and by means of Eqs. (6a,b) yields

$$\frac{\partial (r j_r)}{\partial r} \theta = - G\left(\alpha, \mathrm{Ta}\right) \frac{\mathrm{Ta}\,\varepsilon}{\mathrm{Ro}} \left[\frac{f'(r)}{H} \theta + 2\mathrm{Ta} \frac{F(\alpha)}{\alpha} \right] r. \tag{12}$$

From Eqs. (8) and (12) we see that in the limiting case $\alpha_0 \Rightarrow 0 \left(r_s \Rightarrow 1\right)$ the radial total flux becomes independent of α and the flux is merely a function of the radius r. If we further assume conical end-caps, $f'(r) = \Phi = \text{const}$, the integration of (12) by taking into acount (10b) produces

$$j_r = \frac{\mathrm{Ta}\,\varepsilon}{r\mathrm{Ro}}\left[\frac{\Phi}{3H}\left(1 - r^3\right) + \frac{\mathrm{Ta}}{\theta}\left(1 - r^2\right)\right]. \tag{13}$$

Then, the concentration is given by integrating Eq. (7b) together with the initial condition (10a) and one obtains

$$\alpha = \alpha_0 \exp\left(-2\mathrm{Ta}\,\varepsilon\, t/\mathrm{Ro}\right). \tag{14}$$

In the case of batch settling we are interested in calculating the settling rate and the total settling time. Thus, we have to study the propagation of kinematic shocks.

2. Kinematic-shock relations

Discontinuities separate the mixture from the purified liquid and from the sediment. By applying the conservation law in a frame of reference in which the shock is at rest the propagation velocity W_0 of a discontinuity separating the mixture with the concetration $\alpha(t)$ from the purified liquid at the radius r_0 is

$$W_0 = \mathrm{d}r_0/\mathrm{d}t = j_{r0} + \left(j_{21_{r0}}/\alpha\right). \tag{15}$$

For the propagation velocity W_s of a shock that separates the suspension layer with concentration α_s ($\alpha_s \sim 0.66$) at the radius r_S one derives

$$W_s = \mathrm{d}r_s/\mathrm{d}t = -j_{21_{rs}}/\left(\alpha_s - \alpha\right). \tag{16}$$

The total settling time is then given when the two discontinuities intersect. From Eq. (15) one obtains for small initial concentrations the total settling time

$$t^* = \int_{R_i}^{1} \frac{r_0\,\mathrm{d}r_0}{-\frac{\alpha}{3H} r_0^3 + \left(1 - \frac{\mathrm{Ta}}{\theta}\right) r_0^2 + \frac{\mathrm{Ta}}{\theta} + \frac{a}{3H}}. \tag{17}$$

Fig. 2 shows the settling time t^* as a function of Ta for different compartment angles Θ.

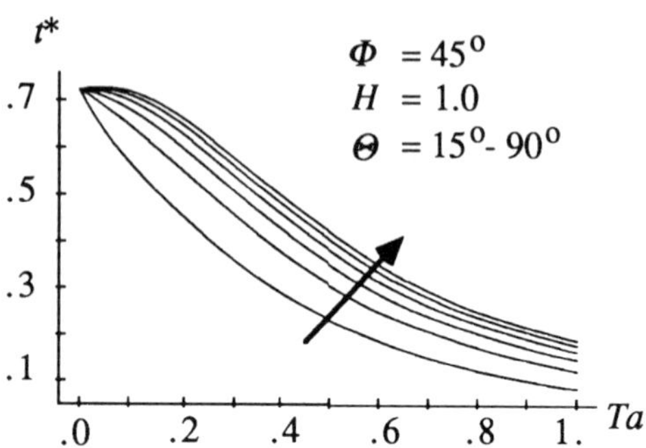

Figure 2. Theoretically predicted total settling time *vs*. Taylor number.

References

Amberg, G., Dahlkild, A.A., Bark, F.H. and Henningson, D.S. (1986). On time-dependent settling of a dilute suspension in a rotating conical channel, *J Fluid Mech.* **166**, 473-502.

Anestis, G. and Schneider, W. (1983). Application of the theory of kinematic-waves to the centrifugation of suspensions, *Ing. Arch.* **53**, 399-407.

Greenspan, H.P. (1983). On centrifugal separation of a mixture, *J. Fluid Mech.* **127**, 91-101.

Greenspan, H.P. and Ungarish, M. (1985). On the enhancement of centrifugal separation, *J. Fluid Mech.* **157**, 359-373.

Schaflinger, U. (1987). Enhanced centrifugal separation with finite Rossby numbers in cylinders with compartment walls. *Chem. Eng. Science* **42**, 1197-1205.

Schaflinger, U. and Stibi, H. (1987). On centrifugal separation of suspensions in cylindrical vessles, *Acta Mech.* **67**, 163-181.

Schaflinger, U., Köppl, A. and Filipczak, G. (1986). Sedimentation in cylindrical centrifuges with compartments, *Ing. Arch.* **56**, 321-331.

Schneider, W. (1982). Kinematic-wave theory of sedimentation beneath inclined walls, *J. Fluid Mech.* **120**, 323-346.

U. Schaflinger
Institut für Strömungslehre und
Wärmeübertragung
Technical University of Vienna
Wiedner Hauptstraβe 7
A-1040 Wien
AUSTRIA.

N.T. KOVATCHEVA AND A.D. POLYANIN

Convective diffusion of particles in a viscous liquid with a variable diffusion coefficient

The convective diffusion problem for particles in viscous liquid flow is a classical problem in applied mathematics. Theoretical and experimental investigations are concerned with the rate at which some diffusible quantity is transferred from a surface. At the same time the authors have assumed that the diffusivity is constant (for example, see([1]-[3]). There is now experimental evidence ([4]-[7]) that the changes in the concentration of the diffusible quantity lead to a change in the diffusion coefficient.

Here, our purpose is to investigate theoretically the two-dimensional steady-state diffusion problem for particles (drops, bubbles and solid particles) with a viscous liquid when the diffusivity $\mathcal{D}$ is an arbitrary function of the concentration C, i.e. $\mathcal{D} = \mathcal{D}(C)$. We also discuss the case of hyperbolic dependence of $\mathcal{D}$ on C when the diffusion equation has a solution in closed form and simple formulae for the diffusion flux and the average Sherwood number (Sh) may be obtained. Asymptotic results for large Péclet numbers (Pe) are considered.

The concentration of the diffusible quantity in the fluid satisfies the equation

$$\mathbf{V}.\nabla C = \mathcal{D}(C)\, \nabla^2 C, \tag{1}$$

where $\mathbf{V}$ is the local fluid velocity relative to axes such that the particle has zero translational velocity. The boundary conditions are assumed to be

$$\begin{aligned} C &= C_1 \quad \text{at the particle surface} \\ C &= C_0 \quad \text{as } r \to \infty, \end{aligned} \tag{2}$$

and r is the distance from the particle surface. The concentrations in each phase are considered to be uniform, C_1 and C_0 are constants and $C_1 \neq C_0$.

We use dimensionless orthogonal curvilinear coordinates ξ, η, χ which are fixed to the particle. The η coordinate is parallel to the surface and the ξ coordinate is normal to the surface. The concentration and the velocity vector do not depend on the χ coordinate.

Eq. (1), subject to the thin boundary layer approximation (Pe >> 1), then becomes

$$\frac{V_\xi}{\sqrt{\overset{s}{g}_{\xi\xi}}}\frac{\partial \tilde{c}}{\partial \xi}+\frac{V_\eta}{\sqrt{\overset{s}{g}_{\eta\eta}}}\frac{\partial \tilde{c}}{\partial \eta}=\frac{1}{\text{Pe}}\frac{1}{\overset{s}{g}_{\xi\xi}}\frac{\partial}{\partial \xi}f(\tilde{c})\frac{\partial \tilde{c}}{\partial \xi} \tag{3}$$

with the boundary conditions

$$\xi = 0,\ \tilde{c} = 1;\quad \xi \to \infty,\ \tilde{c} \to 0. \tag{4}$$

The dimensionless concentration, diffusivity and Péclet number are

$$\tilde{c} = \frac{C_0 - C}{C_0 - C_1},\ f = f(\tilde{c}) = \frac{\mathcal{D}(C)}{\mathcal{D}(C_0)},\ \text{Pe} = \frac{LU}{\mathcal{D}(C_0)}.$$

The components $\overset{s}{g}_{\xi\xi}$, $\overset{s}{g}_{\eta\eta}$ and $\overset{s}{g}_{\chi\chi}$ of the metrical tensor are taken on the particle surface (at $\xi = 0$).

For this flow we may define a stream function, ψ, as follows:

$$V_\xi = -\frac{1}{\sqrt{\overset{s}{g}_{\eta\eta}\overset{s}{g}_{\chi\chi}}}\frac{\partial \psi}{\partial \eta},\quad V_\eta = \frac{1}{\sqrt{\overset{s}{g}_{\eta\eta}\overset{s}{g}_{\chi\chi}}}\frac{\partial \psi}{\partial \xi}. \tag{5}$$

Near the particle surface (when $\xi \to 0$) ψ may be approximated by $\psi = \xi^n \varphi(\eta)$ and n is equal to 1, 2 or 3, depending on the form of the flow as well as on the particle consistence and geometry (see [3]).

By introducing the new variables

$$\zeta = \text{Pe}^{\frac{1}{n+1}}\psi^{\frac{1}{n}},\quad \tau = \frac{1}{n}\int_0^\eta \left(\frac{\overset{s}{g}_{\eta\eta}\overset{s}{g}_{\chi\chi}}{\overset{s}{g}_{\xi\xi}}\right)^{\frac{1}{2}}\left[\varphi(\eta)\right]^{\frac{1}{n}}\, d\eta \tag{7}$$

the problem (3), (4) is reduced to

$$\frac{\partial \tilde{c}}{\partial \tau} = \zeta^{1-n}\frac{\partial}{\partial \zeta}f(\tilde{c})\frac{\partial \tilde{c}}{\partial \zeta}, \tag{8}$$

$$\tau = 0,\ \tilde{c} = 0;\ \zeta = 0,\ \tilde{c} = 1;\ \zeta \to \infty,\ \tilde{c} \to 0. \tag{9}$$

In order to solve Eq. (8) we introduce the similarity variable

$$z = \zeta \tau^{-\frac{1}{n+1}}, \tag{10}$$

which leads to the ordinary differential equation

$$(n+1)\left[f(\tilde{c})\tilde{c}'_z\right]'_z + z^n \tilde{c}'_z = 0 \tag{11}$$

and

$$z = 0,\ \tilde{c} = 1;\ z \to \infty,\ \tilde{c} \to 0. \tag{12}$$

For example some unsteady problems of the nonlinear heat transfer and filtration may be reduced to Eq. (11) when $n = 1$. Eq. (11) has a solution in closed form for $f(\tilde{c}) = (1 + \alpha\tilde{c} + \beta\tilde{c}^2)^{-1}$ and $n = 1$ (see [8]) and the problem (11), (12) has been solved numerically for $f(\tilde{c}) = \exp(\alpha\tilde{c})$, $f(\tilde{c}) = 1 + \alpha\tilde{c}$ and $n = 1$.

Now we consider the hyperbolic dependence

$$\mathcal{D} = \frac{\mathcal{D}_0}{1 + \mathrm{K}c} \tag{13}$$

and an arbitrary value of the parameter n. The constants $\mathcal{D}_0$ and K are, respectively, the diffusivity for infinite dilution and an empirical constant. By using the dimensionless quantities we have

$$f(\tilde{c}) = \frac{1}{1 + \kappa\tilde{c}}, \qquad \text{where} \qquad \kappa = \frac{\mathrm{K}(c_1 - c_0)}{1 + \mathrm{K}c_0}. \tag{14}$$

After a detailed analysis of Eq. (11) with the boundary conditions (12) and after some calculations and evaluations it is possible to obtain the solutoin of Eq. (11) in closed form. This mathematical procedure is lengthy and may be omitted here.

It must be noted that in the case of hyperbolic dependence (14) a simple formula for the average Sherwood number is derived. Thus we introduce the coefficient of nonlinearity

$$\alpha = \alpha(\kappa, n) = \bar{j}/\bar{j}_0 \tag{15}$$

as a characteristic quantity of the nonlinearity of the problem, where $\bar{j}$ and $\bar{j}_0$ are the normalized diffusion flux for $f(\tilde{c})$ given where $\bar{j}$ and $\bar{j}_0$ are the normalized diffusion flux for $f(\tilde{c})$ given by (14) and for constant diffusion coefficient $f = 1$, respectively. The dimensionless local diffusion flux on the particle surface (drop, bubble, solid particle) is

$$j = -\frac{1}{\sqrt{g^s_{\xi\xi}}}\left(f\frac{\partial\tilde{c}}{\partial\xi}\right)_{\xi=0} = -\left(\frac{f}{\sqrt{g^s_{\xi\xi}}}\frac{\partial\tilde{c}}{\partial z}\frac{\partial z}{\partial\zeta}\frac{\partial\zeta}{\partial\xi}\right)_{\xi=0} \tag{16}$$

$$= \mathrm{Pe}^{\nu}\frac{\varphi^{1/n}\tau^{-\nu}}{\sqrt{g^s_{\xi\xi}}}\left(-f\tilde{c}_z'\right)_{z=0} = \mathrm{Pe}^{\nu}\frac{\varphi^{1/n}\tau^{-\nu}}{\sqrt{g^s_{\xi\xi}}}\bar{j} = \alpha\left(\mathrm{Pe}^{\nu}\frac{\varphi^{1/n}\tau^{-\nu}}{\sqrt{g^s_{\xi\xi}}}\bar{j}_0\right)$$

and we have the simple relationship

$$j = \alpha\, j_0, \tag{17}$$

where j_0 is the dimensionless local diffusion flux for $f = 1$ (or $\kappa = 0$). By integrating (16) over the particle surface we obtain the average Sherwood number as follows:

$$\mathrm{Sh}(f) = \alpha\, \mathrm{Sh}\,(1). \tag{18}$$

This formula may be applied for an arbitrary function f if α is given by the formula

$$\alpha = \nu^{1-2\nu}\,\Gamma(\nu)\left[-f(\tilde{c})\,\tilde{c}'_z\right]_{z=0},\quad \nu = \frac{1}{n+1} \tag{19}$$

and $\tilde{c}$ is the solution of the problem (11), (12), Γ is the gamma function.

References

[1] Levich, V.G. *Physiochemical Hydrodynamics*, Prentice-Hall, New York, 1962.

[2] Clift, R., Grace, J.R. and Weber, M.E. *Bubbles, drops and particles,* Academic Press, New York, San Francisco and London, 1978.

[3] Gupalo, Yu.P., Polyanin, A.D. and Ryazantsev, Yu.S. *Mass and Heat Transfer of Reacting Particles with a Flow* (in Russian), Nauka, Moscow, 1985.

[4] Dil'man, V.V. and Polyanin, A.D. *Methods of Model Equations and Analogies in the Chemical Engineering* (in Russian), Khimiya, Moscow, 1988.

[5] Bretshnayder, S. *Properties of Gases and Liquids (Engineering Methods)* (in Russian), Khimiya, Leningrad, 1986.

[6] Rid, R., Prausnits, D. and Sherwood, T. *Properties of Gases and Liquids* (in Russian), Khimiya, Leningrad, 1982.

[7] Sherwood, T., Pigford, R. and Wilke, C. *Mass Trasnfer* (in Russian), Khimiya, Moscow, 1982.

[8] Fujita, M. The exact pattern of a concentration-dependent diffusion in a semi-infinite medium. *Textile Res. J.* **22**, 757-760 (1952).

N.T. Kovatcheva
Institute of Mechanics
Bulgarian Academy of Sciences
P.O. Box 373 1090 Sofia,
BULGARIA

A.D. Polyanin
Institute for Problems in Mechanics
USSR Academy of Sciences
Vernadsky Avenue 101 117526 Moscow,
USSR

Part IV
Material Science and Thermodynamic Aspects of Continuum Mechanics

I. MÜLLER

The structure of extended thermodynamics

Foreword

In my contribution to STAMM 8 I presented an account of the structure of extended thermodynamics together with an application to relativistic gases which nicely illustrates the restrictive power of the constitutive theory of extended thermodynamics. That application is not contained in the present work, because it had been published before, see [1], and [2].

Instead, here I have tried to give a streamlined account of the mathematical structure of extended thermodynamics. Much credit for the revelation of this structure must go to Ruggeri and his co-workers, e.g. see [3].

1. Structure of extended thermodynamics

The objective of thermodynamics is the determination of N fields which we combine in the N-vector $\mathbf{u}(x^\delta)$. x^δ are space-time coordinates of an event with $x^0 = t$, $x^a = (x^1, x^2, x^3)$.

The necessary field equations are based on N equations of balance of the form

$$\mathbf{F}^\alpha{}_{,\alpha} = \mathbf{f}\,. \tag{1.1}$$

The comma denotes partial differentiation. The components of $\mathbf{F}^0$ may be called densities and the components of $\mathbf{F}^a$ fluxes and we refer to $\mathbf{F}^\alpha$ as four-dimensional fluxes. Their divergence in space-time is equal to the production vector $\mathbf{f}$.

In order to give field equations for the fields $\mathbf{u}$, the balance equations must be supplemented by constitutive equations. Such constitutive equations relate the fluxes $\mathbf{F}^\alpha$ and the products $\mathbf{f}$ to the fields $\mathbf{u}$ in a materially dependent manner. In extended thermodynamics the constitutive equations are of the form

$$\mathbf{F}^\alpha = \hat{\mathbf{F}}^\alpha(\mathbf{u}) \quad \text{and} \quad \mathbf{f} = \hat{\mathbf{f}}(\mathbf{u}) \tag{1.2}$$

so that the fluxes $\mathbf{F}^\alpha$ and productions $\mathbf{f}$ at one event depend only on the values of the fields $\mathbf{u}$ at that event. We say that the constitutive relations are local in space-time.

If the constitutive functions $\hat{\mathbf{F}}^\alpha$ and $\hat{\mathbf{f}}$ were known explicitly, we could eliminate $\mathbf{F}^\alpha$ and $\mathbf{f}$ between (1.1) and (1.2) and obtain an explicit set of field equations.

Every solution of these field equations is called a *thermodynamic process*.

2. Constitutive theory

In reality of course the constitutive functions are not known and it is the task of the constitutive theory to determine these functions, or at least to reduce their generality. The tools of the constitutive theory are certain universal principles which we rely on from experience. The most important three among these principles are

(i) the entropy principle,
(ii) the requirement of convexity and causality,
(iii) the principle of relativity (in non-relativistic theories this principle is sometimes known as the principle of material objectivity).

The most important part of the entropy principle is the entropy inequality

$$h^{\alpha}{}_{,\,\alpha} \leq 0, \tag{2.1}$$

which must hold for all thermodynamic processes. Here $-h^{\alpha}$ is the entropy flux vector which is a constitutive quantity such that in extended thermodynamics we have

$$h^{\alpha} = \hat{h}^{\alpha}(\mathbf{u}). \tag{2.2}$$

The requirement of convexity and causality demands that the field equations be symmetric hyperbolic. This requirement ensures the well-posedness of Cauchy problems for the field equations and it ensures that all wave speeds are finite. If the entropy principle is accepted as stated above, the requirement of convexity and causality is equivalent to the statement that

$$-\hat{h}^{0}(\mathbf{u}) \text{ is a concave function, i.e. } \frac{\partial^2 \hat{h}^0}{\partial \mathbf{u}\,\partial \mathbf{u}} \sim \text{pos. def.} \tag{2.3}$$

The principle of relativity states that the field equations have the same form in all frames. In particular this implies that the constitutive functions must be invariant under transformations of frame. (In the non-relativistic case the statement of this principle is a little more complicated, because the balance equations are usually different in inertial and non-inertial frames).

3. Exploitation of the entropy principle

We recall that the entropy inequality must hold for all thermodynamic processes, i.e. solutions of the field equations. In a manner of speaking the field equations provide constraints for the fields which must satisfy the entropy inequality. It is possible to use Lagrange multipliers $\mathbf{\Lambda}$ to eliminate such constraints. Indeed, the new inequality

$$h^{\alpha}{}_{,\alpha} - \mathbf{\Lambda}\,(\mathbf{F}^{\alpha}{}_{,\alpha} - \mathbf{f}) \leq 0 \tag{3.1}$$

must hold for *all* fields $\mathbf{u}$. The Lagrange multipliers are functions of $\mathbf{u}$. For proof of the existence and of the properties of Lagrange multipliers see [4].

We rewrite (3.1) in the form

$$(h^{\alpha} - \mathbf{\Lambda}\,\mathbf{F}^{\alpha})_{,\alpha} + \mathbf{F}^{\alpha}\,\mathbf{\Lambda}_{,\alpha} + \mathbf{\Lambda}\,\mathbf{f} \leq 0 \tag{3.2}$$

and introduce a new vector

$$h'^{\alpha} = -\,h^{\alpha} + \mathbf{\Lambda}\,\mathbf{F}^{\alpha}, \tag{3.3}$$

which we call the vector potential for reasons that will soon become apparent. Thus (3.2) reads

$$-\,h'^{\alpha}{}_{,\alpha} + \mathbf{F}^{\alpha}\,\mathbf{\Lambda}_{,\alpha} + \mathbf{\Lambda}\,\mathbf{f} \leq 0. \tag{3.4}$$

In the next step we introduce the Lagrange multipliers $\mathbf{\Lambda}$ as new variables instead of $\mathbf{u}$. The requirement of convexity and causality ensures that there is a one-to-one correspondence between the two sets, but we postpone proof of that statement. With the $\mathbf{\Lambda}$s as variables $\mathbf{F}^{\alpha}, \mathbf{f}$ and h^{α}, hence h'^{α} are all functions of $\mathbf{\Lambda}$

$$\mathbf{F}^{\alpha} = \bar{\mathbf{F}}(\mathbf{\Lambda}),\ \mathbf{f} = \bar{\mathbf{f}}(\mathbf{\Lambda}),\ h^{\alpha} = \bar{h}^{\alpha}(\mathbf{\Lambda}),\ h'^{\alpha} = \bar{h}'^{\alpha}(\mathbf{\Lambda}). \tag{3.5}$$

Therefore, by use of the chain rule, we may write (3.4) in the form

$$\left(-\frac{\partial h'^{\alpha}}{\partial \mathbf{\Lambda}} + \mathbf{F}^{\alpha}\right)\mathbf{\Lambda}_{,\alpha} + \mathbf{\Lambda}\cdot\mathbf{f} \leq 0. \tag{3.6}$$

This inequality must now hold for all fields $\mathbf{\Lambda}$. Therefore it must hold in particular for all gradients $\mathbf{\Lambda}_{,\alpha}$ at an event. And since the left-hand side of the inequality is explicitly linear in $\mathbf{\Lambda}_{,\alpha}$, the factor of that gradient must vanish lest the inequality be violated for some choice of $\mathbf{\Lambda}_{,\alpha}$. Thus we conclude

$$\mathbf{F}^{\alpha} = \frac{\partial h'^{\alpha}}{\partial \mathbf{\Lambda}} \quad \text{and with (3.3)} \quad h^{\alpha} = -\,h'^{\alpha} + \mathbf{\Lambda}\frac{\partial h'^{\alpha}}{\partial \mathbf{\Lambda}}. \tag{3.7}$$

It follows that *the constitutive qualities* $\mathbf{F}^{\alpha}$ *and* h^{α} *can be derived from a single vector*, viz. h'^{α}. This justifies the name 'vector potential' for h'^{α}.

Of course there is a residual inequality, viz.

$$\mathbf{\Lambda}\,\mathbf{f} \leq 0. \tag{3.8}$$

The relations (3.7) and (3.8) exhaust the consequences of the entropy principle and now we proceed to investigate the significance of these results for our primary constitutive quantities $\mathbf{F}^{\alpha}$ and $\mathbf{f}$. First $\mathbf{F}^{\alpha}$: The integrability condition for h'^{α} implied by (3.7) states that the four $(N \times N)$-matrices

$$\frac{\partial \mathbf{F}^{\alpha}}{\partial \mathbf{\Lambda}} \text{ be symmetric.} \tag{3.9}$$

These are obviously many severe restrictions on the functions $\mathbf{F}^{\alpha}(\mathbf{\Lambda})$. Next $\mathbf{f}$: $\mathbf{f}$ is less severely restricted, because all we know is that it must satisfy the inequality (3.8).

There is, however, a serious blemish on these results, because they hold for the Lagrange-multipliers as variables and the Lagrange multipliers – *a priori* – have no physical significance. Before we obtain 'useful' results, i.e. results for $\mathbf{F}^{\alpha}$ and $\mathbf{f}$ as functions of the chosen physically significant variables $\mathbf{u}$, we must reverse the switch of variables form $\mathbf{u}$ to $\mathbf{\Lambda}$. This is a most cumbersome task which will not be described in this paper. The reason is that the arguments strongly depend on the tensorial character of the variables $\mathbf{u}$ or $\mathbf{\Lambda}$ in space-time, which is different for different materials. Therefore such argumetns do not fit well into this description of the generic mathematical structure of extended thermodynamics.

I conclude the present section with the proof of the above statement that the switch of variables from $\mathbf{u}$ to $\mathbf{\Lambda}$ is always possible, cumbersome in detail as it may be.

Note that the inequality (3.1) may be written in the form

$$\left(\frac{\partial h^{\alpha}}{\partial \mathbf{u}} - \mathbf{\Lambda}\,\frac{\partial \mathbf{F}^{\alpha}}{\partial \mathbf{u}}\right)\mathbf{u}_{,\alpha} + \mathbf{\Lambda}\,\mathbf{f} \leq 0 \tag{3.10}$$

which must hold for all fields $\mathbf{u}$. By an argument akin to the one that led from (3.6) to (3.7) we obtain

$$\frac{\partial h^{\alpha}}{\partial \mathbf{u}} = \mathbf{\Lambda}\frac{\partial \mathbf{F}^{\alpha}}{\partial \mathbf{u}}. \tag{3.11}$$

Now let us choose – without essential loss of generality – the fields $\mathbf{u}$ to be equal to the densities $\mp^{0}$. Hence it follows from (3.11) for $\alpha = 0$

$$\frac{\partial h^{0}}{\partial \mathbf{u}} = \mathbf{\Lambda} \tag{3.12}$$

and by differentiation with respect to $\mathbf{u}$

$$\frac{\partial^2 h^0}{\partial \mathbf{u} \partial \mathbf{u}} = \frac{\partial \mathbf{\Lambda}}{\partial \mathbf{u}}. \tag{3.13}$$

Now by the requirement (2.3) of convexity and causality the left-hand side of (3.13) is negative definite. Therefore the Jacobian of the transformation $\mathbf{u} \Leftrightarrow \mathbf{\Lambda}$ is non-zero and the change of variables is possible.

4. Symmetric hyperbolic field equations

In this section I propose to show that the field equations (1.1) and (1.2) or $(3.5)_{1,2}$, form a symmetric hyperbolic system, if their solution satisfies the entropy inequality (2.1) with (2.2), or $(3.5)_3$, in which $-\hat{h}^0(\mathbf{u})$ is a concave function.

Indeed, we combine (1.1) and (3.5) to give

$$\frac{\partial F^\alpha}{\partial \mathbf{\Lambda}} \mathbf{\Lambda}_{,\alpha} = \mathbf{f} \text{ or by } (3.7)_1 \ \frac{\partial^2 h'^\alpha}{\partial \mathbf{\Lambda} \partial \mathbf{\Lambda}} \mathbf{\Lambda}_{,\alpha} = \mathbf{f}\,. \tag{4.1}$$

Therefore the system of equations is symmetric, since the Hessian matrices $\partial^2 h'^\alpha/\partial\mathbf{\Lambda}\partial\mathbf{\Lambda}$ are, of course symmetric. And the system is symmetric hyperbolic, if $\partial^2 h'^0/\partial\mathbf{\Lambda}\partial\mathbf{\Lambda}$ is positive definite so that h'^0 is a convex function of the components of $\mathbf{\Lambda}$. This latter condition, however, is satisfied, because of (2.3). For proof we assume - again without essential loss of generality - that $\mathbf{u} = \mathbf{F}^0$ holds. By $(3.7)_1$, (3.12) and (2.3) we then have

$$\frac{\partial^2 h'^0}{\partial \mathbf{\Lambda} \partial \mathbf{\Lambda}} \delta\mathbf{\Lambda}\, \delta\mathbf{\Lambda} = \delta\mathbf{u}\, \delta\mathbf{\Lambda} = \frac{\partial^2 h^0}{\partial \mathbf{u} \partial \mathbf{u}} \delta\mathbf{u}\, \delta\mathbf{u} > 0. \tag{4.2}$$

The reason why we focus attention on symmetric hyperbolic systems lies in the convenient and desirable properties of existence, uniqueness and well-posedness of their solutions. We note that the symmetric hyperbolic character has several ingredients which are all satisfied in extended thermodynamics, viz.

field equations of balance type,
local constitutive relations,
existence of an entropy inequality,
convexity of the entropy density.

This is quite satisfactory.

And yet, one might have more far-reaching ambitions. Indeed, there is a persistent suspicion among thermodynamicists that the entropy principle should not be postulated *a priori*, but rather that it should follow from a reasonable and well-motivated requirement for the solution of the field equations.

In thermostatics Carathéodory [5] has realized such a program: For the solution of the adiabatic energy equation he postulated inaccessibility of some states adjacent to the initial state and, starting from there, he derived the concept of entropy and the entropy inequality.

In thermodynamics an analogous argument is not yet known. In my paper 'Entropy in non-equilibrium - a challenge to mathematicians' [6] I have tried to advertise the view that such an argument is needed. So far no mathematician has seriously taken up that challenge.

5. Equilibrium

Whatever the choice of balance equations (1.1) may be in different thermodynamic theories, they invariably contain the *conservation laws* of mass, momentum and energy. The components of $\mathbf{F}^0$ corresponding to those conservation laws are the densities of mass, momentum and energy and we take those to be the first five components of $\mathbf{F}^0$. The first five components of $\mathbf{f}$ are then equal to zero, of course. The remaining components of $\mathbf{f}$ form an $(N - 5)$-vector which we shall denote by $\bar{\mathbf{f}}$. The residual inequality (3.8) thus reads

$$\bar{\Lambda}\bar{\mathbf{f}} \leq 0, \tag{5.1}$$

where $\bar{\Lambda}$ is an $(N - 5)$-vector consisting of the components Λ_6 through Λ_N.

Equilibrium is defined as a process in which the productions $\bar{\mathbf{f}}(\mathbf{u})$ all vanish. Therefore the residual inequality (3.8), whose left-hand side represents the positive entropy production, has its minimum, namely zero in equilibrium. This implies - under the assumption that the constitutive equations $\bar{\mathbf{f}} = \bar{\mathbf{f}}(\mathbf{u})$ may be inverted - that the Lagrange multipliers $\bar{\Lambda}$ must vanish in equilibrium

$$\bar{\Lambda}|_E = 0. \tag{5.2}$$

6. Absolute temperature and chemical potential

We now proceed to determine the values of the five non-vanishing Lagrange multipliers in equilibrium. In order to cut the argument short we do this by reference of the Gibbs equation of thermostatics which reads

$$\mathrm{d}\eta = \frac{1}{T}(\mathrm{d}\varepsilon + p d(1/\rho)) \tag{6.1}$$

where p is the pressure and T is the absolute temperature; ε and η are the specific values of internal energy and equilibrium entropy respectively. In particular, we conclude from (6.1) that the equilibrium entropy is independent of the velocity v_i.

Seeing that the first five components of $\mathbf{F}^0$ are the densities of mass momentum and energy we choose the conventional notation and denote them by

$$\rho, \rho v_i, \quad \rho e = \rho\,(\frac{1}{2}v^2 + \varepsilon), \tag{6.2}$$

where ρ is the mass density, v_i is the velocity and e is the specific value of the energy, kinetic and internal. The corresponding Lagrange multipliers will be denoted by

$$\Lambda^{\rho}, \ \Lambda^{\rho v_i}, \ \Lambda^{\rho e}. \tag{6.3}$$

By (5.2) those are the only non-vanishing Lagrange multipliers. Without essential loss of generality we may now take the densities (6.2) to be the first five components of the vector $\mathbf{F}^0$. Thus in equilibrium the equations (3.11) for $\alpha = 0$ may be combined to give

$$d(\rho\eta) = -\Lambda_E^{\rho}\,d\rho - \Lambda_E^{\rho v_i}\,d(\rho v_i) - \Lambda_E^{\rho e}\,d(\rho\,(\frac{1}{2}\,v^2 + \varepsilon)), \tag{6.4}$$

where the equilibrium entropy density $-h_E^0$ has been denoted by $\rho\eta$. A slight rearrangement of this expression and comparison with the Gibbs equation (6.1) gives

$$\begin{aligned}
\Lambda_E^{\rho} &= +\frac{1}{T}\,(\varepsilon + \frac{1}{2}\,v^2 - T\eta + \frac{p}{\rho}),\\
\Lambda_E^{\rho v_i} &= +\frac{1}{T}v_i,\\
\Lambda_E^{\rho e} &= -\,\frac{1}{T}.
\end{aligned} \tag{6.5}$$

The equilibrium values of the Lagrange multipliers have thus been determined: All but the first five of them vanish and those five are related by (6.5) to measurable quantities, viz. the temperature T, the velocity v_i, and the chemical potential $g = \varepsilon - T\eta + p/\rho$.

Actually the foregoing analysis holds for a non-relativisitc theory; it is different in several subtle ways for relativistic theories.

7. On the difficulty in eliminating the Lagrange multipliers

We recall that the conditions (3.9) indicate strong restrictions on the constitutive functions. However, there is no immediately obvious way to evaluate the conditions to obtain restrictions on the functions $\mathbf{F}^\alpha(\mathbf{u})$, because (3.9) concerns the functions $\mathbf{F}^\alpha(\mathbf{\Lambda})$. We need to have $\mathbf{\Lambda} = \mathbf{\Lambda}(\mathbf{u})$ before the conditions (3.9) become useful.

In all successful special cases the determination of the relation $\mathbf{\Lambda} = \mathbf{\Lambda}(\mathbf{u})$ has made use of the principle of relativity. As a consequence of that principle the balance equations (1.1) must be tensor equations in space-time and the constitutive functions (1.2), (2.2), or (3.5) must be isotropic functions with respect to space-time transformations.

Therefore the actual calculation of the relation $\mathbf{\Lambda} = \mathbf{\Lambda}(\mathbf{u})$ must be preceded by a specification of the tensorial character - in space-time - of the components of $\mathbf{u}$ and $\mathbf{F}^\alpha$. This is different for different classes of materials and therefore the generic structure of extended thermodynamics cannot be maintained beyond this point. I conclude this paper with a summary of special cases that have been treated in the literature.

Non-relativisitic mon-atomic gases (see [7] or [8])

Variables: mass-density F
momentum density F_i
momentum flux density F_{ij}
energy flux density $\frac{1}{2} F_{ijj}$

Balance laws:

conservation of mass $$\frac{\partial F}{\partial t} + \frac{\partial F_i}{\partial x_i} = 0$$

conservation of momentum $$\frac{\partial F_i}{\partial t} + \frac{\partial F_{ij}}{\partial x_j} = 0$$

balance of momentum flux $$\frac{\partial F_{ij}}{\partial t} + \frac{\partial F_{ijk}}{\partial x_k} = f_{<ij>}$$

balance of energy flux $$\frac{\partial F_{ijj}}{\partial t} + \frac{\partial F_{ijjk}}{\partial x_k} = f_{ijj}.$$

Relativisitic mon-atomic gases (see [1] and [2]

Variables: particle flux A^α
stress-energy-momentum tensor $A^{\alpha\beta}$

Balance laws: conservation of particle number $A^{\alpha}{}_{,\alpha} = 0$
conservation of energy-momentum $A^{\alpha\beta}{}_{,\beta} = 0$
balance of fluxes $A^{\alpha\beta\gamma}{}_{,\gamma} = I^{\alpha\beta}$.

Non-relativisitic molecular gases (see [9])

Variables: mass density F
momentum density F_i
momentum flux density F_{ij}
translational energy flux density $\frac{1}{2} F_{ijj}$
intrinsic energy density ρg
intrinsic energy flux density G_i

Balance laws: conservation of mass $$\frac{\partial F}{\partial t} + \frac{\partial F_i}{\partial x_i} = 0$$

conservation of momentum $$\frac{\partial F_i}{\partial t} + \frac{\partial F_{ij}}{\partial x_j} = 0$$

balance of momentum flux $$\frac{\partial F_{ij}}{\partial t} + \frac{\partial F_{ijk}}{\partial x_k} = f_{ij}$$

balance of translational energy flux $$\frac{\partial F_{ijj}}{\partial t} + \frac{\partial F_{ijjk}}{\partial x_u} = f_{ijj}.$$

balance of intrinsic energy $$\frac{\partial \rho g}{\partial t} + \frac{\partial G_i}{\partial x_i} = -\tfrac{1}{2} f_{ii}$$

balance of intrinsic energy flux $$\frac{\partial G_i}{\partial t} + \frac{\partial G_{ij}}{\partial x_j} = f_i \,.$$

The reader who is interested enough to check the references given to the above specific cases will find out that the restrictive principles of extended thermodynamics are powerful enough to determine all constitutive functions for the densities and fluxes, at least in irreversible processes close to equilibrium. The constitutive functions for the productions can be related to transport coefficients like heat-conductivity or the viscosities.

References

[1] Liu, I.-Shih, Müller, I., Ruggeri, T. Relativistic thermodynamics of gases. *Annals of Physics* **169** (1986).

[2] Müller, I. *Relativistic Extended Thermodynamics*, London Math. Soc. Lecture Notes **122**, Cambridge University Press (1987).

[3] Ruggeri, T. *Relativistic Extended Thermodynamics: General Assumptions and Mathematical Procedure*, Springer Lecture Notes in Mathematics **1385** (1989).

[4] Liu, I.-Shih, On the use of Lagrange multipliers for the exploitation of the entropy inequality. *Arch. Rational Mech. Anal.* **46** (1972).

[5] Carathéodory, C. Untersuchungen über die Grundlagen der Thermodynamik. *Math. Annalen* **67** (1909).

[6] Müller, I. *Entropy in Non-Equilibrium - A Challenge to Mathematicians. Trends in Applications of Pure Mathematics to Mechanics II*, Pitman, London (1979).

[7] Liu, I.-Shih and Müller, I. Extended thermodynamics of classical and degenerate gases, *Arch. Rational Mech. Anal.* **83** (1983).

[8] Müller, I. *Thermodynamics*, Pitman, London (1985).

[9] Kremer, G.M. Extended thermodynamics of molecular ideal gases. *Cont. Mech. Thermodyn.* **1** (1989).

I. Müller
TU-Berlin
Hermann-Föttinger-Institute für Thermo- und Fluiddynamics
Str. des 17. Juni 135
D-1000 Berlin 12
GERMANY

A.D. PIERCE

Irreversible thermodynamics formulation of sound generation by heat in liquids with internal relaxation

Abstract

When heat is added to a liquid, as by absorption of electromagnetic radiation, one of the outcomes is the generation of an acoustic wave that can possibly be detected at large radial distances. The present paper attempts a rational formulation following the general methodology of Meixner, de Groot, and others of the governing equations for this sound generation when the fluid remains in a single phase, but does not necessarily expand or contract during slow addition of an infinitesimal amount of heat. Possible applications would include generation of sound by very short laser pulses or by short microwave pulses in pure water at 4°C. The formulation is based on the postulated existence of an instantaneous specific entropy function that depends on internal energy, specific volume, and one or more internal variables.

Introduction

When electromagnetic energy is suddenly transmitted to a fluid in a localized region, the absorption of the radiation heats the fluid, and this heating can generate sound. This phenomenon, in connection with the heating of fluids by laser beam radiation, has been the subject of many papers. A good review of the early literature is given by Lyamshev and Sedov [1]. Related work on the generation of sound by microwave radiation has been reported by Nasoni *et al.* [2].

A basic quantitative theory that applies to such phenomena can be found in a 1955 report by Chu [3] and was independently derived or stated in articles by Ingard [4], by Westervelt and Larson [5], and by Bunkin and Komissarov [6]. A review of such derivations, with an extension to allow for the presence of thermal conduction and viscosity, has been given in a previous paper by the present author [7]. This theory, however, is developed within the context of equilibrium thermodynamics and can possibly give grossly erroneous results if the fluid neither expands nor contracts when heated. Such would be the case, for example, with pure water at 4°C. In oral presentations [8, 9], at meetings of the Acoustical Society of America in 1982 and 1983, the present author suggested an improved theory that included internal relaxation. This theory has circulated in preliminary form in the form of handouts and preprints, but has not yet been published in an archival form. The present paper attempts to give a concise and clear derivation of that theory.

Implications of equilibrium thermodynamics

To introduce the terminology and notation of the present paper and for convenience of referral, the Bunkin and Komissarov version [6] of the earlier theory of sound generation by heat is first briefly restated. As far as practicable, the notation is adapted from that of Ingo Müller's text [10].

One lets Q denote the heat added per unit time and per unit volume to the fluid by external means (as by the absorption of electromagnetic radiation). With the neglect of thermal conduction and of viscous heating and with the assumption that the heat addition is a reversible process, the entropy η per unit mass in the fluid increases in accord with the manner of its definition and with the laws of equilibrium thermodynamics as

$$\rho T \frac{D\eta}{Dt} = Q, \qquad (1)$$

where T is the absolute temperature and the density ρ is the mass per unit volume. The time derivative that appears here is the time derivative following the fluid motion.

The fluid is presumed to be in a single thermodynamic phase, so one can express the density ρ

$$\rho = \rho(p, \eta) \qquad (2)$$

as a function of the total pressure p and the entropy η per unit volume. The differential of this relationship can be expressed as

$$d\rho = \left\{ \frac{\partial \rho}{\partial p} \right\}_{\text{fixed } \eta} dp + \left\{ \frac{\partial \rho}{\partial \eta} \right\}_{\text{fixed } p} d\eta \,. \qquad (3)$$

The coefficients here can be expressed, by various thermodynamic relationships, in terms of standard thermodynamic properties, so that the time derivative of ρ can be expressed, with the aid of Eq. (1), as

$$\frac{D\rho}{Dt} = \frac{1}{c^2}\frac{Dp}{Dt} - \frac{\beta}{c_p}\rho T \frac{D\eta}{Dt} = \frac{1}{c^2}\frac{Dp}{Dt} - \frac{\beta}{c_p} Q \,, \qquad (4)$$

where c is the speed of sound, c_p is the coefficient of specific heat at constant pressure, and

$$\beta = \frac{1}{\text{volume}} \left\{ \frac{\partial \,\text{volume}}{\partial T} \right\}_{\text{fixed } p} = -\frac{1}{\rho} \left\{ \frac{\partial \rho}{\partial T} \right\}_{\text{fixed } p} \tag{5}$$

is the coefficient of thermal expansion.

When Eq. (4) is inserted into the differential equation expressing the conservation of mass, one obtains

$$-\rho \nabla \cdot \mathbf{v} = \frac{1}{c^2} \frac{Dp}{Dt} - \frac{\beta}{c_p} Q\,, \tag{6}$$

where $\mathbf{v}$ is the fluid velocity. In addition to this equation, one uses the Euler equation of motion for a fluid, which has the form

$$\rho \frac{D\mathbf{v}}{Dt} = -\nabla p. \tag{7}$$

Linearization of these two equations about an ambient state and manipulation subsequently yields the inhomogeneous acoustic wave equation

$$\nabla^2 p' - \frac{1}{c^2} \frac{\partial^2}{\partial t^2} p' = -\frac{\beta}{c_p} \frac{\partial Q}{\partial t}. \tag{8}$$

Here the ambient state is assumed homogeneous and nonmoving. The acoustic pressure p' which appears here as the dependent variable is the deviation of the total pressure from its ambient value.

The inhomogeneous wave equation stated above has implications in regard to the dependence of waveforms on the properties of the ambient medium. Suppose one does a sequence of similar experiments in which the spatial and temporal dependence of the rate Q of heat addition per unit volume is the same for each experiment and in which the heat addition is the only cause of acoustic wave generation. Then, the acoustic pressure p' at any given fixed point can be regarded as of the form

$$p'(t) = \frac{\beta}{c_p} F(t), \tag{9}$$

where the function $F(t)$ depends on the overall geometry of the experiment, on the sound speed c, and on the function Q, but is independent of both the coefficient β of thermal expansion and the specific heat coefficient c_p. It follows that there will be no acoustic signal if β is 0.

Suppose that the successive experiments differ only in the ambient temperature T_0 of the medium. Suppose also, as in invariably the case, that the sound speed c has only slight dependence on temperature, but that β may depend strongly on temperature. Then the function $F(t)$ that appears in Eq. (9) will be very nearly the same for each experiment, and the received waveforms will all have essentially the same shape. They will differ primarily only in amplitude or possibly in sign, with this difference being explained because of the linear dependence of the waveform on β/c_p and because of the temperature dependence of this coefficient.

Review of pertinent experiments

There appears now to be ample experimental evidence that the implications described above break down for experiments carried out in pure water near ambient temperatures of 4°C. At atmospheric pressure, such water has a density maximum at an ambient temperature $T_{\max\rho}$ which corresponds to 4°C, so β vanishes at this temperature, in accord with Eq. (5). In the vicinity of this temperature, the ratio β/c_p varies approximately linearly with temperature in accord with the relation

$$\beta/c_p \approx K[T - T_{\max\rho}] \tag{10}$$

where K is positive and of the order of 2×10^{-9} J/(kg·K). The quantity β/c_p goes from negative to positive through zero when the temperature increases from below to above $T_{\min\rho}$. Thus the simple theory would predict waveforms in successive experiments with amplitudes that vary with T as in Eq. (10).

Experiments that show such a prediction is not satisfied near 4°C have been reported by Tam and Patel [11], by Hunter *et al.* [12], and most recently by Liew and Bowen [13]. The former two papers were concerned with sound generation by absorption of laser light; the third was concerned with sound generation by absorption of microwave radiation. A detailed critique and summary of the data reported by these authors is beyond the scope of the present paper, but it is fair to state that the predictions of Eqs. (9) and (10) were not met at 4°C. Since there was a signal received at a temperature when the equilibrium thermodynamic theory predicted no such signal, the authors refer to such a signal as a nonthermal signal. The following sentence fragments lifted from a preprint of the paper by Liew and Bowen summarize the features of the received waveforms:

'A closer look at the waveform of the signals at temperatures around 4°C showed the presence of a nonthermal component, which had a waveform similar to the derivative of the waveform at room temperature'.

'[When one studies the temperature evolution of the first positive peak found in the 1°C waveform, one finds] the positive peak shifts to the right gradually as temeprature increases...'

'[The waveforms received at any given temperature is oscillatory, and when these for a sequence of different temperatures are juxtaposed, one can perceive a net phase change (corresponding to a change in sign) of π which takes place as the temperature is increased from 1 °C to 15°C.] ... a phase change of $\pi/2$ occurs at 4°C. The phase is completely reversed at higher temperature'.

Irreversible thermodynamics formulation

A formulation consistent with the feature of the data as described above ensues from the internal variable theory of irreversible thermodynamics developed by Meixner [14], de Groot [15], Prigogine [16] and others. Internal variables a_1, a_2, etc., enter into the formulation in that the instantaneous entropy η per unit volume depends on these quantities as well as on the internal energy e per unit mass and on the specific volume ρ^{-1}. For simplicity, the following derivation presumes that only one internal variable need be explicitly taken into account. Thus one writes

$$\eta = \eta(e, \rho^{-1}, a) \tag{11}$$

as the equation of state. This in turn has the differential relation

$$T\,\mathrm{d}\eta = \mathrm{d}e + p\mathrm{d}\rho^{-1} + T\Lambda^{a}\,\mathrm{d}a \tag{12}$$

with T identified as the instantaneous temperature and p identified as the instantaneous pressure. The quantity Λ^{α} is defined by this equation. Since the absolute temperature is always positive, the relation (11) can be inverted to yield the energy e per unit mass in a functional relationship of the form

$$e = e(\eta, \rho^{-1}, a) \tag{13}$$

and the differential relation (12) consequently allows one to derive expressions of the form

$$p = p(\eta, \rho^{-1}, a) \tag{14a}$$

$$T = T(\eta, \rho^{-1}, a) \tag{14b}$$

$$\Lambda^{\alpha} = \Lambda^{\alpha}(\eta, \rho^{-1}, a). \tag{14c}$$

The pressure that appears in the equations above is related to the local stress tensor in the fluid by the relation

$$\frac{1}{3}\sum_{i} \sigma_{ii} = -p + \mu_B\,\nabla\cdot\mathbf{v} \tag{15}$$

in accord with Stokes's general hypothesis for a Newtonian fluid. The quantity μ_B is the bulk viscosity. Thus the Navier–Stokes equation takes the form

$$\rho \frac{D\mathbf{v}}{Dt} = -\nabla p + \nabla(\mu_B \nabla\cdot\mathbf{v}) + \sum_{ij} \mathbf{e}_i \frac{\partial(\mu\varphi_{ij})}{\partial x_j} \tag{16}$$

where μ is the shear viscosity and

$$\varphi_{ij} = \frac{\partial v_i}{\partial x_j} + \frac{\partial v_j}{\partial x_i} - \frac{2}{3}\nabla\cdot\mathbf{v}\delta_{ij} \tag{17}$$

is the rate of shear tensor.

Similarly, the temperature T that appears in Eqs. (12) and (14b) is that which enters into the Fourier law of heat conduction,

$$\text{heat flux vector} = -\kappa\nabla T, \tag{18}$$

where κ is the coefficient of thermal conductivity, so the energy conservation equation becomes

$$\rho \frac{De}{Dt} = -p\nabla\cdot\mathbf{v} + \mu_B(\nabla\cdot\mathbf{v})^2 + \tfrac{1}{2}\sum_{i,j} \varphi_{ij}^2 + \nabla\cdot(\kappa\nabla T) + Q, \tag{19}$$

where Q is the rate at which energy is added per unit volume by electromagnetic energy absorption. Note that the forms of Eqs. (16) and (19) are independent of the existence of an internal variable.

For acoustical applications, the most natural thermodynamic variables with which to work are the pressure p and the entropy η per unit mass. Since the enthalpy per unit mass,

$$h = e + p\rho^{-1} \tag{20}$$

has the differential

$$dh = Td\eta + \rho^{-1}dp - T\Lambda^\alpha da, \tag{21}$$

it is the natural thermodynamic potential for acoustical applications. With the aid of the mass conservation relation and of Eq. (20), the energy conservation equation (19) is recast into the form

$$\rho \frac{Dh}{Dt} = \frac{Dp}{Dt} + \mu_B(\nabla\cdot\mathbf{v})^2 + \tfrac{1}{2}\sum_{i,j} \varphi_{ij}^2 + \nabla\cdot(\kappa\nabla T) + Q. \tag{22}$$

When the fluid is in local thermodynamic equilibrium, the internal variable a is not arbitrary but takes on the value

$$a = a_{\rm eq}(e, \rho^{-1}), \tag{23}$$

which maximizes the entropy η per unit mass for fixed internal energy e per unit mass and for fixed volume ρ^{-1} per unit mass, so that

$$\Lambda^{\alpha}(e, \rho^{-1}, a_{\rm eq}) = 0 \tag{24a}$$

$$\frac{\partial \Lambda^a}{\partial a} \leq 0 \text{ when } a = a_{\rm eq}(e, \rho^{-1}). \tag{24b}$$

[Here Λ^a is regarded as a function of e, ρ^{-1}, and a.] Thus the second law of thermodynamics takes the form

$$\eta(e, \rho^{-1}, a) \leq \eta(e, \rho^{-1}, a_{\rm eq}). \tag{25}$$

A more general statement, which is based on the conceptual subdivision of a large mass into two or more (spatially disjoint) masses and on the additiveness of the entropies, internal energies, and volumes of the constituent masses, is that the equation of state near an equilibrium state be such that

$$\delta T \delta\eta - \delta p \delta\rho^{-1} - \delta\Lambda^a \delta a \geq 0. \tag{26}$$

[The derivation is straightforward, but lengthy, and similar to that given by Landau and Lifshitz [17], only with the explicit consideration of internal variables.] Suppose a given choice $\{p, \eta, a\}$ corresponds to a equilibrium state such that T, ρ^{-1}, and Λ^a are the corresponding temperature, specific volume, and internal variable 'affinity', the latter being zero in accord with Eq. (24a). Then for a slightly different, possibly nonequilibrium, state $\{p + \delta p,\ \eta + \delta\eta,\ \Lambda^a + \delta\Lambda^a\}$, the other thermodynamic variables take on slightly different values $T + \delta T$, $\rho^{-1} + \delta\rho^{-1}$, and $\Lambda^a + \delta\Lambda^a$. If the equation of state is in accord with the second law of thermodynamics, then these increments must satisfy the inequality of Eq. (26), one consequence of which is that, when Λ^a is regarded as a function of p, η, and a,

$$\frac{\partial \Lambda^a}{\partial a} \leq 0 \text{ when } a = a_{\rm eq}(p, \eta) \tag{27}$$

where $a_{\rm eq}(p, \eta)$ is the value of a for which $\Lambda^a(p, \eta, a) = 0$. Consequently, one can infer from Eq. (21) that, in the vicinity of an equilibrium state corresponding to given p and η,

$$h(p, \eta, a) \geq h_{\mathrm{eq}}(p, \eta) \tag{28}$$

or to second order,

$$h(p, \eta, a) \approx h_{\mathrm{eq}}(p, \eta) + \tfrac{1}{2}K\left[a - a_{\mathrm{eq}}(p, \eta)\right]^2 \tag{29}$$

where K is a function of p and η, but independent of a.

This above approximate form of the equation of state is adequate for linear acoustics, in which only first-order deviations of the thermodynamic variables from a given (equilibrium) ambient state are taken into account. The specific volume, for example, is derived from $h(p, \eta, a)$ via the relation

$$\rho^{-1} = \frac{\partial h}{\partial p}, \tag{30}$$

so the acoustic perturbation to this quantity caused by deviations p', χ', and a' of the corresponding quantities from their ambient values, is

$$\left(\rho^{-1}\right)' = -\frac{1}{\rho^2 c^2}p' + \frac{\beta T}{\rho c_p}\eta' - K\frac{\partial a_{\mathrm{eq}}}{\partial p}\left[a' - \frac{\partial a_{\mathrm{eq}}}{\partial p}p' - \frac{\partial a_{\mathrm{eq}}}{\partial \eta}\eta'\right]. \tag{31}$$

Here K and all other unprimed quantities are evaluated at the ambient state. The indicated thermodynamic coefficients are those appropriate to equilibrium thermodynamics. Similarly, the acoustic perturbations to the enthalpy h per unit mass and to the temperature T arc

$$h' = T_0\eta' + \frac{1}{\rho_0}p' \tag{32}$$

$$T' = \frac{T\beta}{\rho c_p}p' + \frac{T}{c_p}\eta' - K\frac{\partial a_{\mathrm{eq}}}{\partial \eta}\left[a' - \frac{\partial a_{\mathrm{eq}}}{\partial p}p' - \frac{\partial a_{\mathrm{eq}}}{\partial \eta}\eta'\right]. \tag{33}$$

With the identifications just given, the linearized versions of the conservation of mass equation, the Navier–Stokes equation, and the energy conservation equation become

$$\frac{1}{\rho c^2}\frac{\partial p'}{\partial t} + \nabla\cdot\mathbf{v}' = \frac{\beta T}{c_p}\frac{\partial \eta'}{\partial t} - \rho K\frac{\partial a_{\mathrm{eq}}}{\partial p}\left[\frac{\partial a'}{\partial t} - \frac{\partial a_{\mathrm{eq}}}{\partial p}\frac{\partial p'}{\partial t} - \frac{\partial a_{\mathrm{eq}}}{\partial \eta}\frac{\partial \eta'}{\partial t}\right] \tag{34}$$

$$\rho \frac{\partial \mathbf{v}'}{\partial t} = - \nabla p' + \left[\mu_B + \frac{4}{3}\mu\right] \nabla(\nabla \cdot \mathbf{v}') - \mu\nabla \times (\nabla \times \mathbf{v}') \tag{35}$$

$$\rho T \frac{\partial \eta'}{\partial t} - \kappa \nabla^2 T' + Q. \tag{36}$$

To these equations one appends a relaxation equation. The basic assumptions are that Da/Dt depends only on the instantaneous state, that it vanishes when the fluid is in thermodynamic equilibrium, and that its sign is such that the tendency is to restore equilibrium. Thus, in the acoustic approximation, the relaxation equation must be of the form

$$\frac{\partial a'}{\partial t} = - \frac{1}{\tau} \left[a' - \frac{\partial a_{\text{eq}}}{\partial p} p' - \frac{\partial a_{\text{eq}}}{\partial \eta} \eta' \right], \tag{37}$$

where the relaxation time τ depends only on the ambient state. With the substitutions (which define r', Δc, and $\Delta\beta$)

$$\left[\rho^2 K \frac{\partial a_{\text{eq}}}{\partial p} \right] a' = r' \tag{38a}$$

$$\rho^2 K \left(\frac{\partial a_{\text{eq}}}{\partial p} \right)^2 = \frac{2[\Delta c]}{c^3} \tag{38b}$$

$$\rho^2 K \frac{\partial a_{\text{eq}}}{\partial p} \frac{\partial a_{\text{eq}}}{\partial \eta} = \frac{\rho[\Delta \beta] T}{c_p}, \tag{38c}$$

the relaxation equation can equivalently be written

$$\frac{\partial r'}{\partial t} = - \frac{1}{\tau} \left[r' - \frac{2[\Delta c]}{c^3} p' - \frac{\rho[\Delta \beta] T}{c_p} \eta' \right]. \tag{39}$$

This change of notation also allows the mass conservation relation (34) to be rewritten

$$\frac{1}{c^2} \frac{\partial p'}{\partial t} + \rho \nabla \cdot \mathbf{v}' = \frac{\rho \beta T}{c_p} \frac{\partial \eta'}{\partial t} - \left[\frac{\partial r'}{\partial t} - \frac{2[\Delta c]}{c^3} \frac{\partial p'}{\partial t} - \frac{\rho[\Delta \beta] T}{c_p} \frac{\partial \eta'}{\partial t} \right] \tag{40}$$

The reasons for the choice of the notation, δc and $\delta\beta$, are evident in this equation.

Derivation of an appropriate approximate wave equation proceeds with the neglect of terms of second order in Δc, $\Delta\beta$, and κ, so that the energy conservation equation (36) becomes

$$\rho T \frac{\partial \eta'}{\partial t} = \frac{T\beta}{\rho c_p}\kappa\nabla^2 p' + \frac{T}{c_p}\kappa^2\eta' + Q. \tag{41}$$

It is also assumed that the relaxation time τ is sufficiently short compared to any characteristic time for the energy deposition that the relaxtion equation (39), rewritten as

$$r' = \frac{2[\Delta c]}{c_0^3}p' + \frac{\rho_0[\Delta\beta]T_0}{c_p}\eta' - \tau\frac{\partial r'}{\partial t}, \tag{42}$$

can be solved by iteration, so that the mass conservation equation (4) becomes

$$\frac{1}{c^2}\frac{\partial p'}{\partial t} + \rho\nabla\cdot\mathbf{v}' = \frac{\rho\beta T}{c_p}\frac{\partial\eta'}{\partial t} + \frac{2[\Delta c]\,\tau}{c^3}\frac{\partial^2 p'}{\partial t^2} + \frac{\rho[\Delta\beta]T\tau}{c_p}\frac{\partial^2\eta'}{\partial t^2}. \tag{43}$$

Also, the divergence of the Navier–Stokes equation (35) yields

$$\left[\rho\frac{\partial}{\partial t} - [\mu_B + \tfrac{4}{3}\mu]\nabla^2\right](\nabla\cdot\mathbf{v}') = -\nabla^2 p', \tag{44}$$

so one obtains the partial differential equation

$$\nabla^2 p' - \left[\frac{\partial}{\partial t} - \frac{1}{\rho}[\mu_B + \frac{4}{3}\mu]\nabla^2\right]\left[\frac{1}{c^2}\frac{\partial p'}{\partial t} - \frac{\rho\beta T}{c_p}\frac{\partial\eta'}{\partial t}\right.$$

$$\left. + \frac{2[\Delta c]\tau}{c^3}\frac{\partial^3 p'}{\partial t^3} + \frac{\rho[\Delta\beta]T\tau}{c_p}\frac{\partial^3\eta'}{\partial t^3}\right] = 0. \tag{45}$$

Solution of the coupled partial differential equations, (41) and (45), can be expressed [in a manner similar to that done previously by the author for the case when no relaxation processes are present [7]] as a sum of disturbances in the entropy (ent) and acoustic (ac) modes, so that one writes

$$p' = p_{\mathrm{ac}} + p_{\mathrm{ent}}. \tag{46}$$

Because of space limitations, the analysis is omitted here. The entropy mode is limited to the immediate vicinity of the heated region and obeys the thermal diffusion equation. The acoustic mode, however, propagates away from the heated region as an acoustic wave. Apart from an attenuation term, its governing partial equation is what results from Eqs. (41) and (45) when the thermal conductivity and viscosity are ignored. The result with the attenuation term is

$$\nabla^2 p_{ac} - \frac{1}{c^2}\frac{\partial^2 p_{ac}}{\partial t^2} + \frac{[\mu_B]}{\rho c^4}\frac{\partial^3 p_{ac}}{\partial t^3} = -\frac{\beta}{c_p}\frac{\partial Q}{\partial t} - \frac{\Delta\beta}{c_p}\tau\frac{\partial^2 Q}{\partial t^2}, \tag{47}$$

where

$$[\mu_B]_{\text{eff}} = 2\rho c\,[\Delta c]\,\tau + \mu_B + \frac{4}{3}\mu + (\gamma - 1)\frac{\kappa}{c_p} \tag{48}$$

is an effective bulk modulus, with

$$\gamma - 1 = c^2\beta^2 T / c_p\,. \tag{49}$$

For the various experiments mentioned in the previous paper, the influence of the absorption term in the wave equation (44) is of minor importance. What is important is the added second source term on the right side, which does not vanish when $\beta = 0$. The general prediction (with the neglect of absorption) which replaces Eq. (9) is that the pressure waveforms are of the general form

$$p_{\text{ac}} = \left\{\frac{\beta}{c_p}F(t)\right\} + \left\{\frac{[\Delta\beta]\tau}{c_p}F'(t)\right\}. \tag{50}$$

Recent work by Liew and Bowen [13] confirms this prediction and gives

$$T\,[\Delta\beta]\tau = 2\times 10^{-10}\ \text{s}. \tag{51}$$

Concluding remark

Any single- or multiple-relaxation process will predict a non-zero bulk viscosity and the anamalous generation of sound by heat addition at 4°C in water. What is remarkable is that the 4°C waveforms are explained by a theory with only one adjustable parameter, this being the quantity $T\,[\Delta\beta]\tau$.

Acknowledgements

The work reported here was supported by the Office of Naval Research (USA) and by the William E. Leonhard endowment to Pennsylvania State University.

References

[1] L.M. Lyamshev and L.V. Sedov, Optical generation of sound in a liquid: thermal mechanism (review), *Akust. Zh.* **27**, 5-29 (1981) [*Sov. Phys. Acoust.* **27**, 4-18 (1981).]

[2] R.L. Nasoni, S.C. Liew, P.G. Halverson and T. Bowen, Thermoacoustic images generated by a 2450 MHz portable source and applicator, *1985 IEEE Ultrasonic Symposium Proceedings*, 899-904 (1985).

[3] Boa-Teh Chu, *Pressure Waves Generated by Addition of Heat in a Gaseous Medium*, NACA Technical Note 3411 (National Advisory Committee for Aeronautics, Washington, June 1955).

[4] U. Ingard, Acoustics, in *Handbook of Physics* (ed. E.U. Condon and H. Odishaw), Ch. 8, McGraw-Hill (1958).

[5] P.J. Westervelt and R.S. Larson, Laser-excited broadside array, *J. Acoust. Soc. Am.* **54**, 121-122 (1973).

[6] F.V. Bunkin and V.M. Komissarov, Optical excitation of sound waves, *Akust. Zh.* **19**, 305-320 (May-June 1973); *Soviet Phys. - Acoust.* **19**(3), 203-211 (Nov.-Dec. 1973).

[7] A.D. Pierce, The inhomogeneous wave equation of thermoacoustics, *Flow of Real Fluids* (ed. G.E.A. Meier and F. Obermeier), Volume 235 of Lecture Notes in Physics, Springer-Verlag, Berlin, 1985, pp. 92-99.

[8] A.D. Pierce, Nonlinear thermoacoustics and electrostriction explanations of acoustic signatures in 4°C water caused by laser pulses and proton beams, *J. Acoust. Soc. Am. Suppl.* **72**(1), S13 (Fall 1982).

[9] A.D. Pierce, Irreversible thermodynamics formulation of multiple relaxation processes, with application to laser-generated sound in 4°C water, *J. Acoust. Soc. Am. Suppl.* **73**(1), S50 (Spring 1983).

[10] I. Müller, *Thermodynamics*, Pitman, Boston, 1985.

[11] A.C. Tam and C.K.N. Patel, Optical absorptions of light and heavy water by laser optoacoustic spectroscopy, *Applied Optics* **18** (19), 3348-3358 (1 October 1979).

[12] S.D. Hunter, W.V. Jones, and D.J. Malbrough, Nonthermal acoustic signals from absorption of a cylindrical laser beam in water, *J. Acoust. Soc. Am.* **69**, 1563-1567 (1981).

[13] S.C. Liew and T. Bowen, Detection of nonthermal microwave-induced photoacoustic signals in water, preprint of paper for 6th International Topical Meeting on Photoacoustic and Photothermal Phenomena, Baltimore (1989), communicated privately by S.C. Liew to A.D. Pierce.

[14] J. Meixner, Flows of fluid media with internal transformations and bulk

viscosity, *Z. Phys.* **131**, 456-469 (1952); J. Meixner and H.G. Reik, Thermodynamik der irreversiblen Prozesse, in *Handbuch der Physik*, Vol III/2, Prinzipien der Thermodynamik und Statistik, edited by S. Flügge (Springer-Verlag, Berlin, 1959), pp. 413-523.

[15] S.R. de Groot, *Thermodynamics of Irreversible Processes*, North-Holland, Amsterdam, 1961; S.R. de Groot and P. Mazur, *Nonequilibrium Thermodynamics*, North-Holland, Amsterdam, 1962).

[16] I. Prigogine, *Introduction to Thermodynamics of Irreversible Processes*, 3rd ed., Interscience, New York, 1967.

[17] L.D. Landau and E.M. Lifshitz, *Statistical Physics*, Pergamon, London, 1959, pp. 57-63. See also J.H. Keenan, *Thermodynamics*, M.I.T. Press, Cambridge, 1970, pp. 395-428.

A.D. Pierce
The Pennsylvania State University
Graduate Program in Acoustics and
Department of Mechanical Engineering
157 Hammond Building
University Park
Pennsylvania 16802
USA

J.A. NOHEL

Shearing motions of viscoelastic fluids satisfying non-monotone constitutive relations

Extended abstract

Viscoelastic materials with fading memory, e.g. polymers, suspensions and emulsions, exhibit behaviour that is intermediate between the nonlinear hyperbolic response of purely elastic materials and the strongly diffusive, parabolic response of viscous fluids; they incorporate a subtle dissipative mechanism induced by effects of the fading memory (see extensive discussion and survey of mathematical results in [1], [2]).

The purpose of this talk, primarily based on recent joint work with D. Malkus and B. Plohr (see [3]-[5]), is to discuss novel phenomena in dynamic shearing flows of non-Newtonian fluids that are important to advanced materials engineering and process design. Examples of such materials are high-strength polymers and lubricant additives used in spinning of synthetic fibres and injection moulding. The materials are often highly elastic and very viscous; consequently, their dynamic response involves multiple time-scales. Understanding the qualitative and quantitative behaviour of the equatins of motion coupled with various constitutive assumptions has proved to be of significant physical, mathematical, and computational interest. The phenomena discussed below appear to be relevant to material instabilities that can disrupt polymer processing.

One striking phenomenon, called 'spurt', was observed by Vinogradov *et al.* [6] in the flow of monodispersive polyisoprenes through capillaries. They found that under quasistatic loading, the volumetric flow rate increased dramatically at a critical stress that was independent of molecular weight. Until recently, spurt had been overlooked or dismissed by rheologists because no plausible mechanism was known to explain it in the context of steady flows that are linearly stable. Spurt was associated with failure of the polymer to adhere to the wall and was lumped together with instabilities such as 'slip', 'apparent slip', 'melt fracture' (see [7], [8]). Computational difficulties in steady flows were generally attributed to a change of type of the governing system at a high Weissenberg number resulting in a loss of evolutionarity that could lead to a Hadamard instability.

By contrast, we show that satisfactory modelling and explanation of spurt and related phenomena require studying the full dynamics of the equations of motion and constitutive equations. The common feature of constitutive models that exhibit spurt is a non-monotonic relation between the steady shear stress and strain rate. This allows jumps in the steady strain rate to form when the driving pressure gradient exceeds a critical value; such jumps correspond to the sudden increase in volumetric flow rate observed in the experiments of Vinogradov *et al.* [6]. Hunter and Slemrod [9] studied the qualitative behaviour of these jumps in an elementary, one-dimensional viscoelastic model of rate type and they predicted shape memory and hysteresis effects related to spurt. A salient feature of their model is linear instability and loss of evolutionarity in a region of state space.

The equations we study in [3]-[5] to model such flows derive from a three-dimensional constitutive equation (see [10], [11]). The total shear stress is decomposed into a polymer contribution, evolving in accordance with a differential constitutive relation with a single relaxation time and a Newtonian viscosity contribution; the flows can also be modelled by a system based on a differential constitutive law with two widely spaced relaxation times but no Newtonian viscosity contribution. The governing one-dimensional system of quasilinear partial differential equations is evolutionary and well posed globally in time for smooth initial data of arbitrary size (see [12]), and can possess discontinuous steady states.

We present results of effective numerical methods developed in [3], [4], [13] for simulating transient, one-dimensional shear flows in a slit die at higher Weissenberg (Deborah) number and very low Reynolds number (appropriate for highly elastic and very viscous fluids). When restricted to flows of monodisperse polyisoprenes, results of numerical simulation using these methods agreed qualitatively and quantitatively with results of the Vinogradov loading experiment exhibiting spurt. Furthermore, numerical calculations also exhibited other phenomena: latency, and when loading is followed by unloading, shape memory, hysteresis, numerical stress oscillations, normal stress oscillations, and molecular weight dependence of hysteressis; the latter phenomena require additional testing and verification in rheological experiment.

The numerical work suggested a particular scaling appropriate for highly elastic and very viscous fluids; in particular, for the Vinogradov material, the ratio of Reynolds number to Deborah number is several orders of magnitude smaller than the Newtonian viscosity component. Setting the former (i.e. the inertial term) equal to zero and integrating the momentum equation reduces the governing system of partial differnetial equations to a quadratic system of two ordinary differential equations, the dynamics of which is determined completely in [5] by phase plane analysis and by determining suitable invariant regions using an appropriate Liapunov function. We also justify the use of Newtonian viscosity to model the effects of relaxation times which are very much shorter than the fundamental time constant of the macromolecules, and we determine conditions under which the character of the phase portrait of a more realistic, two relaxation-time model without Newtonian viscosity matches the phase portrait of the simpler model with Newtonian viscosity. Results of this analysis are then used to explain for both models not only the occurrence of spurt, but also the phenomena of shape memory, hysteresis, latency, and of other effects that have been observed in the numerical simulations. We believe that our results are not limited to the two specific models discussed above. Indeed, numerical simulation in [4] using yet a different constitutive law with a single relaxation time and a Newtonian viscosity also yields spurt solutions.

It is a challenge to prove that solutions of the full system of governing PDE's also exhibit these phenomena. In a collaboration with R. Pego and A. Tzavaras, we have undertaken a deeper analytical study of the full unsteady shear-flow problem keeping the inertial term. In [12], we identify nonlinearly stable discontinuous steady states of a model initial-boundary value problem in one space dimension for incompressible, isothermal shear flow of a non-Newtonian fluid between parallel plates, driven by a constant pressure gradient. The non-Newtonian contribution to the shear stress is assumed to satisfy a single, much simpler differential constitutve law. The model incorporates several features of the more complex problem described above (e.g., for a particular choice, the two systems have the same steady states). Again the key property is a non-monotone relation between the total steady shear stress and steady shear strain rate that results in steady states having, in general, discontinuities in the strain rate. We explain why every solution tends to a steady state as $t \to \infty$, and we establish a simple criterion for determining which steady states are non-linearly asymptotically stable. For the more complex problem described above, research in progress indicates that discontinuous steady states that take their values on increasing parts of the steady total shear stress vs. strain rate curve are stable (perhaps not asymptotically), provided the ratio of Reynolds number to Deborah number is sufficiently small.

References

[1] M. Renardy, W. Hrusa and J. Nohel, *Mathematical Problems in Viscoelasticity*, Pitman Monographs and Surveys in Pure and Applied Mathematics V. 35, Longman Scientific and Technical, Essex, 1987.

[2] W. Hrusa, J. Nohel and M. Renardy, Initial value problems in viscoelasticity, *Appl. Mech. Rev.* **41** (1988), 371-378.

[3] D. Malkus, J. Nohel and B. Plohr, Time-dependent shear flow of a non-Newtonian fluid, in *Current Progress in Hyperbolic Problems: Riemann Problems and Computations* (Bowdoin, 1988), ed. B. Lindquist, Amer. Math. Soc., Providence, 1989. Contemporary Mathematics 91-110.

[4] D. Malkus, J. Nohel and B. Plohr, Dynamics of shear flow of a non-Newtonain fluid, *J. Comput. Phys.* **87** (1990), 464-487.

[5] D. Malkus, J. Nohel and B. Plohr, Analysis of new phenomena in shear flow of non-Newtonian fluids, *SIAM J. Appl. Math.*, 1990 (accepted). Also CMS Technical Summary Report No. 90-6.

[6] G. Vinogradov, A. Malkin, Yu. Yanovskii, E. Borisenkova, B. Yarlykov and G. Berezhnaya, 'Viscoelastic properties and flow of narrow distribution polybutadienes and polyisoprenes, *J. Polymer Sci.*, A-2 **10** (1972), 1961-1084.

[7] M. Denn, Issues in viscoelastic fluid dynamics, *Annual Reviews of Fluid Mechanics*, **22** (1990), 13-34.

[8] J. Parson, *Mechanics of Polymer Processing*, Elsevier Applied Science, London, 1985.

[9] J. Hunter and M. Slemrod, Viscoelastic fluid flow exhibiting hysteretic phase changes, *Phys. Fluids* **26** (1983), 2345-2351.

[10] M. Johnson and D. Segalman, A model for viscoelastic behavior which allows non-affine deformation, *J. Non-Newtonian Fluid Mech.* **2** (1977), 255-270.

[11] J. Oldroyd, Non-Newtonian effects in steady motion of some idealized elastico-viscous liquids, *Proc. Roy. Soc. London* **A 245** (1958), 278-297.

[12] J. Nohel, R. Pego and A. Tzavaras, Stability of discontinuous shearing motions of non-Newtonian fluids, *Proc. Royal Soc. Edinburgh*, **115 A** (1990), 39-59.

[13] R. Kolkka, D. Malkus, M. Hansen, G. Ierley and R. Worthing, Spurt Phenomena of the Johnson-Segalman fluid and related models, *J. Non-Newtonian Fluid Mech.* **29** (1988), 303-325.

J.A. Nohel
Center for the Mathematical Sciences
University of Wisconsin-Madison
610 Walnut Street
Madison, Wisconsin 53705
USA

K. WILMANSKI

On plasticity of multicomponent metallic alloys - evolution of texture

Abstract

The paper is devoted to the derivation of a new equation for the texture evolution in multi-phase polycrystals. This equation has a purely kinematical character, which means that it does not require any constitutive relations of its own, additional to those present in the continuum model anyway.

1. Introduction remarks

The construction of a continuous model of elastic-plastic processes in a multi-component metallic polycrystal should account for three basic properties of such materials:

- different plastic reaction of each component on the same state of stresses,
- presence of an extensive interface region with material properties different from each phase, at least in the case of substantially different phases,
- changes of crystallographic properties due to the development of textures induced by plastic deformation.

These properties indicate that the model of such processes should be based on the description of a continuum with microstructure, the latter changing during the process.

In this paper, we present only one part of the model, reflecting the changes of texture. The full presentation, based on the texture description, contained in this paper, can be found in the forthcoming paper [1].

In contrast to all previous continuous models of crystallographic structure and its changes in the process of deformation, we attempt here to construct a purely kinematical model. It means that the evolution of texture follows solely from the motion of the body and it does not require any material (constitutive) assumptions of its own.

We assume the body to consist of M crystallographically distinguishable phases. We consider a certain microdomain V_M, whose dimensions are small enough to assume the macroscopic representation of the motion of V_M as the motion of a point of continuum and large enough to contain a large number ${}^{\mu}N$ of grains of the μ-phase.

Within each grain the processes are assumed to be homogeneous. This assumption, though very restrictive, is corrected by the existence of an additional 'phase', connected with interfaces. Details of this problem can be found again in [1] and shall not be presented in this paper.

In the next section, we present the basic notions of the multi-component continuum, which are needed for the formulation of the evolution equation of texture.

Section 3 contains the derivation of the main result of this paper, namely of equation (3.14).

2. Basic notions

We assume the domain of microstructure V_M to consist of disjoint domains ${}^{\mu}M^{(i)}$, $i = 1, \dots, {}^{\mu}N$, $\mu = 1, \dots, M$, each of them being connected with a certain crystallographic phase μ. The deformation of such a domain, which we call the i-grain of μ-phase, is described by the mapping ${}^{\mu}\mathbf{F}$ of the tangent space at a given point $\mathbf{Z}$ of ${}^{\mu}M^{(i)}$ into the three-dimensional vector space V^3

$$ {}^{\mu}\boldsymbol{F} : T_Z\, {}^{\mu}M^{(i)} \rightarrow V^3, \ \mathbf{Z} \in {}^{\mu}M^{(i)}. \tag{2.1} $$

This mapping is supposed to be the same for all points $\mathbf{Z} \in {}^{\mu}M^{(i)}$ of a chosen grain ${}^{\mu}M^{(i)}$ (microhomogeneity). Such mappings are called *local configurations* or *deformation gradients*, even though we do not assume the integrability of ${}^{\mu}\boldsymbol{F}$ and, hence, it does not have to be a gradient of anything. The deformation gradient ${}^{\mu}\boldsymbol{F}$ changes from one grain to another, even within the same μ-phase, and, for this reason, should be also marked by the index i. To simplify the notation, we shall not do so, as long as it does not lead to any confusion within the framework of this work.

The fundamental assumption, concerning the mapping ${}^{\mu}\mathbf{F}$, is its multiplicative decomposition

$$ {}^{\mu}\boldsymbol{F} = \mathbf{F}^{e}\, {}^{\mu}\boldsymbol{F}^{p}, \tag{2.2} $$

where $\mathbf{F}^{e}$ is constant over V_M and denotes the *macroscopic elastic* deformation gradient, and ${}^{\mu}\boldsymbol{F}^{p}$ varies from grain to grain and denotes the *microscopic plastic* deformation gradient. The latter is supposed to be caused by slip-mechanisms alone. In this paper, we do not need any detatils of this part of the model.

According to the general polar decomposition theorem, we have

$$ \begin{gathered} {}^{\mu}\boldsymbol{F} = {}^{\mu}\boldsymbol{R}\, {}^{\mu}\boldsymbol{U}, \ {}^{\mu}\boldsymbol{R}^{T} = {}^{\mu}\boldsymbol{R}^{-1}, \ {}^{\mu}\boldsymbol{U}^{T} = {}^{\mu}\mathbf{U}, \\ \mathbf{F}^{e} = \mathbf{R}^{e}\, \mathbf{U}^{e}, \ \mathbf{R}^{eT} = \mathbf{R}^{e-1}, \ \mathbf{U}^{eT} = \mathbf{U}^{e}, \\ {}^{\mu}\boldsymbol{F}^{p} = {}^{\mu}\boldsymbol{R}^{p}\, {}^{\mu}\boldsymbol{U}^{p}, \ {}^{\mu}\boldsymbol{R}^{pT} = {}^{\mu}\boldsymbol{R}^{p-1}, \ {}^{\mu}\boldsymbol{U}^{pT} = {}^{\mu}\boldsymbol{U}^{p}. \end{gathered} \tag{2.3} $$

It is obvious that, in general, the full rotation, described by the orthogonal tensor ${}^{\mu}\boldsymbol{R}$, is not the superposition of the elastic rotation $\mathbf{R}^{e}$ and the plastic rotation ${}^{\mu}\boldsymbol{R}^{p}$. It is indeed the case only in absence of the elastic deformation, described by the stretch tensor $\mathbf{U}^{e}$:

$$\mathbf{U}^{e} = \mathbf{1} \Rightarrow {}^{\mu}\boldsymbol{R} = \mathbf{R}^{e}\,{}^{\mu}\boldsymbol{R}^{p}. \tag{2.4}$$

Simultaneously it should be stressed that both ${}^{\mu}\boldsymbol{R}$ and $\mathbf{R}^{e}$ depend on the choice of the reference frame in the configuration space, while ${}^{\mu}\boldsymbol{R}^{p}$ does not depend on this choice. We have for the rigid rotation of such a frame:

$$\left.\begin{array}{l} \overset{*}{\mathbf{x}} = \mathbf{O}\,\mathbf{x}, \\ \mathbf{O} = \mathbf{O}(t),\ \mathbf{O}^{T} = \mathbf{O}^{-1} \end{array}\right\} \Rightarrow {}^{\mu}\overset{*}{\boldsymbol{R}} = \mathbf{O}\,{}^{\mu}\boldsymbol{R},\ \overset{*}{\mathbf{R}}{}^{e} = \mathbf{O}\,\mathbf{R}^{e},\ {}^{\mu}\overset{*}{\boldsymbol{R}}{}^{p} = {}^{\mu}\boldsymbol{R}^{p}. \tag{2.5}$$

The connected spins transform as follows:

$$\begin{aligned} ({}^{\mu}\dot{\boldsymbol{R}}\,{}^{\mu}\boldsymbol{R}^{T})^{*} &= \mathbf{O}\,{}^{\mu}\dot{\boldsymbol{R}}\,{}^{\mu}\boldsymbol{R}^{T}\,\mathbf{O}^{T} + \dot{\mathbf{O}}\,\mathbf{O}^{T}, \\ (\dot{\mathbf{R}}^{e}\,\mathbf{R}^{eT})^{*} &= \mathbf{O}\,\dot{\mathbf{R}}^{e}\,\mathbf{R}^{eT}\,\mathbf{O}^{T} + \dot{\mathbf{O}}\,\mathbf{O}^{T}, \\ ({}^{\mu}\dot{\boldsymbol{R}}^{p}\,{}^{\mu}\boldsymbol{R}^{pT})^{*} &= {}^{\mu}\dot{\boldsymbol{R}}^{p}\,{}^{\mu}\boldsymbol{R}^{pT}, \end{aligned} \tag{2.6}$$

the latter being defined by the constitutive relation within the structural plasticity (e.g. [1]), the other two following from the external kinematical conditions, imposed on the motion of the body.

We need further the following spin-like object:

$${}^{\mu}\boldsymbol{\Omega} := {}^{\mu}\boldsymbol{R}^{T}(\dot{\mathbf{R}}^{e}\,\mathbf{R}^{eT} - {}^{\mu}\dot{\boldsymbol{R}}\,{}^{\mu}\boldsymbol{R}^{T})\,{}^{\mu}\boldsymbol{R} = -\,{}^{\mu}\boldsymbol{\Omega}^{T}. \tag{2.7}$$

It is easy to see that the above *material relative spin* is invariant with respect to transformation (2.5)

$${}^{\mu}\boldsymbol{\Omega}^{*} = {}^{\mu}\boldsymbol{\Omega}. \tag{2.8}$$

3. Texture

One of the fundamental problems of structural plasticity of polycrystals arises due to the change of crystallographic orientation of grains caused by plastic deformation. We present here a kinematic model of evolution of textures, reflecting these changes, for the multi-phase polycrystal, considered in this work.

Let us first define the notion of a texture in the reference state. With this aim, for a given domain of microstructure, we introduce a local reference basis $\{\mathbf{K}_A\}$, A = 1,2,3 and the basis $\{{}^{\mu}\mathbf{K}_A^{(i)}\}$, A = 1, 2, 3 for each grain i of the μ-phase, The vectors of this basis define the crystallograpic directions, to which we refer the initial geometry of slip systems. The *initial texture* of the μ-phase is then defined as the set of proper orthogonal transformations of $\{\mathbf{K}_A\}$ into $\{{}^{\mu}\mathbf{K}_A^{(i)}\}$

$$ {}^{\mu}\mathbf{K}_A^{(i)} = {}^{\mu}\mathbf{\Pi}^{(i)}\mathbf{K}_A, \ \mathbf{\Pi}^{(i)} \in o(3), \quad i = 1, \dots, {}^{\mu}N. \tag{3.1} $$

$o(3)$ being the space of orthogonal tensors.

Apart form the local orientation of grain lattice, it is important to know the distribution of crystallographic directions within the domain of microstructure V_M. This is achieved by the following number density distribution

$$ n_n^{\text{o}}(\mathbf{\Pi}) = \sum_{i=1}^{{}^{\mu}N} \delta(\mathbf{\Pi} - {}^{\mu}\mathbf{\Pi}^{(i)}), \ \ \forall\ \mathbf{\Pi} \in o(3), \tag{3.2} $$

defined for all proper orthogonal tensors $\mathbf{\Pi}$. Obviously

$$ \int_{o(3)} n_n^{\text{o}}(\mathbf{\Pi})\, \mathrm{d}\nu(\mathbf{\Pi}) = {}^{\mu}N, \tag{3.3} $$

$\nu(\mathbf{\Pi})$ being a volume measure in the $o(3)$-space

The number density has a simple physical interpretation. For a chosen $\mathbf{\Pi}$ and a chosen neighbourhood of $\mathbf{\Pi}$, say $\Delta\mathbf{\Pi} \subset o(3)$, the integral

$$ \int_{\Delta\mathbf{\Pi}} n_n^{\text{o}}(\mathbf{\Pi})\, \mathrm{d}\nu(\mathbf{\Pi}) = {}^{\mu}N(\mathbf{\Pi}, \Delta\,\mathbf{\Pi}) \tag{3.4} $$

defines the number of grains, whose orientation lies in $\Delta\mathbf{\Pi}$-neighbourhood of $\mathbf{\Pi}$.

In practical applications, the exact number density distribution (3.2) is replaced by a certain *smooth function*, which follows from (3.2) by a so-called *coarse-graining* procedure [1]. We skip these considerations in this paper.

We proceed now to define the time evolution of the texture. Let us notice first that ${}^{\mu}\mathbf{K}_A^{(i)}$-vectors should remain unchanged under the plastic deformation, if we base the description of the deformation process on the slip model of plasticity:

$$ ({}^{\mu}\boldsymbol{F}^{\mathrm{p}} - \mathbf{1})\, {}^{\mu}\mathbf{K}_A^{(i)} = 0. \tag{3.5} $$

It means that the current image ${}^{\mu}\mathbf{k}_A^{(i)}$ of those vectors is given by the formula

$$ {}^{\mu}\mathbf{k}_A^{(i)} = \mathbf{F}^{\mathrm{e}}\, {}^{\mu}\mathbf{K}_A^{(i)}. \tag{3.6} $$

Simultaneously, the vectors $\mathbf{K}_A$ of the reference basis should transform for the μ-phase in the following way:

$$ {}^{\mu}\mathbf{k}_A = {}^{\mu}\boldsymbol{F}\, \mathbf{K}_A. \tag{3.7} $$

We connect now these spatial images, eliminating simultaneously the influence of local rigid rotations

$$ ({}^{\mu}\boldsymbol{R}^{\mathrm{T}}\, {}^{\mu}\mathbf{k}_A^{(i)}) = {}^{\mu}\hat{\pi}^{(i)}\, ({}^{\mu}\boldsymbol{R}^{\mathrm{T}}\, {}^{\mu}\boldsymbol{F}\, \mathbf{K}_A) \equiv ({}^{\mu}\hat{\pi}^{(i)}\, {}^{\mu}\boldsymbol{U})\mathbf{K}_A. \tag{3.8} $$

The tensor

$$ {}^{\mu}\pi^{(i)} := {}^{\mu}\hat{\pi}^{(i)}\, {}^{\mu}\boldsymbol{U} \tag{3.9} $$

describes the transformation of reference vectors $\mathbf{K}_A$ into crystallographic directions $({}^{\mu}\boldsymbol{R}^{\mathrm{T}}\, {}^{\mu}\mathbf{k}_A^{(i)})$ in the objective manner, i.e. the tensor ${}^{\mu}\pi^{(i)}$ does not change under the time-dependent orthogonal transformation of spatial coordinates (2.5).

Bearing in mind relations (3.1), (3.6) and (3.8), we obtain

$$ {}^{\mu}\pi^{(i)} = {}^{\mu}\boldsymbol{R}^{\mathrm{T}}\, \mathbf{F}^{\mathrm{e}}\, {}^{\mu}\Pi^{(i)}. \tag{3.10} $$

In the case of metals, the changes of crystallographic directions, caused by the elastic stretch $\mathbf{U}^{\mathrm{e}}$, are negligible. For this reason, relation (3.10) can be replaced by the following *approximation*

$$ {}^{\mu}\pi^{(i)} \approx ({}^{\mu}\boldsymbol{R}^{\mathrm{T}}\, \mathbf{R}^{\mathrm{e}})\, {}^{\mu}\Pi^{(i)}. \tag{3.11} $$

The tensor ${}^{\mu}\pi^{(i)}$, defined by relation (3.11), is certainly orthogonal. According to the above remarks, it describes the rotation of crystallographic directions with respect to the *material* direction of the polycrystal. Hence, we may consider (3.11) to be the definition of the *spatial image of the texture* with the obvious relation

$$ {}^{\mu}\mathbf{k}_A^{(i)} = {}^{\mu}\boldsymbol{R}\, {}^{\mu}\pi^{(i)}\, \mathbf{K}_A. \tag{3.12} $$

The above definition gives rise to the natural definition of the spatial number density distribution for the μ-phase

$$ {}^{\mu}n(\pi, t) = \sum_{i=1}^{{}^{\mu}N} \delta(\pi - {}^{\mu}\pi^{(i)}) \quad \forall\, \pi \in o(3). \tag{3.13} $$

We have made here the tacit assumption that the deformation process does not yield phase transformations, i.e. the redistribution of crystallographic directions in the domain of microstructure is caused by rotations but not by the change of the total number of grains in a chosen phase.

Definition (3.13) yields immediately the following differential equation:

$$ \frac{\partial {}^{\mu}f}{\partial t} - \operatorname{tr} {}^{\mu}\Omega \left(\frac{\partial {}^{\mu}f}{\partial \pi} \pi^{T} \right) = 0, \quad {}^{\mu}f := \frac{{}^{\mu}n}{{}^{\mu}N}, \tag{3.14} $$

where we have used definition (2.7), and

$$ {}^{\mu}f(\pi, t = 0) = \frac{1}{{}^{\mu}N}\, {}^{\mu}\overset{\circ}{n} $$

(π), (3.15)

are the initial conditions.

In the case of the exact number density (3.13), equation (3.14) is just the identity. It is not so if we smear out the distribution ${}^{\mu}n$ by the *course-graining* procedure. In the latter case, we assume equation (3.14) also to hold true. Then the smooth solutions ${}^{\mu}f(\pi, t)$ of this equation *define* the distribution functions ${}^{\mu}f$ on $o(3)$-space for each instant of time and these functions no longer coincide with (3.13).

The evolution equation (3.14) for textures replaces rather speculative considerations, appearing in the literature on this subject (e.g. [2]). It replaces, for instance, one of the oldest theories of the texture evolution, which was based on the assumption of the diffusion of crystallographic directions on an appropriate sphere (e.g. [3]).

In contrast to those theories, the model of texture development, presented above, has a purely kinematic character. It does not require any further material parameters except of those, which are anyway present in the model of the deformation process. The solution of equation (3.14) requires only the initial condition (3.15) to be given. The data for this condition are easily available and measured in metallurgical experiments since a long time.

References

[1] K. Wilmanski, Macroscopic thoery of evolution of deformation textures, *Int. J. Plasticity* (in prepreation).

[2] T.D. Shermegor, Elasticity theory of microheterogeneous media (in Russian), Nauka, Moscow, 1977., Nauka, Moscow, 1977.

[3] J. Cook, The elastic constants of an isotropic matrix reinforced with imperfectly oriented fibres, *Brit. J. Appl. Phys.*, **1**, ser. 2, 799, 1968.

K. Wilmanski
Technische Universität
Hamburg-Harburg
Arbeitsbereich Meerestechnik II
Postfach 90 14 03
BRD-2100 Hamburg 90
GERMANY

S. MINAGAWA

On the force exerted on dislocations in a body by an external electric field

1. Introduction

This paper aims to answer the question: 'Does the external electric field interact with dislocations in the body?' The answer to this question is affirmative, as far as there exists a field of electric displacement which varies from point to point in the body.

The formula for the force between an electric field and a loop of discrete dislocation will be derived by the method of Peach and Koehler [1]. It will be concluded that the force on an aggregate of dislocations is given by the substitution of the 'asymmetric' Maxwell stress, composed of the external electric field and the field of electric displacement, for the ordinary stress in the tensorial formulation for the Peach-Koehler force (see [2, 3]).

2. Discrete dislocations

We consider a dislocation loop with the Burgers vector $\mathbf{b}$. The positive sense of description of the dislocation line, that of the Burgers circuit linking it, and the plus and minus sides of a dislocation surface are given as shown in Fig. 1. We shall give the dislocation by translating the plus side of the surface through the distance $\mathbf{b}$, while the minus side remains fixed.

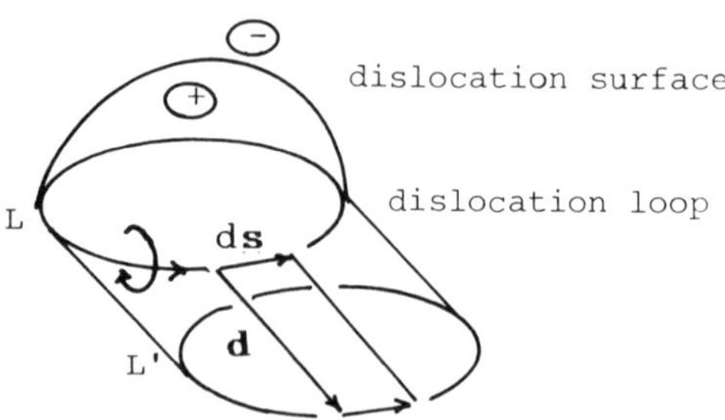

Figure 1. Sign convention and the translation of dislocation loop

Let $\mathbf{E}$ be the external electric field. Assuming that this electric field is stationary, we have

$$\mathbf{E} = - \nabla\varphi, \tag{1}$$

where φ is the electric scalar potential. We shall assume that a field of electric displacement, $\mathbf{D}$, is given beforehand in the body. When the interaction between an electric field and a dislocation is involved, $\mathbf{D}$ becomes the field induced by the given electric field. If the interaction between dislocations in a piezoelectric crystal is dealt with, the $\mathbf{E}$ and $\mathbf{D}$ are the fields stimulated by those dislocations in the crystal (see [4, 5]).

Let ds be a vector line element belonging to the dislocation loop. If the dislocation loop is displaced parallel from L to L′ by a short vectorial distance $\mathbf{d}$, ds draws a surface given by $\mathrm{d}\mathbf{S} = \mathrm{ds} \times \mathbf{d}$. The material in the plus side of this surface is translated by $\mathbf{b}$, and accordingly the value of the electric displacement is changed by $-\mathbf{b}\cdot\nabla\mathbf{D}$, while in the minus side they remain fixed. Therefore, there appears a discontinuous jump in the field of electric displacement at this surface.

Its component, normal to the surface, is equal to the amount of charges spread over the surface (see, e.g., [6]). In other words, the surface charge on this surface is increased by $-(\mathbf{b}\cdot\nabla\mathbf{D}) \cdot \mathrm{d}\mathbf{S}$, as the translation of the loop is carried out. Since this increase of the surface charge is achieved under the action of the electric field E, the work required to do this is given by $\delta W = \varphi(\mathbf{b}\cdot\nabla\mathbf{D}) \cdot \mathrm{d}\mathbf{S}$.

We have $(\ ) \cdot (\mathrm{ds} \times \mathbf{d}) = \{(\ ) \times \mathrm{ds}\} \cdot \mathbf{d}$, and the force exerted on ds is given by $-\partial w/\partial \mathbf{d}$, therefore the total force exerted on the dislocation loop L is given by the line integral belonging to the dislocation loop, such as

$$\mathbf{F} = -\oint_L \varphi(\mathbf{b}\cdot\nabla\mathbf{D}) \times \mathrm{ds}. \tag{2}$$

From the above discussion, if the field of electric displacement is constant throughout the body, no force is exerted on dislocations by an external electric field.

3. Continuous distribution of dislocations

We assume an orthogonal Cartesian coordinate system, with respect to which the position of a point is stated as x_i. Throughout this paper, $i, j, k, \ldots$ take 1, 2 or 3, Einstein's summation convention is used for indices appearing twice in one expression, and the comma followed by an index means the derivative with regard to the corresponding space coordinate.

In tensor notations, the last equation becomes

$$F_j = - \varepsilon_{jpq}\, b_k \oint_L \varphi D_{p,k}\, ds_q, \tag{3}$$

where ε_{ijk} is Eddington's permutation symbol.

We substitute the volume integral $\int_\Omega [\ldots] \alpha_{qk}\, dV$ for the line integral $b_k \oint_L [\ldots] ds_q$ in equation (3) to give the force exerted on an aggregate of dislocations, such as

$$F_j = -\varepsilon_{jpq} \int_\Omega \varphi \alpha_{qk} D_{p,k}\, dV, \tag{4}$$

where the volume integral is calculated over region Ω which is occupied by the aggregate of dislocations, and α_{ij} is the dislocation density tensor, which vanishes at the boundary of Ω, because we are assuming that no dislocations pierce the latter.

By the method of integration by parts, and by using the assumption that α_{ij} vanishes at the boundary, equation (4) is transformed into

$$F_j = \varepsilon_{jpq} \int_\Omega (\varphi \alpha_{qk})_{,k} D_p\, dV \tag{5}$$

and by substituting from equation (1) and by using $\alpha_{ij,j} = 0$ in Ω,

$$F_j = -\varepsilon_{jpq} \int_\Omega \alpha_{qk} E_k D_p\, dV. \tag{6}$$

The last equation implies the force exerted on an aggregate of dislocations by an external electric field.

4. Remarks

The Peach–Koehler force exerted on an aggregate of dislocations has been given in [2, 3] by

$$F_n^S = 2 \int s_{jni}\, \sigma_{ij}\, dV, \tag{7}$$

where σ_{ij} is the external stress field, and

$$s_{jni} = \tfrac{1}{2} \varepsilon_{jnp} \alpha_{pi}. \tag{8}$$

We combine the last two equations to obtain

$$F_j^S = -\varepsilon_{jpq} \int_\Omega \alpha_{qk} \sigma_{kp}\, dV. \tag{9}$$

The right–hand side of equation (6) is given by the substitution of the $E_k D_p$ for σ_{kp} in equation (9).

The **E** and **D** may be indepenent of each other, and therefore $E_k D_p$ is in general asymmetric. Calling $E_k D_p$ the 'asymmetric' Maxwell stress, we arrive at the conclusion that the force exerted on an aggregate of dislocations by an external electric field, under the existence of a field of electric displacement, is given by the substitution of the asymmetric Maxwell stress, composed of those fields, for the ordinary stress in the formula for the Peach-Koehler force.

References

[1] M. Peach and I.S. Koehler, The force exerted on dislocations and the stress fields produced by them. *Phys. Rev.*, **80** (1950), 436-439.

[2] S. Minagawa, *On the force on distributed dislocations caused by stresses.* RAAG Research Notes, 3rd Ser., No. 143, 1969.

[3] S. Minagawa, On the force exerted upon continuously distributed dislocations in a Cosserat continuum. *Phys. Stat. Sol.*, **39** (1970), 217-222.

[4] S. Minagawa and K. Shintani, Dislocation dynamics in anisotropic piezoelectric crystals. *Phil. Mag.*, A, **51** (1985), 277-285.

[5] S. Minagawa and K. Shintani, Dislocation dynamics in anisotropic piezoelectric crystals; II. Additional Examples. *Phil. Mag.*, **56** (1987), 343-352.

[6] J.A. Stratton, *Electromagnetic Theory*, McGraw-Hill, New York/London, 1941.

S. Minagawa
University of Electro-Communications
Department of Mechanical and Control Engineering
Chofu, Tokyo 182
JAPAN

J.P. NOWACKI

Dislocation dynamics in elastic dielectrics

1. Introduction

In recent years, due to the technological applications of piezoelectric materials in experimental devices (resonators, filters, sensors, delay lines,...) the problem of elastic dielectrics has received considerable attention from researchers.

The investigations were initially involved in the framework of the Voigt's linear theory of piezoelectricity. The first approach to the problem of piezoelectric field generated by the dislocation line was made by Merten [1]-[4]. Using the approximate theory (neglecting the piezoelectric stress and assuming the elastic isotropy), the stress field around the dislocation has been calculated. It has been shown that the distortion of elastic strain field caused by inverse piezoelectric effect represents a correction of only a few per cent to the primary elastic strain. Merten also obtained rather unrealistic results that the electric potential field does depend on direction, but not on the distance from the dislocation line. Next Kosevich *et al*. [5], Saada [6] and Faivre and Saada [7] extended the analysis to avoid Merten's approximation. They generalized the exact plane-strain solution given by Eshelby *et al*. [8] for an infinite straight dislocation in an anisotropic medium, to include the piezoelectric effect. The application of this exact method shows that the electric potential due to dislocation depends not only on direction but also logarithmically on distance. Following this fundamental consideration, the detailed calculations of the magnitudes and configurations of piezoelectric field around rectilinear dislocations which may form in crystals having quartzite structure has been made by Smirnowa [9]. For example, it has been shown that the screw dislocation lying in the basal plane and dislocations oriented along the c axis have no piezoelectric field. On the other hand, the piezoelectric field of edge dislocation lying in the basal plane, with Burger's vector also in the same plane, has a considerable radial component, and this dislocation may behave as a positively charged filament. The most recent study of piezoelectric field surrounding a static dislocation by Huston and Walker [10] was devoted to scrutinizing previous results by means of machine computation. It is also worth mentioning the paper of Booyens *et al* [11], although it uses Merten's approximation and its main interest is in semiconductors.

In recent years Minagawa [12] introduced the general theory of moving dislocations and disclinations in thermo-piezoelectric continua. This approach is based on the modelling of defects by the initial plastic distortion. The final expressions for thermo-piezoelectric fields are obtained in terms of the defect density and defect-flux tensor with the aid of Green's functions and dynamic Green's potential. Recently Minagawa and Shintani [13, 14] studied the electric and mechanical fields of moving dislocations by the method of continuously distributed dislocations. The formal solution in the most general form is shown by means of the convolution integral in space-time. Next the piezoelectric field of a uniformly moving infinite straight dislocation is considered and the integrals are calculated completely. The several examples have been evaluated by numerical computation.

In the last three years a series of papers, concerned with description of a piezoelectric field of dislocation, in the frame of the theory of dielectrics with polarization gradient [15], has been published by Nowacki and Hsieh [16] and Nowacki [17]-[19]. According to the classical theory of elastic dielectrics no coupling exists between the mechanical displacement and the polarization. However, in the frame of extended theory, such coupling effects exist even in material of highest symmetry, i.e. centrosymmetric, isotropic. The continuum equations of this linear theory are the correct, long wave limits of the difference equations of the dynamic lattice formulation. The crystal is based on the shell model and the relations between polarization gradient, shell-shell and core-shell interaction of lattice theories of crystal have all been indicated. The theory of elastic dielectrics with polarization gradient accommodated observed and experimentally measured phenomena (which are not accounted for by other linear theories describing electrically polarized material in interaction with an electrostatic field) such as electromechanical interaction in symmetric materials, capacitance of thin dielectric film, surface energy of polarization, deformation and optical activity in quartz. The theory of dislocation, based on the generalization of Weingarten's theorem, as well as the theory based on Burger's condition, has been given. There are two ways of introducing the concept of dislocation to the theory of elastic dielectrics. The first of them, due to Volterra, consists in the representation of the dislocation as an arbitrary surface resting on the dislocation line, on which the displacement vector of the points of the medium has a constant discontinuity equal to b. So, the dislocation is a surface defect. However, in reality, in any simply connected region which does not contain the boundary of the discontinuity surface, the medium remains a perfect elastic dielectric medium deformed in a continuous manner; thus, the essential source of the deformation is exactly the boundary of the cut, i.e. the dislocation line. However, if one wants to describe the dislocation as a line defect, the displacement description must be given up. In this paper we concentrate on the calculation of elastic and electric fields generated by a moving line, based on the second equivalent concept of the description of a dislocation, namely on the Burgers condition.

2. Governing equations

The system of basic linear equations for an elastic dielectric including polarization gradient effects occupying the domain V and bounded by surface S is given by

(i) The field equation

$$\sigma_{ij,i} = \rho \ddot{u}_j, \quad \sigma_{ij} = \sigma_{ji},$$

$$E_{ij,i} + E_j^{\mathrm{L}} - E_{,j}^{\mathrm{MS}} = 0,$$

$$\varepsilon_0 \varphi_{,ii} + P_{i,i} = 0 \ \text{in} \ V,$$

$$\varphi_{,ii} = 0 \ \text{in} \ V'. \tag{2.1}$$

(ii) The boundary conditions

$$\sigma_{ij} n_i = k_j,$$

$$E_{ij} n_i = S_j,$$

$$(P_i - \varepsilon_0 \| \varphi_{,i} \|) = 0 \ \text{on} \ S. \tag{2.2}$$

(iii) The constitutive equations

$$\sigma_{ij} = c_{ijkl} \beta_{kl} + f_{kij} P_k + d_{klij} P_{lk},$$

$$E_{ij} = d_{ijkl} \beta_{kl} + j_{kij} P_k + b_{ijkl} P_{lk} + b_0,$$

$$E_j^L = f_{jkl} \beta_{kl} + a_{jk} P_k + j_{jkl} P_{lk}. \tag{2.3}$$

(iv) The kinematic relations

$$E_i^{\mathrm{MS}} = -\varphi_{,i}, \ S_{ij} = \tfrac{1}{2}(u_{i,j} + u_{j,i}), \ P_{kl} = P_{l,k}, \tag{2.4}$$

where σ_{ij}, E_{ij} and S_{ij} designate the components of the stress tensor, electric tensor and strain tensor, respectively; u_i, P_i, $E_i^{\mathrm{L}}, E_i^{\mathrm{MS}}$ and n_i are components of displacement vector, polarization vector, local electric vector, Maxwell self-field vector and the exterior unit normal vector, respectively; φ, $\| \varphi_{,i} \|$ denote Maxwell potential, jump in $\varphi_{,i}$ across S; k_i and S_j are surface traction and surface electric force, while V' stands for outer vacuum and ε_0 its permittivity.

Substitution of the constitutive equations and kinematic relations into the field equations yields Navier's equations of dielectrics

$$C_{ji}\, u_i + D_{ji}\, P_i = \rho \ddot{u}_j,$$

$$D_{ji}\, u_i + B_{ji}\, P_i - \varphi_{,j} = 0,$$

$$-\varepsilon_0\, \varphi_{,ii} + P_{i,i} = 0, \tag{2.5}$$

where the following differential operators have been defined:

$$C_{ji} = C_{jpis}\, \nabla_p \nabla_s = [c_{12}\, \varphi_{jp}\, \delta_{is} + c_{44}\, (\delta_{ps}\, \delta_{ij} + \delta_{js}\, \delta_{pi})]\, \nabla_p \nabla_s,$$

$$D_{ji} = D_{jpis}\, \nabla_p \nabla_s = [d_{12}\, \delta_{jp}\, \delta_{is} + d_{44}\, (\delta_{ps}\, \delta_{ij} + \delta_{js}\, \delta_{pi})]\, \nabla_p \nabla_s,$$

$$B_{ji} = B_{jpis}\, \nabla_p \nabla_s - a\delta_{ij} = [b_{12}\, \delta_{jp}\, \delta_{is} + (b_{44} + b_{77})\delta_{ps}\, \delta_{ij}$$

$$+ (b_{44} - b_{77})\delta_{js}\, \delta_{pi}]\, \nabla_p \nabla_s - a\delta_{ij}.$$

3. The theory of dislocation line

The point of departure is the mathematical description of the Burgers condition for distortion around the dislocation line. Then one presents a method of derivation, exactly from the Burgers condition, the expression for the distortion field, the gradient of polarization field, the electric field of a perfect medium containing a dislocation. The required fields should possess the following properties; first they should satisfy the Burgers conditions, and secondly outside dislocation they should constitute the fields of an unloaded dielectric medium. If the dislocation is a closed line or its ends are in infinity, then it can be described by

$$\int_C \beta_{ki}\, dl_k = b_i\,, \tag{3.1}$$

where β_{ki} is the distortion field, C is a curve encircling once the dislocation line, and b denotes the Burgers vector. Replacing the line integral by the surface integral over the arbitrary surface S, resting on C, using Stokes' theorem, one obtains

$$b_i = \int_C \beta_{ki}\, dl_k = \int_S dS_l\, \varepsilon_{mlk}\, \beta_{ki,m}, \tag{3.2}$$

where the integrand has a singularity of the Dirac delta type. This singularity is due to puncture of S by the dislocation line and it takes the form

$$\varepsilon_{mpk}\, \beta_{ki,m} = b_i \int_D d\xi_p\, \delta_3(\mathbf{x} - \xi(l, t)) = \alpha_{pi}\,, \tag{3.3}$$

where α_{pi} stands for the tensor of density of dislocation and D describes the dislocation line. This equation implies that there does not exist a displacement field $\mathbf{u}$ such that $u_{k,i} = \beta_{ik}$ at all points of the medium, for then one would have

$\varepsilon_{mlk}\beta_{ki,m} = \varepsilon_{mlk}u_{i,km} = 0$, which contradicts (3.3). However such a field can be constructed in the simply connected region which contain no dislocations and this field obey Navier's equation (2.5). Differentiating these equations with respect to x_k, one obtains the homogeneous system of equations for β_{ki}, P_{ki} and $\varphi_{,i}$ in an infinite domain without the dislocation line. If a dislocation line is to occur, then β_{ki}, P_{ki}, and $\varphi_{,i}$ should satisfy the nonhomogeneous system of equations

$$C_{ji}\beta_{ki} + D_{ij}P_{ik} = -A_{jk} + \rho\frac{\partial^2\beta_{ij}}{\partial t^2},$$

$$D_{ji}\beta_{ki} + B_{ij}P_{ik} - \varphi_{,jk} = -B_{jk},$$

$$-\varepsilon_0\varphi_{,iik} + P_{ik,i} = 0 \tag{3.4}$$

where the tensor A_{jk} and B_{jk} differ from zero only on dislocation line.

Next, using the convolution theorem, one can represent the solution of the above system for β_{ij}, P_{ij}, E_i in the form

$$\beta_{kn}(\mathbf{x}, t) = \int dt' \int [A_{ik}(\mathbf{x}', t)\, {}^{f}G_{ni}(\mathbf{x} - \mathbf{x}', t - t')$$

$$+ B_{ik}(\mathbf{x}', t')\, {}^{f}\Pi_{n\,i}(\mathbf{x} - \mathbf{x}', t - t')]\, dV(\mathbf{x}').$$

$$P_{nk}(\mathbf{x}, t) = \int dt' \int [A_{ik}(\mathbf{x}', t)\, {}^{E}G_{ni}(\mathbf{x} - \mathbf{x}', t - t')$$

$$+ B_{ik}(\mathbf{x}' - t')\, {}^{E}\Pi_{ni}(\mathbf{x} - \mathbf{x}', t - t')]\, dV(\mathbf{x}').$$

$$-E_k(\mathbf{x}, t) = \int dt' \int [A_{ik}(\mathbf{x}', t)\, {}^{\rho}G_i(\mathbf{x} - \mathbf{x}', t - t')$$

$$+ B_{ik}(\mathbf{x}', t')\, {}^{\rho}\Pi_i(\mathbf{x} - \mathbf{x}', t - t')]\, dV(\mathbf{x}'). \tag{3.5}$$

Next, using the definition of the Green's G_{ij} and Π_{ij} function and the divergence theorem, one obtains

$$\alpha_{li}(\mathbf{x}, t) = \int dt' \int dV(\mathbf{x}')\delta_{ir}\,\delta_3(\mathbf{x} - \mathbf{x}')\delta(t - t')\alpha_{lr}(\mathbf{x}', t')$$

$$= \int dt' \int dV(\mathbf{x}')\,[C_{rpns}\,{}^{f}G_{in,ps}(\mathbf{x} - \mathbf{x}', t - t') + D_{rpns}\,{}^{f}\Pi_{ni,ps}(\mathbf{x} - \mathbf{x}', t - t')$$

$$-\rho\frac{\partial^2 f}{\partial t^2}\, G_{ir}(\mathbf{x} - \mathbf{x}', t - t')]\,\alpha_{lr}(\mathbf{x}', t')$$

$$= \int \mathrm{d}t' \int \mathrm{d}V(\mathbf{x}') [C_{rpns} fG_{in}(\mathbf{x}-\mathbf{x}', t-t') + D_{rpns} f\Pi_{ni}(\mathbf{x}-\mathbf{x}', t-t')]$$

$$\times \alpha_{lr,p's'}(\mathbf{x}', t') - \int \mathrm{d}t' \int \mathrm{d}V(\mathbf{x}') fG_{ir}(\mathbf{x}-\mathbf{x}', t-t')] \rho \frac{\partial^2}{\partial t'^2} \alpha_{lr}(\mathbf{x}', t'). \qquad (3.6)$$

Here we integrated by parts and we used the relations

$$\frac{\partial G_{ik}}{\partial x_j} = -\frac{\partial G_{ik}}{\partial x_j'}, \quad \frac{\partial G_{ik}}{\partial t} = -\frac{\partial G_{ik}}{\partial t'}.$$

Let us consider the first integral of (3.6)

$$\int \mathrm{d}t' \int \mathrm{d}V(\mathbf{x}') [C_{rpns} fG_{in}(\mathbf{x}-\mathbf{x}', t-t') + D_{rpns} f\Pi_{ni}(\mathbf{x}-\mathbf{x}', t-t')] \alpha_{lr,p's'}(\mathbf{x}', t')$$

$$= \varepsilon_{klm} \frac{\partial}{\partial x_m'} \varepsilon_{kba} \int \mathrm{d}V [C_{rbns} fG_{ni}(\mathbf{x}-\mathbf{x}', t-') + D_{rbns} f\Pi_{ni}(\mathbf{x}-\mathbf{x}', t-t')] \alpha_{ar,s'}(\mathbf{x}', t').$$

We have used here the identity

$$C_{rpns}\alpha_{lr,ps} - C_{rlns}\alpha_{pr,ps} = \varepsilon_{klm}\varepsilon_{kba} C_{rbns}\alpha_{ar,sm}$$

and we have observed that

$$\alpha_{pr,p} = \varepsilon_{pmk} \beta_{kr,mp} = 0.$$

Let us consider the last integral of (3.6)

$$\int \mathrm{d}t' \int \mathrm{d}V(\mathbf{x}') fG_{ir}(\mathbf{x}-\mathbf{x}', t-t') \rho \frac{\partial^2}{\partial t'^2} \alpha_{lr}(\mathbf{x}', t')$$

$$= \int \mathrm{d}t' \int \mathrm{d}V(\mathbf{x}') \rho fG_{ir}(\mathbf{x}-\mathbf{x}', t-t') \, b_r \frac{\partial^2}{\partial t'^2} \oint_D \mathrm{d}\xi_\ell \, \delta_3(\mathbf{x}' - \xi(\ell, t'))$$

$$= \int \mathrm{d}t' \int \mathrm{d}V(\mathbf{x}') fG_{ir}(\mathbf{x}-\mathbf{x}', t-t') \, b_r \, \rho \frac{\partial}{\partial t'} \oint_D \mathrm{d}\xi_\ell \, \delta_3(\mathbf{x}' - \xi(1\ell, t'))$$

$$-\oint_D d\xi_\ell \xi_k \nabla_k \delta_3 (\mathbf{x}' - \xi(\ell, t'))$$

$$= \int dt' \int dV(\mathbf{x}') fG_{ir}(\mathbf{x} - \mathbf{x}', t - t')] \, b_r \, \rho \frac{\partial}{\partial t'} \{ \oint_D [\xi_\ell \, \delta_3 (\mathbf{x}' - \xi(1\ell, t'))]$$

$$+ \oint_D d\xi_k \xi_\ell \nabla_k \delta_3(\mathbf{x}' - \xi(1\ell, t')) - \oint_D d\xi_{1\ell} \xi_k \nabla_k \delta_3(\mathbf{x}' - \xi(1\ell, t'))\}$$

$$= \int dt' \int dV(\mathbf{x}') \, \rho \, fG_{ir}(\mathbf{x}-\mathbf{x}', t-t') \{ b_r \frac{\partial}{\partial t'} \varepsilon_{imk} \nabla_m \varepsilon_{kba} \oint_D d\xi_a \xi_b \, \delta_3(\mathbf{x}' - \xi(\ell, t'))\}.$$

We have used here the identity

$$\varepsilon_{lmk} \varepsilon_{abk} = \delta_{la} \, \delta_{mb} - \delta_{lb} \, \delta_{ma}.$$

Finally, adding the results, we obtain

$$\alpha_{li}(\mathbf{x}, t) = \varepsilon_{klm} \frac{\partial}{\partial x'_m} \varepsilon_{kba} \int dt' \int dV \{ [C_{rbns} fG_{ni} (\mathbf{x} - \mathbf{x}', t - t')$$

$$+ D_{rbns} f\Pi_{ni}(\mathbf{x} - \mathbf{x}', t - t')] \alpha_{ar,s'}(\mathbf{x}', t')$$

$$+ fG_{ir}(\mathbf{x} - \mathbf{x}', t - t') \{ b_r \, \rho \frac{\partial}{\partial t'} \oint_D d\xi_a \xi_b \, \delta_3(\mathbf{x}' - \xi(l, t'))\}. \tag{3.7}$$

Comparing (3.7) with (3.6), one arrives at

$$\beta_{kn}(\mathbf{x}, t)$$

$$= \varepsilon_{kba} \int dt' \int dV(\mathbf{x}) \{ [C_{rbns} fG_{ni}(\mathbf{x} - \mathbf{x}', t - t')$$

$$+ D_{rbns} f\Pi_{ni}(\mathbf{x} - \mathbf{x}', t - t')] \alpha_{ar,s'}(\mathbf{x}', t')$$

$$+ fG_{ir}(\mathbf{x} - \mathbf{x}', t - t') \, b_r \, \rho \frac{\partial}{\partial t'} \oint_D d\xi_a \xi_b \, \delta_3(\mathbf{x}' - \xi(l, t'))\}. \tag{3.8}$$

And finally, comparing (3.8) with (3.5), we obtain

$$A_{ik}(\mathbf{x}', t') = \varepsilon_{kba} \, [C_{rbis} \, \alpha_{ar,s'}(\mathbf{x}', t')$$

$$+ b_i \, \rho \frac{\partial}{\partial t'} \oint_D d\xi_a \xi_b \, \delta_3(\mathbf{x}' - \xi(l, t'))]$$

$$B_{nk}(\mathbf{x}', t') = \varepsilon_{kba} D_{rbns} \, \alpha_{ar,s'}(\mathbf{x}', t'). \tag{3.9}$$

Now, using the definition of the dislocation density tensor, one can write the final form of the solution for an electric field:

$$\beta_{kn}(\mathbf{x}, t) = \varepsilon_{k\,b\,a=}\int dt' \int dV(\mathbf{x}')\{[C_{r\,b\,i\,s}\,\alpha_{a\,r,\,s'}(\mathbf{x}', t')$$

$$+ b_i\,\rho \frac{\partial}{\partial t'} \oint_D d\xi_a \xi_b\, \delta_3(\mathbf{x}' - \xi(l, t'))]\, {}^f G_{in}(\mathbf{x} - \mathbf{x}', t - t')$$

$$+ D_{r\,b\,i\,s}\, {}^f\Pi_{in}(\mathbf{x} - \mathbf{x}', t - t')]\, \alpha_{a\,r,\,s'}(\mathbf{x}', t'),$$

$$P_{nk}(\mathbf{x}, t) = \varepsilon_{k\,b\,a}\int dt' \int dV(\mathbf{x}')\,[C_{r\,b\,i\,s}\,\alpha_{a\,r,\,s'}(\mathbf{x}', t')$$

$$+ b_i\,\rho \frac{\partial}{\partial t'} \oint_D d\xi_a \xi_b\, \delta_3(\mathbf{x}' - \xi(l, t')]\, {}^E G_{in}(\mathbf{x} - \mathbf{x}', t - t')$$

$$+ D_{r\,b\,i\,s}\, {}^E\Pi_{in}(\mathbf{x} - \mathbf{x}', t - t')]\, \alpha_{a\,r,\,s'}(\mathbf{x}', t'),$$

$$-E_k\,(\mathbf{x}, t) = \varepsilon_{k\,b\,a}\int dt' \int dV(\mathbf{x}')\{\,[C_{r\,b\,i\,s}\,\alpha_{a\,r,\,s'}(\mathbf{x}', t')$$

$$+ b_i\,\rho\, \frac{\partial}{\partial t'} \oint_D d\xi_a \xi_b\, \delta_3(\mathbf{x}' - \xi(l, t'))]\, {}^\rho G_i\,(\mathbf{x} - \mathbf{x}', t - t')$$

$$+ D_{r\,b\,i\,s}\, {}^\rho\Pi_i\,(\mathbf{x} - \mathbf{x}', t - t')]\, \alpha_{a\,r,\,s'}\,(\mathbf{x}', t'). \tag{3.10}$$

The above formulae completely describes the behaviour of the fields generated by moving dislocation line and can be used as a starting point for numerical solution for the given geometry of dislocation line.

References

[1] Merten, L. *Phys. Kondens. Mater.* **2**, 53 (1964).
[2] Merten, L. *Phys. Kondens. Mater.* **2** 66 (1964).
[3] Merten, L. *Z. Naturforsch.* **192**, 788 (1964).
[4] Merten, L. *Z. Naturforsch.* **192**, 1161 (1964).
[5] Kosevich, A.M., Pastur, L.A., Fel'dman, E.P., Sov. Phys.-Crystallogr. 12, 797 (1968).
[6] Saada, G., *Phys. Status Solidi* (B) **44**, 717 (1971).
[7] Faivre, G. and Saada, G., *Phys. Status Solidi* **52**, 127 (1972).
[8] Eshelby, J.D., Read, W.T. and Shockley, W., *Acta Metall.* **1**, 251 (1953).
[9] Smirnowa, I.S., *Sov. Phys. -Solid State* **15**, 1543 (1974).
[10] Huston, A.R. and Walker, L.R. *J. Appl. Phys.*, **50**, 6247 (1979).
[11] Booyens, H., Vermaak, J.S. and Proto, G.R., *J. Appl. Phys.*, **48**, 3008 (1977).
[12] Minagawa, S., *Phys. Stat. Sol.* (B), **124**, 565 (1984).

[13] Minagawa, S. and Shintani, K. *Phil. Mag.* A **51**, 277 (1985).

[14] Minagawa, S., and Shintani, K. *Phil. Mag.* A, **56**, 3343 (1987).

[15] Mindlin, R.D., *J. Elasticity* **2**, 217 (1972).

[16] Nowacki, J.P. and Hsieh, R.K.T. *Int. J. Engng. Sci*,. **24** (10), 1655 (1986).

[17] Nowacki, J.P. In *Deformable Solids and Structures*, ed. Yamamoto, Y. and Miya, K., North-Holland, 1987.

[18] Nowacki, J.P., *J. Tech. Phys.* **29** (1), 99 (1988).

[19] Nowacki, J.P. In *Applied Electromagnetics in Materials*, ed. Miya, K., Pergamon Press, 1989.

J.P. Nowacki
Institute of Fundamental Technological Research
Polish Academy of Sciences
Swietokrzyska 21
00-049 Warsaw
POLAND

H. ZORSKI

Dynamics of dipoles

Introduction

The knowledge of the motion of dipoles in external fields is important for many theories in mechanics and physics. It constitutes (or rather should constitute) an important ingredient of the theory of polarization of material media, in particular in the determination of the dependence of the polarization vector on the vector of the electrical field, on the basis of elementary models, derived either statistically or by means of a constitutive relation.

In this paper the dipole is regarded as a pair of point charges subject to the following constraint: the distance between them is a known function of time. Rather than regard the elementary model as an oscillator (a simple but hardly rational model), called sometimes the Drude model of polarization, we prefer to regard it as a rotator. The consequence is roughly speaking that the simplest equation is now the pendulum equation with the coefficient proportional to the external field, not the oscillator equation. Thus, one of the principal differences between the models is that in the case of the rotator the resonance is parametric.

We confine ourselves in this paper to the basic equations and some exact qualitative results. Approximate solutions will be published elsewhere.

1. Derivation of the equations of motion

We begin by introducing our model of the dipole. To this end, consider two particles P^+ and P^- placed at $\mathbf{x}^+$ and $\mathbf{x}^-$, with masses and charges m^+, q^+ and m^-, q^-, respectively (Fig. 1). The quantity $\boldsymbol{\mu} = \mathbf{x}^+ - \mathbf{x}^-$ will be called the dipole moment. Denoting by $\mathbf{x}$ the position of the centre of mass, $m^+ \mathbf{x}^+ + m^- \mathbf{x}^- = m\mathbf{x}$, $m = m^+ + m^-$, we have

$$\mathbf{x}^+ = \mathbf{x} + \frac{m^-}{m}\boldsymbol{\mu}, \quad \mathbf{x}^- = \mathbf{x} - \frac{m^+}{m}\boldsymbol{\mu}. \tag{1.1}$$

Rather than introducing an interaction potential between the two particles constituting the dipole (cf. [1]), we assume the constraint

$$|\boldsymbol{\mu}| = l(t) \tag{1.2}$$

where $l(t)$ is a prescribed length. (1.2) is a holonomic constraint in the dynamics of the two particles P^+ and P^-. Let now these two particles be subject to the Lorentz forces due to the electric filed (the time variable is suppressed, generally $\mathbf{E} = \mathbf{E}(\mathbf{x}, t)$)

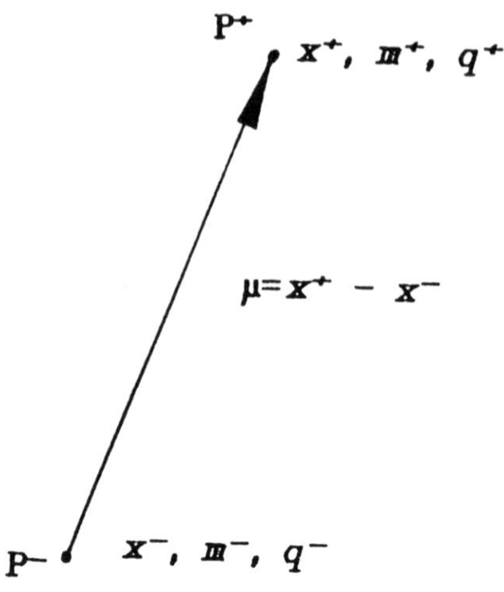

Figure 1.

$$\mathbf{E}(\mathbf{x}) = - \nabla\varphi(\mathbf{x}); \tag{1.3}$$

then the Lagrangian with the multiplier λ has the form

$$L = \tfrac{1}{2}(m^+ \dot{\mathbf{x}}^{+2} + m^- \dot{\mathbf{x}}^{-2}) - [q^+ \varphi(\mathbf{x}^+) + q^- \varphi(\mathbf{x}^-)] + \lambda\,[\,|\,\mathbf{x}^+ - \mathbf{x}^-|^2 - l^2(t)\,], \tag{1.4}$$

and the conventional procedure yields the equations of motion

$$m^+ \ddot{\mathbf{x}}^+ = q^+ \mathbf{E}(\mathbf{x}^+) + 2\lambda\,(\mathbf{x}^+ - \mathbf{x}^-)$$

$$m^- \ddot{\mathbf{x}}^- = q^- \mathbf{E}(\mathbf{x}^-) - 2\lambda\,(\mathbf{x}^+ - \mathbf{x}^-). \tag{1.5}$$

We note that the reaction forces are now central, but in general not necessarily potential. Adding and subtracting (1.5), we obtain

$$m\ddot{\mathbf{x}} = q^+ \mathbf{E}(\mathbf{x}^+) + q^- \mathbf{E}(\mathbf{x}^-)$$

$$\ddot{\boldsymbol{\mu}} = \Big[\frac{q^+}{m^+}\mathbf{E}(\mathbf{x}^+) - \frac{q^-}{m^-}\mathbf{E}(\mathbf{x}^-)\Big] + 2\lambda\,\frac{m}{m^+m^-}\,\boldsymbol{\mu}. \tag{1.6}$$

Scalar-multiplying the rotation equation by $\boldsymbol{\mu}$, in view of (1.2) we obtain the expression for λ:

$$\lambda = \frac{m^+ m^-}{2l^2 m}\, \mu \cdot \Big[\mu - \Big(\frac{q^+}{m^+}\mathbf{E}(\mathbf{x}^+) - \frac{q^-}{m^-}\mathbf{E}(\mathbf{x}^-)\Big)\Big]$$

and substituting back into the system (1.6) we have

$$m\,\ddot{\mathbf{x}} = q^+\,\mathbf{E}(\mathbf{x}^+) + q^-\,\mathbf{E}(\mathbf{x}^-)$$

$$\Big(\mathbf{I} \cdot \frac{\mu\mu}{l^2}\Big) - \Big\{\ddot{\mu} - \Big[\frac{q^+}{m^+}\mathbf{E}(\mathbf{x}^+) - \frac{q^-}{m^-}\mathbf{E}(\mathbf{x}^-)\Big]\Big\} = 0 \tag{1.7}$$

$$|\,\mu\,| = l(t).$$

Since

$$\Big(\mathbf{I} - \frac{\mu\mu}{l^2}\Big) \cdot A = -\frac{1}{l^2}\,\mu \times (\mu \times A),$$

the rotation equation can be written in the form

$$\mu \times \Big\{\ddot{\mu} - \Big[\frac{q^+}{m^+}\mathbf{E}(\mathbf{x}^+) - \frac{q^-}{m^-}\mathbf{E}(\mathbf{x}^-)\Big]\Big\} = 0. \tag{1.7$'$}$$

We note that the rotation equation multiplied scarlarly by μ yields an identity, due to the fact that the inertia about the dipole axis is zero. The operator $I - \mu\mu/l^2 = P$, where I is the unit matrix is projective, i.e. $P^2 = P$. Thus (1.7) is a system of 6 equations with 6 unknowns.

So far nothing has been said about the magnitude of $l(t)$; now we assume that it is small compared with the variation in space of the electric field (or any other dimension which may appear in the problem), i.e.

$$\frac{1}{l} >> \frac{|\nabla \mathbf{E}|}{|\mathbf{E}|},$$

and then $\mathbf{E}(\mathbf{x}^+)$ and $\mathbf{E}(\mathbf{x}^-)$ can be expanded into Taylor series about the centre of gravity of the dipole; retaining only the first two terms

$$\mathbf{E}(\mathbf{x}^+) = [1 + \frac{m^-}{m}(\mu\cdot\nabla)]\,\mathbf{E}(\mathbf{x}),\ \ \mathbf{E}(\mathbf{x}^-) = [1 - \frac{m^+}{m}(\mu\cdot\nabla]\,\mathbf{E}(\mathbf{x}),$$

whence the equations of motion (1.7) take the form

$$m\,\ddot{\mathbf{x}} = q^+\,[K_0 + \frac{m^-K_1}{m}(\mu\cdot\nabla)]\,\mathbf{E}(\mathbf{x})$$

$$\mu \times \left\{\mu - \frac{q^+}{m^+}[K_1 + \frac{m^+}{m}K_2\,(\mu\cdot\nabla)]\,\mathbf{E}(\mathbf{x})\right\} = 0, \tag{1.8}$$

where

$$K_0 = 1 + \frac{q^-}{q^+},\ K_1 = 1 - \frac{q^-}{q^+}\frac{m^+}{m^-},\ K_2 = \frac{m^-}{m^+}\left[1 + \frac{q^-}{q^+}\left(\frac{m^+}{m^-}\right)^2\right]$$

are dimensionless constants. In the particular case of a neutral dipole $q^- = -q^+$, $K_0 = 0$, and if we also assume that $m^+ = m^-$, then $K_1 = 2$, $K_2 = 0$ and

$$m\ddot{\mathbf{x}} = q^+\,(\mu\cdot\nabla)\mathbf{E}$$

$$\mu \times (\ddot{\mu} - \frac{2\,q^+}{m^+}\mathbf{E}) = 0; \tag{1.9}$$

as before we have to bear in mind that there is the constraint

$$|\,\mu\,| = l(t).$$

2. Properties of the derived system of equations

We first consider the rotation equation; denoting for a moment $\mathbf{r} = \mathbf{x}^+ - \mathbf{x}^-$, we can decompose the velocity $\dot{\mathbf{r}}$ into its angular part and the velocity along $\mathbf{r}$:

$$\dot{\mathbf{r}} = \omega \times \gamma + \frac{\mathbf{r}}{|\mathbf{r}|}\left(\frac{\mathbf{r}}{|\mathbf{r}|}\cdot\dot{\mathbf{r}}\right),$$

i.e. in our notation

$$\dot{\mu} = \omega \times \mu + \gamma\mu \tag{2.1}$$

where $\gamma = \dot{l}/l$; thus, in view of (2.1),

$$\omega = \frac{1}{l^2}\,\mu \times \dot{\mu} \tag{2.2}$$

and

$$|\dot{\mu}|^2 = l^2(|\omega|^2 + \gamma^2)$$

In order to replace the second-order rotation equation (1.8) by a system of first-order equations we differentiate (2.1) and we substitute for $\ddot{\mu}$ into (1.8). Thus

$$\mu \times \left\{\dot{\omega} + 2\gamma\omega) - \frac{1}{l^2}\frac{\overset{+}{q}}{\overset{+}{m}}\,\mu \times [\,K_1 + \frac{\overset{+}{m}}{m}\,K_2\,(\mu\cdot\nabla)]\mathbf{E}\right\} = 0,$$

and since $\omega\cdot\mu = 0$, taking into account (2.1) we find that $\dot{\omega}\cdot\mu = 0$, and it follows that the vector product in both sides of the above equation can be omitted. Hence, we arrive at the required system

$$\dot{\mu} = \omega \times \mu + \gamma\mu, \quad |\mu| = l(t)$$

$$\dot{\omega} + 2\gamma\omega = \frac{\overset{+}{q}K_1}{\overset{+}{m}\,l^2}\,\mu \times \mathbf{E}^*, \quad \mu\cdot\omega = 0. \tag{2.3}$$

where

$$\mathbf{E}^* = [1 + \frac{\overset{+}{m}}{m}\frac{K_2}{K_1}(\mu\cdot\nabla)]\mathbf{E}.$$

Observe that multiplying the first equation (2.3) by ω, the second by μ and adding

$$\omega\cdot\mu = \omega(0)\mu(0)\exp\,(-\int_0^t \gamma\,dt).$$

It is therefore sufficient to assume that $\boldsymbol{\omega}(0)\cdot\boldsymbol{\mu}(0) = 0$ and then $\boldsymbol{\omega}\cdot\boldsymbol{\mu} = 0$ follows from the equations

$$\dot{\boldsymbol{\mu}} = \boldsymbol{\omega} \times \boldsymbol{\mu} + \gamma\boldsymbol{\mu}, \quad |\boldsymbol{\mu}| = l(t)$$

$$\dot{\boldsymbol{\omega}} + 2\gamma\boldsymbol{\omega} = \frac{q^+ K_1}{m^+ l^2}\, \boldsymbol{\mu} \times \mathbf{E}^*. \tag{2.4}$$

For a prescribed field $\mathbf{E}(\mathbf{x}, t)$ the system (2.4) and the translation equation (1.8) constitute a complicated nonlinear system of ordinary differential equations with variable coefficients; even in the case of absence of translations (e.g. for neutral dipoles, when in view of the smallness of l the second term in the right-hand side of (1.8) can be neglected) there is nonlinearity in the first equation.

We shall now prove that the terms containing γ in (2.4) can be eliminated by changing the time and the dependent variables; in fact, introduce first

$$\tilde{\boldsymbol{\mu}} = \frac{l_0}{l}\boldsymbol{\mu}, \quad \tilde{\boldsymbol{\omega}} = \left(\frac{l}{l_0}\right)^2 \boldsymbol{\omega} \tag{2.5}$$

where $l_0 = l(0)$. Then the system (2.4) takes the form

$$\dot{\tilde{\boldsymbol{\mu}}} = \left(\frac{l_0}{l}\right)^2 \tilde{\boldsymbol{\omega}} \times \tilde{\boldsymbol{\mu}}$$

$$\dot{\tilde{\boldsymbol{\omega}}} = \frac{q^+ K_1}{m^+}\, \frac{l}{l_0}\, \tilde{\boldsymbol{\mu}} \times \mathbf{E}^*. \tag{2.6}$$

(2.6) is simpler than (2.4) since there are now no terms containing γ. It can be further simplified by eliminating the time-dependent coefficient in the first equation (2.6). To this end we define a new time variable by the monotonic transformation

$$\tilde{t} = \int_0^t \left(\frac{l_0}{l}\right)^2 \mathrm{d}t. \tag{2.7}$$

Then after simple transformations we obtain

$$\frac{d\tilde{\mu}}{d\tilde{t}} = \tilde{\omega} \times \tilde{\mu}$$

$$\frac{d\tilde{\omega}}{d\tilde{t}} = \frac{q^+ K_1}{m^+ l_0^2} \tilde{\mu} \times \tilde{\mathbf{E}}, \tag{2.8}$$

where

$$\tilde{\mathbf{E}} = \left(\frac{l}{l_0}\right)^3 \mathbf{E}^*$$

is a new field. Observe that it follows form Eq. (2.8) that $|\tilde{\mu}| =$ const. and $\tilde{\omega} \cdot \tilde{\mu} =$ const. We assume hereafter $\tilde{\mu}(0) = l_0$ in accordance with (2.5) and $\tilde{\omega}(0) \cdot \tilde{\mu}(0) = 0$. Unfortunately, the new time $\tilde{t}$ leads to complicated terms in the translation Eq. (1.8).

Note that now

$$|\tilde{\omega}| = \frac{1}{l_0^2} \left|\frac{d\tilde{\mu}}{d\tilde{t}}\right|^2. \tag{2.9}$$

There is a useful form of a second-order equation for μ, which will be used later. To derive it we differentiate the first Eq. (2.8) with respect to $\tilde{t}$ and we substitute for $d\tilde{\omega}/dt$. Thus

$$\frac{d^2\tilde{\mu}}{d\tilde{t}^2} = \frac{q^+ K_1}{m^+ l_0^2} (\tilde{\mu} \times \tilde{\mathbf{E}}) \times \tilde{\mu} + \tilde{\omega} \times (\tilde{\omega} \times \tilde{\mu})$$

$$= \frac{q^+ K_1}{m^+ l_0^2} (\tilde{\mu} \times \tilde{\mathbf{E}}) \times \mu - |\tilde{\omega}|^2 \tilde{\mu} \tag{2.10}$$

and taking into account (2.10) we obtain

$$\frac{d^2\tilde{\mu}}{d\tilde{t}^2} + \frac{1}{l_0^2} \left|\frac{d\tilde{\mu}}{d\tilde{t}}\right|^2 \tilde{\mu} = \frac{q^+ K_1}{m^+} \left(\mathbf{I} - \frac{\tilde{\mu}\tilde{\mu}}{l_0^2}\right) \cdot \tilde{\mathbf{E}}. \tag{2.11}$$

Scalar-multiplying (2.11) by $d\tilde{\mu}/d\tilde{t}$ we find the formula for the time derivative of the square of the angular velocity

$$\frac{d}{d\tilde{t}} |\tilde{\omega}|^2 = \frac{2q^+K_1}{m^+} \frac{d\tilde{\mu}}{d\tilde{t}} \cdot \tilde{\mathbf{E}},$$

which is important in the case of small fields; in fact, then $|\tilde{\omega}|^2$ is a slowly varying variable [2].

In absence of the external field, there is the following solution of the homogeneous system (2.4) or (2.8) (or, of course, (2.11)):

$$\mu(t) = \frac{l(t)}{l_0} \left[\mu(0) \cos|\omega_0|\tilde{t} + \frac{1}{|\omega_0|} \omega_0 \times \mu(0) \sin|\omega_0|\tilde{t} \right]$$

$$\omega(t) = \left[\frac{l_0}{l(t)}\right]^2 \omega_0, \tag{2.12}$$

where ω_0 is an arbitrary vector. This simple solution can be used as the zeroth approximation in the case of small external field. In the above solution $\mu(0)$ and $1/|\omega_0|\ \omega_0 \times \mu_0$ can be replaced by arbitrary constant vectors $\mathbf{a}_0$ and $\mathbf{b}_0$, respectively, provided $|\mathbf{a}_0| = |\mathbf{b}_0| = l_0$ and $\mathbf{a}_0 \cdot \mathbf{b}_0 = 0$.

3. The energy balance

To derive the law of energy balance we scalar-multiply Eqs. (1.5) by $\dot{\mathbf{x}}^+$ and $\dot{\mathbf{x}}^-$, respectively, and add them. Hence

$$\dot{K} + q^+\dot{\varphi} = \lambda\, \dot{\overline{l^2}} = 2\lambda l^2 \gamma, \tag{3.1}$$

where

$$q^+\varphi = q^+\varphi(\mathbf{x}^+) + q^-\varphi(\mathbf{x}^-)$$

and the kinetic energy K has the form

$$K = \tfrac{1}{2}(m^+\dot{\mathbf{x}}^{+2} + m^-\dot{\mathbf{x}}^{-2}) = \tfrac{1}{2}(m\,\dot{\mathbf{x}}^2 + \jmath|\dot{\mu}|^2, \quad \jmath = \frac{m^+m^-}{m}$$

or, since $|\dot{\mu}|^2 = l^2(|\omega|^2 + \gamma^2)$,

$$k = \tfrac{1}{2}[m\,\dot{\mathbf{x}}^2 + \mathcal{J}\,l^2\,(|\boldsymbol{\omega}|^2 + \gamma^2)\,]. \tag{3.2}$$

Thus, the kinetic energy is a sum of the translational, rotational and longitudinal kinetic energies.

The time derivative of the total energy $K + q^+\,\varphi$ therefore vanishes when l = const.; we observe, however, that the energy balance (3.1) becomes the energy conservation law when $2l\lambda = -\,\Lambda'(l)$, where $\Lambda(l)$ is the potential of the reaction forces between P^+ and P^-, for then (3.1) takes the form

$$\dot{K} + q^+\,\dot{\varphi}\,\dot{\Lambda} = 0. \tag{3.3}$$

To examine this problem in more detail we return to the expression for λ (p.215) noting that

$$\boldsymbol{\mu}\cdot\ddot{\boldsymbol{\mu}} = \overline{\boldsymbol{\mu}\cdot\dot{\boldsymbol{\mu}}}^{\,\cdot} - |\dot{\boldsymbol{\mu}}|^2 = \tfrac{1}{2}\,\ddot{\overline{l^2}} - |\dot{\boldsymbol{\mu}}|^2 = l^2\,(\dot{\gamma} + \gamma^2 - |\boldsymbol{\omega}|^2). \tag{3.4}$$

Expanding $\mathbf{E}(\mathbf{x}^+)$ and $\mathbf{E}(\mathbf{x}^-)$ about the centre of gravity we obtain, retaining only the first terms,

$$\ddot{l} - |\boldsymbol{\omega}|^2\,l = \frac{q^+\,\mathrm{K}_1}{m^+\,l}\,\boldsymbol{\mu}\cdot\mathbf{E} - \frac{\Lambda'(l)}{\mathcal{J}}. \tag{3.5}$$

In general (3.5) is a differential (ordinary, second-order) equation for l, coupled with Eqs. (1.7) or (1.8): it replaces the constraint $|\boldsymbol{\mu}| - l(t)$. In the absence of an external field the angular velocity $\mathbf{w}(t)$ is given by the formula (2.12), i.e. $|\boldsymbol{\omega}| = (l_0/l)^2\,|\,\omega_0|$ and then, after simple transformations we reduce (3.5) to the form

$$l\,\ddot{l} + \frac{\Lambda'(l)\,l}{\mathcal{J}} = l_0^2\,|\,\omega_0|^2. \tag{3.6}$$

This two-body problem can be solved for varius potentials $\Lambda(l)$.

We note that Eq. (3.5) should actually be written in the form

$$\dot{l}\,[\ddot{l} - |\boldsymbol{\omega}|^2\,l - \frac{q^+\,\mathrm{K}_1}{m^+\,l}\,\boldsymbol{\mu}\cdot\mathbf{E} + \frac{\Lambda'(l)}{\mathcal{T}}] = 0,$$

for then $\dot{K} + q^+\,\dot{\varphi} = 0$; thus, for $l = l_0$, Eq. (3.5) should be disregarded.

4. The case of the external field of constant direction

We now proceed to investigate the case of a linearly polarized external field $\mathbf{E}(\mathbf{x}, t)$, assuming that only the first term in the expansion of this field about the centre of gravity is important. This assumption in our case is equivalent to the assumption that $\mathbf{E}(\mathbf{x}, T)$ depends on the time only. Thus, the rotation equation (1.8) reads

$$\mu \times [\ddot{\mu} - \frac{\overset{+}{q}\,K_1}{\overset{+}{m}}\,\mathbf{E}(t)] = 0, \quad |\mu| = l(t). \tag{1.8'}$$

Set now $\mathbf{E}(t) = \mathbf{e}g(t)$ where $\mathbf{e}$ is a constant unit vector and the scalar $g(t)$ determines the time variation of the considered field. We can write without loss of generality $\mathbf{e} = (0,0,e)$ (the same letter being used for the third component of the vector) and Eq. (1.8′) takes the form

$$\begin{aligned} \mu_\alpha \ddot{\mu}_3 - \ddot{\mu}_\alpha \mu_3 &= \frac{\overset{+}{q}\,K_1}{\overset{+}{m}}\, g(t)\,\mu_\alpha, \quad \alpha = 1,2 \\ \mu_1 \ddot{\mu}_2 - \ddot{\mu}_1 \mu_2 &= 0, \end{aligned} \tag{4.1}$$

where μ_1, μ_2 are the components of μ in the plane perpendicular to the field $\mathbf{E}(t)$ and μ_3 is its component parallel to $\mathbf{E}(t)$. There is a solution of the above equations $\mu_\alpha = 0$, $\alpha = 1,2$, $|\mu_3| = l(t)$, describing the situation in which there is no moment on the dipole due to the Lorentz forces acting on the two particles P^+ and P^-. There are also solutions with either $\mu_2 = 0$ or $\mu_1 = 0$, i.e. a two-dimensional case.

Setting now

$$\begin{aligned} \mu_1(t) = d(t)\cos\varphi(t), \quad \mu_2(t) &= d(t)\sin\varphi(t) \\ d^2 + \mu_3^2 &= l^2(t) \end{aligned} \tag{4.2}$$

we find that the last equation (4.1) is reduced to the simple statement

$$d^2\,\dot{\varphi} = \text{const.} = l_0^2\,\Omega_0,$$

whereas the first two equations are reduced to one, namely

$$\mu_\alpha [\ddot{\mu}_3 - (\dot{\delta} + \delta^2 - \dot{\varphi}^2)\mu_3 - \frac{\overset{+}{q} K_1}{\overset{+}{m}} g(t)] = 0 \tag{4.3}$$

where $\delta = \dot{d}/d$ (we assume of course that $d \neq 0$). As in Section 2 it is convenient to introduce the new variable $q = \mu_3/l$, and the new time

$$\tilde{t} = \int_0^t \left(\frac{l_0}{l}\right)^2 dt.$$

Then Eq. (4.3) takes the form

$$\frac{d^2 q}{d\tilde{t}^2} + \omega^2 q = (1 - q^2)\, E^*(t) \tag{4.4}$$

where

$$E^*(t) = \frac{\overset{+}{q} K_1}{\overset{+}{m}\, l_0} \left(\frac{l_0}{l}\right)^3 E(t)$$

and

$$\omega^2 = \frac{1}{1 - q^2} \left[\Omega_0^2 + \left(\frac{dq}{d\tilde{t}}\right)^2\right]. \tag{4.5}$$

To simplify the notation, we denote by dot the time differentiation with respect to $\tilde{t}$; it should be borne in mind however that the functions $g(t)$ and $l(t)$ are prescribed in the original time t. We are faced therefore with the nonlinear ordinary differential equation

$$\ddot{q} + \omega^2 q = (1 - q^2)\, E^*(t), \quad \omega^2 = \frac{\Omega_0^2 + \dot{q}^2}{1 - q^2}. \tag{4.6}$$

If q is found, then we calculate $d(t)$ from the formula $d^2 = l^2 - \mu_3^2$ and the angle in the plane perpendicular to the field from (4.2).

We note first that in absence of the external field the solution of (4.6) has the form of that for a linear oscillator

$$q = q_0 \cos \omega_0 t + \frac{\dot{q}_0}{\omega_0} \sin \omega_0 t \tag{4.7}$$

but the frequency ω_0 depends on the initial conditions and Ω_0 in accordance with the formula

$$\omega_0^2 = \frac{\Omega_0^2 + \dot{q}_0^2}{1 - q_0^2}. \tag{4.8}$$

Returning to the general case we first observe that (4.6) can be written in the equivalent form

$$\dot{\overline{\frac{1}{\beta}\dot{q}}} + \Omega_0^2 \frac{q}{\beta^3} = E^*\beta, \ \beta = \sqrt{1 - q^2}.$$

Multiplying this equation by $(1/\beta)\dot{q}$ we obtain

$$\dot{\overline{\omega^2}} = 2\,\dot{q}\,E^*. \tag{4.9}$$

As before we note that in the case of small external field, the square of the frequency is a slowly varying quantity. If now E^* = const., (4.9) yields the first integral. Setting $q = \cos\theta$ we obtain

$$\frac{1}{2}\,\frac{\Omega_0^2 + \dot{\theta}^2 \sin^2\theta}{\sin^2\theta} - E^* \cos\theta = \text{const.}$$

For $\Omega_0 = 0$, i.e., $\dot{\varphi}$ = const. and the motion occurs in a fixed plane $\varphi = \varphi_0$; with the appropriate choice of the constant we have

$$\frac{1}{2}\dot{\theta}^2 + E^*(1 - \cos\theta) = H_0 = \frac{1}{2}\dot{\theta}_0^2 + E^*(1 - \cos\theta_0) \tag{4.10}$$

i.e. the first ingegral of the pendulum equation. It can be directly verified that assuming $\Omega_0 = 0$, E^* = const. and $q = \cos\theta$, which is the simplext nontrivial case, (4.6) is in fact the pendulum equation

$$\ddot{\theta} + E^* \sin\theta = 0$$

or assuming $E^* > 0$ and denoting $E^* = \Omega^2$:

$$\ddot{\theta} + \Omega^2 \sin\theta = 0. \tag{4.11}$$

The case $E^* < 0$, $E^* = -\Omega^2$ is reduced to (4.11) by changing the angle θ, namely $\theta = \bar{\theta} + \pi$. Then the stable position for $E^* > 0$ becomes unstable and vice versa.

The dependence of the angle θ (i.e. of $q = \mu_3$) on the external field is of essential importance, since it determines the dependence of the polarization on the applied field. We shall now show that the first derivative $\partial q/\partial\Omega$ suffers discontinuity for one definite value of Ω, namely exactly on the separatrix of the pendulum equation. In fact, we know the solution of Eq. (4.11) in the rotation and oscillation ranges:

$$\text{for } H_0 > 2\Omega^2, \quad q(t, \Omega) = 1 - 2\sinh^2\left(\sqrt{\frac{H_0}{2}}\,(t_0 + t), \kappa\right)$$

where $\kappa^2 = 2\Omega^2/H_0 > 1$,

$$t_0 = \sqrt{\frac{2}{H_0}} \int_0^{\frac{1}{2}\theta_0} \frac{d\beta}{\sqrt{1 - \kappa^2 \sin^2\beta}}$$

and for $H_0 < 2\Omega^2$, $q(t, \Omega) = 1 - 2\kappa^2 \sinh^2(\Omega(t + t_0), \kappa)$ where $\kappa^2 = H_0/2\Omega^2 < 1$,

$$t_0 = \frac{1}{\Omega} \int_0^{\frac{1}{2}\theta_0} \frac{d\beta}{\sqrt{1 - \kappa^2 \sin^2\beta}}.$$

To prove that

$$\frac{\partial}{\partial\Omega} q(t, \Omega)$$

has a discontinuity (at $\Omega = \frac{1}{2}\theta_0$) it is sufficient to consider the special case $\theta_0 = 0$ i.e. $q(0, \Omega) = 1$, $\dot{q}(0, \Omega) = 0$, whence $t_0 = 0$ and $H_0 = \frac{1}{2}\dot{\theta}_0^2$. We have for $\dot{\theta}_0 \neq 0$,

$$\text{for} \qquad H_0 > 2\Omega^2, \ \frac{\partial q}{\partial \Omega} = -\frac{8}{\dot{\theta}_0} \tanh \frac{\dot{\theta}_0 t}{2} \operatorname{sech}^2 \frac{\dot{\theta}_0 t}{2} \left(\frac{\dot{\theta}_0 t}{2} - \tanh \frac{\dot{\theta}_0 t}{2} \right)$$

$$\text{and for} \qquad H_0 < 2\Omega^2, \ \frac{\partial q}{\partial \Omega} = -\frac{8}{\dot{\theta}_0} \tanh^2 \frac{\dot{\theta}_0 t}{2} \operatorname{sech}^2 \frac{\dot{\theta}_0 t}{2} \left(1 - \cosh^2 \frac{\dot{\theta}_0 t}{2} \right)$$

and for difference between the above

$$[\![\frac{\partial q(t,\Omega)}{\partial \Omega}]\!]_{\Omega = \frac{1}{2}\dot{\theta}_0} = -\frac{4}{\Omega} \tan \Omega t \operatorname{sech}^2 \Omega t \left[\Omega t - \tanh \Omega t \left(1 - \operatorname{sech}^2 \Omega t \right) \right]. \quad (4.12)$$

Furthermore

$$\lim_{t \to \infty} [\![\frac{\partial q(t,\Omega)}{\partial \Omega}]\!] = -\frac{8}{\dot{\theta}_0},$$

i.e. the above difference does not vanish even after long time.

The considered position on the separatrix $H_0 = 2\Omega^2$ is unstable for $E^* > 0$, i.e. for the field parallel to $q = \mu_3$, while it is stable for $E^* < 0$ with the same absolute value.

We note that for constant E^* and $\Omega_0 \neq 0$ Eq. (4.9) has a solution in the form of elliptic integrals.

Two remarks are in order here.

1. Eq. (4.6) is the Euler–Lagrange equation, derivable from the Lagrangian

$$L = \frac{1}{2} \frac{1}{1 - q^2} (\dot{q}^2 - \Omega_0^2) + E^* q, \qquad (4.13)$$

the momentum canonical to q is

$$p = \frac{\partial L}{\partial \dot{q}} = \frac{\dot{q}}{1 - q^2}$$

and the Hamiltonian is given by the formula

$$H = \tfrac{1}{2}\omega^2 - E^* q, \quad \omega^2 = \frac{\Omega_0^2 + \dot{q}^2}{1 - q^2}, \quad \dot{H} = -\dot{E}^* q.$$

2. For $\Omega_0 \neq 0$ (4.6) can be reduced to a more standard form of first-order system by introducing in the phase space $q = x_1$, $\dot{q} = x_2$ the elliptic coordinates

$$x_1 = r\cos\theta, \quad x_2 = \Omega_0 r \frac{\sin\theta}{\sqrt{1 - r^2}},$$

$$\det[(x_1, x_2) \leftrightarrow (r, \theta)] = \frac{\Omega_0 r (1 - r^2\cos^2\theta)}{(1 - r^2)^{3/2}}.$$

After simple transformations we obtain for $0 < r < 1$

$$\dot{r} = (1 - r^2)^{3/2} \sin\theta \, \frac{E^*}{\Omega_0}$$

$$\dot{\theta} = -\frac{\Omega_0}{\sqrt{1 - r^2}} + (1 - r^2)^{3/2} \frac{\cos\theta}{r} \frac{E^*}{\Omega_0}. \tag{4.14}$$

For small external fields this is a system with one slow (r) and one fast (θ) variable (see. e.g. [2]). For $\Omega_0 = 0$, setting $q = \cos\theta$ we arrive at the pendulum equation with a variable coefficient

$$\ddot{\theta} + E^*(t)\sin\theta = 0 \tag{4.15}$$

which for $\dot{E}^*(t) \neq 0$ can be solved only approximately.

We shall now deduce the system of equations for μ_3 and l, leading to the conservation of energy, i.e. (3.5) (in the considered particular case) and (4.3). Since $d^2 = l^2 - \mu_3^2$ the expression in the bracket in (4.3) can be calculated and we have

$$\ddot{\mu}_3 + \Big[\frac{\Lambda'}{\mathcal{T} l} + \Big(\frac{l_0}{l}\Big)^4 \frac{\Omega_0^2}{L - \Big(\frac{\mu_3}{l}\Big)^2}\Big]\mu_3 = \frac{\overset{+}{q} K_1}{\overset{+}{m}} E$$

$$\ddot{l} - \frac{l}{1 - \Big(\frac{\mu_3}{l}\Big)^2} \overline{\Big(\frac{\mu_3}{l}\Big)}^{\,\cdot} = \frac{\overset{+}{q} K_1}{\overset{+}{m}\, l} \mu_3 E - \frac{\Lambda'}{\mathcal{T}}. \tag{4.16}$$

Setting naturally $\mu_3 = l \cos\theta$,

$$\frac{1}{l}\Big(\overline{l^2 \dot{\theta}}^{\,\cdot}\Big) + \Big(\frac{l_0}{l}\Big)^4 \frac{\Omega_0^2}{\sin\theta}\, l \cos\theta + \frac{\overset{+}{q} K_1}{\overset{+}{m}} E \sin\theta = 0$$

$$\ddot{l} - l^2 \dot{\theta}^2 + \frac{\Lambda'(l)}{\mathcal{T}} - \frac{\overset{+}{q} K_1}{\overset{+}{m}} E \cos\theta = 0, \tag{4.17}$$

which is the required system.

5. Circularly polarized external field

Consider now the external field

$$E_1 = \frac{\overset{+}{m}}{\overset{+}{q} K_1} e \cos \Omega t, \; E_2 = \frac{\overset{+}{m}}{\overset{+}{q} K_1} e \sin \Omega t, \; E_3 = 0.$$

There is no difficulty in generalizing the results of this section to elliptically polarized fields. Eq. (1.8′) now takes the form

$$\mu_2 \ddot{\mu}_3 - \ddot{\mu}_2 \mu_3 = -\frac{\overset{+}{q} K_1}{\overset{+}{m}} E_2 \mu_3$$

$$\mu_1 \ddot{\mu}_3 - \tilde{\mu}_1 \mu_3 = -\frac{\overset{+}{q} K_1}{\overset{+}{m}} E_1 \mu_3 \tag{5.1}$$

$$\mu_1 \ddot{\mu}_2 \ddot{\mu}_1 \mu_2 = \frac{\overset{+}{q} K_1}{\overset{+}{m}} (\mu_1 E_2 - \mu_2 E_1), \ |\mu| = l.$$

A comparison with Eqs. (4.11) shows that the present case is much more complicated. As in Section 4 we introduce the polar coordinates in the plane

$$\mu_1 = d \cos \varphi, \ \ \mu_2 = d \sin \varphi, \ \ \mu_3^2 + d^2 = l^2.$$

Then the third Eq. (5.1) takes the form

$$\overline{d^2 \dot{\varphi}}^{\cdot} = e\, d \sin (\Omega t - \varphi). \tag{5.2}$$

We note that there is a plane solution, i.e. $\mu_3 = 0$, $d = l$; introducing the new angle $\psi = \varphi - \Omega t$ we reduce (5.2) to the form

$$\ddot{\psi} + 2\gamma\dot{\psi} + \frac{e}{l} \sin \psi = -2\gamma\Omega. \tag{5.3}$$

The term containing $\gamma = \dot{l}/l$ can be removed as before by introducing the new time variable (2.7). In fact

$$\frac{d^2\psi}{d\tilde{t}^2} + \left(\frac{l}{l_0}\right)^4 \frac{e}{l} \sin \psi = -2\gamma \left(\frac{l_0}{l}\right)^4 \Omega. \tag{5.3'}$$

Returning to the general case we find that the first two Eqs. (5.1) are

$$\sin \varphi \left[\ddot{\mu}_3 - \left(\dot{\varphi}^2 - \frac{\ddot{d}}{d}\right) \mu_3\right] = \frac{e}{d} \mu_3 \left[\cos \varphi \sin (\Omega t - \varphi) - \sin \Omega t\right]$$

$$\tag{5.4}$$

$$\cos \varphi \left[\ddot{\mu}_3 -+ \left(\dot{\varphi}^2 - \frac{\ddot{d}}{d}\right) \mu_3\right] = -\frac{e}{d} \mu_3 \left[\sin \varphi \sin (\Omega t - \varphi) + \cos \Omega t\right],$$

where it has been taken into account that $\overline{d^2 \dot{\varphi}}^{\cdot} = ed \sin (\Omega t - \varphi)$. Eqs. (5.4) are not independent. In fact multiplying the first by $\cos \varphi$, the second by $\sin \varphi$, and subtracting, we arrive at an identity. To get an equation for μ_3 we multiply the first (5.4) by $\sin \varphi$, the second by $\cos \varphi$ and add the results. Hence, finally, we have the system

$$\ddot{\mu}_3 + \left(\ddot{\varphi} - \frac{\ddot{d}}{d}\right)\mu_3 + \frac{e}{d}\cos(\Omega t - \varphi)\,\mu_3 = 0$$

$$\overline{d^2\dot{\varphi}}^{\,\cdot} - e\,d\sin(\Omega t - \varphi) = 0 \tag{5.5}$$

for μ_3 and φ.

Assume now that $l = l_0$ and consider the system (2.3)

$$\dot{\mu} = \omega \times \mu$$

$$\dot{\omega} = \frac{\overset{+}{q}K_1}{\overset{+}{m}}\,\mu \times E. \tag{5.6}$$

The external field is now the same as before, with a change in notation, namely

$$E_1 = e\cos\Omega t,\ E_2 = e\sin\Omega t,\ E_3 = 0. \tag{5.7}$$

We choose now (5.6) rather than (1.8′) for an easier comparison with the optical Bloch equations. It turns out that the system (5.6) has simple periodic solutions. We seek it in the form $\boldsymbol{\omega} = (AE_1, AE_2, \omega_3)$ where A and ω_3 are constant. Then the system (5.6) written in full is

$$\dot{\mu}_1 = \omega_2\,\mu_3 - \omega_3\,\mu_2 = Ae\sin\Omega t\,\mu_3 - \omega_3\,\mu_2$$

$$\dot{\mu}_2 = -\,\omega_1\,\mu_3 + \omega_3\,\mu_1 = -\,Ae\cos\Omega t\,\mu_3 + \omega_3\,\mu_1$$

$$\dot{\mu}_3 = \omega_1\,\mu_2 - \omega_2\,\mu_1 = Ae(\cos\Omega t\,\mu_2 - \sin\Omega t\,\mu_1)$$

$$\dot{\omega}_1 = -\,\Omega\,Ae\sin\Omega t = -\frac{\overset{+}{q}K_1}{\overset{+}{m}\,l_0^2}e\sin\Omega t\,\mu_3 \tag{5.8}$$

$$\dot{\omega}_2 = \Omega\,A\cos\Omega t = \frac{\overset{+}{q}K_1}{\overset{+}{m}\,l_0^2}e\cos\Omega t\,\mu_3$$

$$\dot{\omega}_3 = \mu_1\sin\Omega t - \mu_2\cos\Omega t = 0.$$

It follows from (5.8^1) and (5.8^2) that μ_3 is constant and $A = \dfrac{\overset{+}{q} K_1}{\overset{+}{m} l_0^2} \dfrac{\mu_3}{\Omega}$. Then ($5.8^3$) and ($5.7^3$) are the same and

$$\mu_1 = B \cos \Omega t, \ \mu_2 = B \sin \Omega t,$$

where B is constant. Substituting these expressions into (5.7^1) and (5.7^2) we find that they are satisfied provided

$$B = \frac{1}{\omega_3 - \Omega} \frac{\overset{+}{q} K_1}{\overset{+}{m} l_0^2} \frac{\mu_3^2 e}{\Omega}.$$

Introducing the notation

$$\bar{\omega} = \frac{\overset{+}{q} K_1}{\overset{+}{m} l_0^2} \frac{e}{\Omega}$$

so that

$$A = \bar{\omega} \frac{\mu_3}{l_0 e}, \ B = \frac{\bar{\omega}}{\omega_3 - \Omega} \frac{\mu_3^2}{l_0}$$

we obtain

$$\omega = (\bar{\omega} \frac{\mu_3}{l_0} \cos \Omega t, \ \bar{\omega} \frac{\mu_3}{l_0} \sin \Omega t, \omega_3)$$

$$\mu = \left(\frac{\bar{\omega}}{\omega_3 - \Omega} \frac{\mu_3}{l_0} \cos \Omega t, \ \frac{\bar{\omega}}{\omega_3 - \Omega} \frac{\mu_3}{l_0} \sin \Omega t, 1 \right) \mu_3.$$

There are two conditions to be satisfied, namely $|\mu| = l_0$ and $\mu \cdot \omega = 0$. Substituting from (5.10) we have

$$\omega_3 = \frac{\bar{\omega}^2}{\Omega}, \ \frac{\mu_3}{l_0} = \pm \sqrt{1 - \left(\frac{\bar{\omega}}{\Omega}\right)^2}, \ \Omega - \omega_3 = \Omega \left(1 - \frac{\bar{\omega}^2}{\Omega^2}\right).$$

Consequently the parameters in the problem must satisfy the condition $\bar{\omega} < \Omega$, i.e.

$$\frac{\overset{+}{q} K_1}{\overset{+}{m} l_0} e < \Omega^2.$$

It follows that $\omega_3 < |\bar{\omega}|$. The only arbitrary parameter in the solution is now l_0. The solution (5.10) can be written in the final form

$$\omega = \left(\pm \sqrt{1 - \frac{\bar{\omega}^2}{\Omega^2}} \cos \Omega t,\ \pm \sqrt{1 - \frac{\bar{\omega}^2}{\Omega^2}} \sin \Omega t,\ \frac{\bar{\omega}}{\Omega} \right) \bar{\omega}$$

$$\mu = \left(- \frac{\bar{\omega}}{\Omega} \cos \Omega t,\ - \frac{\bar{\omega}}{\Omega} \sin \Omega t, \pm \sqrt{1 - \frac{\bar{\omega}^2}{\Omega^2}} \right) l_0. \tag{5.11}$$

There are in fact two solutions in (5.11): with the upper and with the lower sign.

The optical Bloch equations describe the average motion of a dipole μ treated as a two-level quantum system subject to the action of a circularly polarized light

$$E_1 = e \cos \Omega t,\ E_2 = e \sin \Omega t,\ E_3 = 0,$$

i.e. the same as our field. Without going into details of the approximations involved in the derivation we write down the optical Bloch equations in the form [4]

$$\dot{\boldsymbol{\mu}} = \bar{\boldsymbol{\Omega}} \times \boldsymbol{\mu} \tag{5.12}$$

where

$$\bar{\boldsymbol{\Omega}} = - \frac{\gamma}{h} \left(e \cos \Omega t, e \sin \Omega t, - \frac{h\, \omega_0}{\gamma} \right).$$

Thus we are faced with a system of three first-order linear differential equations, with periodic coefficients. There are two free parameters, namely γ and ω_0. It follows from (5.12) that $|\mu|$ is constant, say l_B.

Assuming the solution in the form

$$\mu_1 = B \cos \Omega t,\ \mu_2 = B \sin \Omega t \tag{5.13}$$

(the same as above) we find that Eqs. (5.12) are satisfied provided

$$B = \frac{\gamma e}{h} \frac{\mu_3}{\Omega - \omega_0}$$

and μ_3 is constant. The normalization condition $|\mu| = l_B$ yields

$$\mu_3^2 = l_B^2 \left[1 + \left(\frac{\gamma e}{h} \right)^2 \frac{1}{(\Omega - \omega_0)^2} \right]^{-1}.$$

If we require that μ be orthogonal to the angular velocity $\overline{\boldsymbol{\Omega}}$, $\gamma^2 = \omega_0/e^2\,(\Omega - \omega_0)h^2$ and hence $\Omega > \omega_0$.

The solution of the Bloch equations is identical in form to the solution for μ in (5.11). The solutions are identical provided (we set $x = \frac{\gamma e}{h} \frac{1}{\omega_0 - \Omega}$)

1. for $\mu_3 > 0$ in (5.14)

$$x l_B (1 + x^2)^{-\frac{1}{2}} = \frac{\overline{\omega}}{\Omega} l_0, \quad l_B (1 + x^2)^{-\frac{1}{2}} = \sqrt{1 - \frac{\overline{\omega}^2}{\Omega^2}}\, l_0$$

whence $\omega_0 > \Omega$; it follows then that also $\mu_3 > 0$ in (5.11).

2. for $\mu_3 < 0$ in (5.14)

$$- x l_B (1 + x^2)^{-\frac{1}{2}} = \frac{\overline{\omega}}{\Omega} l_0, \quad l_B (1 + x^2)^{-\frac{1}{2}} = \sqrt{1 - \frac{\overline{\omega}^2}{\Omega^2}}\, l_0$$

whence $\omega_0 < \Omega$; then $\mu_3 < 0$ in (5.11).

Both cases can be described by one set of relations between the constants, namely

$$|x|\, l_B (1 + x^2)^{-\frac{1}{2}} = \frac{\overline{\omega}}{\Omega} l_0, \quad l_B (1 + x^2)^{-\frac{1}{2}} = \sqrt{1 - \frac{\overline{\omega}^2}{\Omega^2}}\, l_0 .$$

The first of the above relations does not in fact contain l_0. for

$$\bar{\omega} = \frac{q^+ K_1}{m^+ l_0} \frac{e}{\Omega}.$$

If therefore the number $q^+ K_1/m^+$ is assumed to be known, this relation is a restriction on the constants entering the Bloch equations.

6. Lagrangians for the rotation equation

We have already presented a Lagrangian for the rotation equation in Section 4, for the case of a constant direction electric field, in which the problem was reduced to a scalar equation for μ_3. Now we shall examine the general case, recalling that the problem is described by Eq. (2.11), where $|\tilde{\mu}| = l_0$. Writing $\tilde{\mu}/l_0 = q$, $|q| = 1$, we obtain

$$\ddot{q} + |\dot{q}|^2 q = (I - \mathbf{q}\,\mathbf{q}) \cdot \bar{\mathbf{E}} \tag{6.1}$$

where

$$\bar{\mathbf{E}} = \frac{q^+ \mathrm{K}_1}{m^+ l_0} \left(\frac{l}{l_0}\right)^3 \mathbf{E},$$

provided we neglect as before the gradient of the electric field. The differentiation with respect to $\tilde{t}$ is now denoted by dot. First we observe that the normalization condition $|\mathbf{q}| = 1$ follows from Eq. (6.1). In fact, scalar-multiplying by $\mathbf{q}$ we get

$$\frac{1}{2}\,\ddot{\overline{|q|^2}} - (1 - |\mathbf{q}|^2)\,|\dot{q}|^2 = 0. \tag{6.2}$$

Regarding $|\dot{q}|^2$ as a known function of time we may regard (6.2) as an ordinary equation of second order for $|\mathbf{q}|^2$. Thus, assuming that $|\dot{\mathbf{q}}|^2$ is sufficiently smooth for the uniqueness of Eq. (6.2) and bearing in mind that $|\mathbf{q}| = 1$ is a solution, we conclude that our assertion is true, provided the initial conditions

$$|\mathbf{q}|^2(0) = 1, \quad \dot{\overline{|q|^2}}(0) = 0$$

are satisfied.

The problem of constructing Lagrangians for equations of Newtonian mechanics has been treated extensively in [4], there occur, however, in our case pecularities due to the presence of the constraint $|\mathbf{q}| = 1$. We shall not be interested in constructing all possible Lagrangians, confining ourselves to some particular cases of interest.

Consider the Lagrangian

$$L = \alpha(|\mathbf{q}|)|\dot{\mathbf{q}}|^2 + \beta(|\mathbf{q}|)\,\mathbf{q}\cdot\mathbf{E} \tag{6.3}$$

where $\alpha(|\mathbf{q}|)$ and $\beta(|\mathbf{q}|)$ are for the time being arbitrary. We have

$$\mathbf{p} = \frac{\partial L}{\partial \dot{\mathbf{q}}} = 2\alpha\dot{\mathbf{q}}, \quad \frac{\partial L}{\partial \mathbf{q}} = \frac{\alpha'}{|\mathbf{q}|}|\dot{\mathbf{q}}|^2\mathbf{q} + \beta\,(\mathbf{I} + \frac{\beta'}{\beta|\mathbf{q}|}\mathbf{q}\,\mathbf{q})\cdot\bar{\mathbf{E}}$$

and the Euler-Lagrange equation takes the form

$$\ddot{\mathbf{q}} + \frac{\alpha'}{2\alpha}\frac{1}{|\mathbf{q}|}\,\dot{\overline{|q|^2}}\,\dot{\mathbf{q}} - \frac{\alpha'}{2\alpha}\frac{|\dot{\mathbf{q}}|^2}{|\mathbf{q}|}\mathbf{q} = \frac{\beta}{2\alpha}(\mathbf{I} + \frac{\beta'}{\beta|\mathbf{q}|}\mathbf{q}\,\mathbf{q})\cdot\bar{\mathbf{E}}. \tag{6.4}$$

Multiplying this equation scalarly by $\mathbf{q}$ and taking into account that $\mathbf{q}\cdot\ddot{\mathbf{q}} = \frac{1}{2}\ddot{\overline{|\mathbf{q}|^2}} - |\dot{\mathbf{q}}|^2$, we have

$$\frac{1}{2}\ddot{\overline{|q|^2}} - (1 + \frac{\alpha'}{2\alpha}|\mathbf{q}|)|\dot{\mathbf{q}}|^2 + \frac{\alpha'}{\alpha}\frac{1}{|\mathbf{q}|}\left(\frac{1}{2}\dot{\overline{|q|^2}}\right)$$

$$= \frac{\beta}{2\alpha}(1 + \frac{\beta'}{\beta}|\mathbf{q}|)\,\mathbf{q}\cdot\bar{\mathbf{E}}\cdot \tag{6.5}$$

We shall now choose the functions $\alpha(|\mathbf{q}|)$ and $\beta(|\mathbf{q}|)$ in such a manner that Eq. (6.4) is equivalent to (6.1). There are many ways to do it; we take

$$\frac{\alpha'}{2\alpha}|\mathbf{q}| = -1, \quad \frac{\beta'}{\beta}|\mathbf{q}| = -1 \;\text{ i.e. }\; \alpha = \frac{\alpha_0}{|\mathbf{q}|^2}, \quad \beta = \frac{\beta_0}{|\mathbf{q}|}.$$

Then (6.5) becomes an equation for $|\mathbf{q}|^2$, which we denote by x:

$$x\ddot{x} - (\dot{x})^2 = 0. \tag{6.6}$$

Hence $x = x_0\,e^{(\dot{x}_0/x_0)t}$ and it suffices to assume that $\dot{x}_0 = 0$ i.e. $\dot{\overline{|\mathbf{q}|^2}}(0) = 0$ to get

$x = x_0$, say 1. The differential equation (6.4) is now

$$\ddot{q} + \frac{1}{|\mathbf{q}|^2}|\dot{\mathbf{q}}|^2\mathbf{q} - \frac{\dot{\overline{|\mathbf{q}|^2}}}{|\mathbf{q}|^2}\dot{\mathbf{q}} = \frac{\beta_0}{2\alpha_0}|\mathbf{q}|(\mathbf{I} - \frac{\mathbf{q}\mathbf{q}}{|\mathbf{q}|^2})\cdot\bar{\mathbf{E}} \tag{6.7}$$

Multiplying it scalarly by $\mathbf{q}$ we obtain again (6.6). Thus, with the initial condition $|\mathbf{q}|^2(0) = 1$, $\dot{\overline{|q|^2}}(0) = 0$ we may set now $|\mathbf{q}| = 1$. Taking also for simplicity $\frac{\beta_0}{2\alpha_0} = 1$ we arrive at the required Eq. (6.1).

Thus, our Lagrangian (6.3) (multiplied for convenience by $\frac{1}{2}$) is

$$L = \frac{1}{2}\Big(\frac{|\dot{\mathbf{q}}|^2}{|\mathbf{q}|^2} + \frac{2}{|\mathbf{q}|}\mathbf{q}\cdot\bar{\mathbf{E}}\Big); \tag{6.8}$$

the canonical momentum $\mathbf{p} = \frac{\dot{\mathbf{q}}}{|\mathbf{q}|}$ and the Hamiltonian has the form

$$H = \frac{1}{2}\Big(|\mathbf{q}|^2|\mathbf{p}|^2 - \frac{2}{|\mathbf{q}|}\mathbf{q}\cdot\bar{\mathbf{E}}\Big). \tag{6.9}$$

We mentioned before that there are many Lagrangians leading to (6.1). One of the possibilities is the following

$$L = \frac{1}{2}e^{2(1-|\mathbf{q}|)}|\dot{\mathbf{q}}|^2 + e^{1-|\mathbf{q}|}\mathbf{q}\cdot\bar{\mathbf{E}}, \quad \mathbf{p} = e^{2(1-|\mathbf{q}|)}\dot{\mathbf{q}},$$

$$H = \frac{1}{2}e^{-2(1-|\mathbf{q}|)}|\mathbf{p}|^2 + e^{1-|\mathbf{q}|}\mathbf{q}\cdot\bar{\mathbf{E}}$$

or

$$L = (1 - \frac{1}{2}|\mathbf{q}|^2)|\dot{\mathbf{q}}|^2 + \frac{3}{2}(1 - \frac{1}{3}|\mathbf{q}|^2)\mathbf{q}\cdot\bar{\mathbf{E}}, \ \mathbf{p} = 2(1 - \frac{1}{2}|\mathbf{q}|^2)\dot{\mathbf{q}},$$

$$H = \frac{1}{4}\frac{|\mathbf{p}|^2}{1 - \frac{1}{2}|\mathbf{q}|^2} - \frac{3}{2}(1 - \frac{1}{3}|\mathbf{q}|^2)\mathbf{q}\cdot\bar{\mathbf{E}}.$$

As already mentioned, there are many other Lagrangians leading to the same system of equation (6.1).

Acknowledgements

This paper was written during the author's stay in the University of Stuttgart. The author is grateful to the University for hospitality, to the Deutsche Forschungsgemeinschaft for the financial support and to Prof. E. Kröner and Prof. H. Bednarczyk for fruitful discussions.

Preliminary numerical work (to be reported in a forthcoming paper on continuously distributed dipoles) carried out by Mr S. Crienietz and Mr A. Hausmann is also gratefully acknowledged.

References

[1] D.C. Hanna, M.A. Yuratich and D. Cotter, *Nonlinear Optics of Free Atoms and Molecules*, Springer Verlag, 1979.

[2] A.H. Nayfeh, *Perturbation Methods*, Wiley, 1973.

[3] Y.R. Shen, *The Principles of Nonlinear Optics*, Wiley, 1984.

[4] M. Santilli, *Foundations of Theoretical Mechanics. The Inverse Problem of Newtonian Mechanics*, Vol. 1 and 2, Springer Verlag, 1979.

H. Zorski
Institute of Fundamental Technological Research
Polish Academy of Sciences
ul. Swietokrzyska 21
0-0049 Warsaw
POLAND

T.S. VASHAKMADZE

On the mathematical theory of piezoelectrically and electrically conductive elastic plates

Abstract

Some of author's previous results are generalized to the case when the initial problems consist in defining the stress–strain state for piezoelectric and electrical conductive transversally-isotropic plates. The corresponding refined theories and some capacity of the new continuum models equivalent to them are constructed, without using simplifying hypotheses of physical and geometrical character.

Introduction

This report illustrates one way of studying the problems arising by transition from multidimensional boundary value problems of mathematical physics to those of fewer dimensions. The latter are also called mathematical models. Below we define the mathematical model as a boundary value problem or a class of such problems which are in some sense equivalent to the initial (original) problem and in which simplifying hypotheses of physical or geometrical character are not used.

The interest in constructing mathematical models in this sense has been increased recently especially with respect to plate and shell problems following the published work of Vorovich [1].

As it is known, there is a sufficient amount of publications in this direction available but I should like to call your attention to the article of Berdichevski [2] and the sequence of works by Ciarlet, Destuynder and their collaborators (see e.g. [3]–[4]).

The two-dimensional mathematical models corresponding to different refined theories are constructed here [2]–[4] by means of the asymptotic method. In these works some refined theories, admitting a strict description by notions of the theory of functions, are constructed. The important conclusions which I want to underline regarding these works are the following:

(i) The results of [2]–[4] contribute only to asymptotic-type estimation of transition error.

(ii) Each of the methods of construction corresponds to a certain concrete mathematical model, which means that there is no mathematical formal rule common for construction of any of the refined theories.

(iii) In these works, the value characterizing the transition approximation of the initial spatial problem by the mathematical model is represented as a remainder term of the asymptotic series. Thus it is not appropriate for effective estimation of the transition error.

In earlier papers (see e.g., [5]–[6]) we constructed refined theories and some capacity of the continuum new models equivalent to them without using simplifying hypotheses; we also obtained an analytical expression for the remainder vector which characterizes the transition approximation when the initial problem of defining stresses and strains of anisotropic elastic plates with a variable thickness is nonlinear, nonhomogeneous and depends also on the time.

Moreover, we developed these methods for the more general cases: for example, when an elastic multilayer composite plate has cracks and is subjected also to electromagnetic fields. Besides, we remark that from our mathematical models there follows also the nonlinear system of equations of Korteweg–de Vries type.

Refined theories for piezoelectrically and electrically conductive elastic plates

According to the suggestion of my colleague Lazarev we develop these methods ([5]–[6]) for piezoelectric and electrically conductive elastic plates.

For the sake of clarity and simplicity we restrict our consideration here to the linear boundary value problem with constant thickness (see e.g. [87]):

$$\sigma_{ij,j} = f_i, \ \operatorname{div} D = 0, \ \varepsilon_{ij} = 0.5\,(u_{i,j} + u_{j,i}), \ \mathcal{E} = \operatorname{grad} \mathcal{S}, x \in \Omega_h,$$

$$\sigma_3 = (\sigma_{31}, \sigma_{32}, \sigma_{33})^{\mathrm{T}} \,|_{S^{\pm}} = g^{\pm}, \ D_3 \,|_{\overset{\pm}{S_1}} = \overset{\pm}{d_3}, \ \mathcal{S} \,|_{\overset{\pm}{S_2}} = \overset{\pm}{v_0},$$

$$\mathfrak{L}[u; \mathcal{S}] = g, \ x \in S = \partial D \times \,]-h, h[,$$

$$\Omega_h = D(x_1, x_2) \times \,]-h, h[, \ S^{\pm} = D \times \{h^{\pm}\} = \overset{\pm}{S_1} \cup \overset{\pm}{S_2}$$

Here σ, ε, u stand for the stress and strain tensors and for displacement vector respectively, D is an electrical induction vector, $\mathcal{E}$ means a tension of electric field, $-\mathcal{S}$ stands for the electric potential.

Hooke's law generalized for transversally isotropic piezoceramic materials is written as follows (e.g. see [7]):

$$\sigma_{\alpha\alpha} = b_{\alpha\beta}\, \varepsilon_{\beta\beta} + b_{13}\, \varepsilon_{33} - e_{31}\, \mathcal{E}_3,$$

$$\sigma_{33} = b_{13}\, \varepsilon_{\alpha\alpha} + b_{33}\, \varepsilon_{33} - e_{33}\, \mathcal{E}_3,$$

$$\sigma_{\alpha 3} = 2b_{44}\, \varepsilon_{\alpha 3} - e_{15}\, \mathcal{E}_{\alpha}, \sigma_{12} = 2b_{66}\, \varepsilon_{12},$$

and

$$D_\alpha = \ni_{11}\, \varepsilon_\alpha + 2e_{15}\, \mathcal{E}_{\alpha 3},\ D_3 = \ni_{33}\, \mathcal{E}_3 + \mathcal{E}_{31}\, \varepsilon_{\alpha\alpha} + e_{33}\, \varepsilon_{33}\,.$$

Here b are elasticity moduli, the quantities e are piezoelectric constants and $\ni$ are dielectric penetrances.

The use of some results of [5] – [6] for this case will be set forth below.

Equations corresponding to the components of the rotation of normals to the middle plane have the following form

$$\overset{*}{u}_\alpha = -\overset{*}{u}_{3,\alpha} + (2hb_{44})^{-1}(1+2\gamma)(Q_\alpha + e_{15}\,\Phi_{1,\alpha}) + R_\alpha\,[\overset{*}{u}_\alpha;\gamma],$$

where γ is an arbitrary parameter,

$$\Phi_1 = \int_{-h}^{h} S(x_1,x_2,t)\,\mathrm{d}t = \int S = (S,1),\ \overset{*}{u}_\alpha = 3(2h^3)^{-1}\int t u_\alpha,\ \overset{*}{u}_3 = 3(4h^3)^{-1}\int (h^2-t^2)u_3.$$

For bending and twisting moments we have:

$$M_\alpha = -\frac{2h^3}{3}(P_{11}\partial_\alpha^2 + P_{12}\partial_{3-\alpha}^2)\,\overset{*}{u}_3 + (3b_{44})^{-1}h^2(1+2\gamma)\,[e_{15}\,(P_{11}\,\partial_\alpha^2 + P_{12}\partial_{3-\alpha}^2)\Phi_1$$
$$+ (P_{11}-P_{12})Q_{\alpha,\alpha}$$
$$+ (b_{44}P_{13}-P_{12})(\overset{+}{g}_3 - \overset{-}{g}_3) + b_{44}P_{14}(\overset{+}{d}_3 - \overset{-}{d}_3) + P_{12}\int f_3] + R_{\alpha+2}\,[M_\alpha;\gamma],$$

$$M_{12} = -\frac{4h^3}{3}b_{66}\,\overset{*}{u}_{3,12} + \frac{2h^2}{3b_{44}}b_{66}(1+2\gamma)(e_{15}\,\Phi_{1,12} + Q_{\alpha,3-\alpha}) + R_5[M_{12};\gamma].$$

Equations corresponding to shearing forces have the following form:

$$Q_\alpha = h(\overset{+}{g}_\alpha + \overset{-}{g}_\alpha) - \int tf_\alpha + (P_{12}+b_{66})\int f_{3,\alpha}\,\mathrm{d}t - \frac{2h^3}{3}(P_{11}\,\partial_\alpha^2 + (P_{12}+2b_{66})\,\partial_{3-\alpha}^2)\,[\overset{*}{u}_{3,\alpha}$$

$$+ \frac{h^{-1}}{2b_{44}}(1+2\gamma)e_{15}\,\Phi_{1,\alpha}] + \frac{h^2}{3b_{44}}(1+2\gamma)\,[(P_{11}-P_{12}-b_{66})\,\partial_\alpha^2 + b_{66}\,\partial_{3-\alpha}^2)Q_\alpha$$

$$+ b_{44}^{-1}P_{14}(\overset{+}{d}_{3,\alpha} - \overset{-}{d}_{3,\alpha}) + (b_{44}P_{13} - P_{12} - b_{66})(\overset{+}{g}_{3,\alpha} - \overset{-}{g}_{3,\alpha})] + R_{5+\alpha}\,[Q_{\alpha j}\ \gamma].$$

The equation for Reissner's averaged deflection are written down as follows:

$$\frac{2h^3}{3}P_{11}\,\Delta^2 u_3^* = \frac{e_{15}h^2}{3b_{44}}(1+2\gamma)P_{11}\,\Delta^2\Phi_1 + \overset{+}{g}_3 - \overset{-}{g}_3 - \int f_3 + h(\overset{+}{g}_{\alpha,\alpha} + \overset{-}{g}_{\alpha,\alpha}) - \int tf_{\alpha,\alpha}$$

$$+\frac{h^2}{3b_{44}}(1+2\gamma)\,[(b_{44}P_{13}-P_{11})\,\Delta(\overset{+}{g}_3-\overset{-}{g}_3) + b_{44}P_{14}\,\Delta(\overset{+}{d}_3-\overset{-}{d}_3) + P_{11}\int\Delta f_3] + R_8\,[u^*_{3j}\,\gamma].$$

This system was derived for the case when D_3 is given on $S_1^{\pm}$. If it is so, from the equation div $\mathbf{D} = 0$, we deduce

$$(\ni_{11} + b_{44}^{-1}e_{15}^2)\,\Delta\Phi_1 = b_{44}^{-1}e_{15}\,(\overset{+}{g}_3 - \overset{-}{g}_3 - \int f_3) - (\overset{+}{d}_3 - \overset{-}{d}_3) + R_9\,[\Phi_1].$$

In another case, when S is given on $S_2^{\pm}$ surfaces, the equations for u_α^* remain unchanged, whereas the others take different forms. For example, the equation with respect to u_3^* and Φ_1 must be replaced by the following ones:

$$\frac{2h^3}{3}P'_{11}\,\Delta^2 u_3^* = (3b_{44})^{-1}\,e_{15}\,h^2 P'_{11}\,(1+2\gamma)\Delta^2\Phi_1 - P'_{31}\,\Delta\Phi_1$$

$$+\overset{+}{g}_3 - \overset{-}{g}_3 - \int f_3 + h\,(\overset{+}{g}_{\alpha,\alpha} + \overset{-}{g}_{\alpha,\alpha}) - \int tf_{\alpha,\alpha}$$

$$+(3b_{44})^{-1}\,h^2(1+2\gamma)\,[(b_{13}\,b_{33}^{-1}\,b_{44}-P'_{11}-b_{66})\Delta(\overset{+}{g}_3-\overset{-}{g}_3) + (P'_{11}-b_{66})\int\Delta f_3] + R'_8\,[u_3^*;\gamma],$$

$$(\ni_{11} + b_{44}^{-1}e_{15}^2)\Delta\Phi_1 = b_{44}^{-1}\,e_{15}(\overset{+}{g}_3 - \overset{-}{g}_3 - \int f_3) - \frac{1}{2h}[(\ni_{11} + \frac{e_{15}^2}{b_{44}})\Delta\,(\overset{+}{v}_0 + \overset{-}{v}_0)$$

$$+\frac{e_{15}}{b_{44}}(\overset{+}{g}_{\alpha,\alpha} + \overset{-}{g}_{\alpha,\alpha})] + R'_9\,[\Phi_1].$$

The above system without remainder terms R gives the system of differential equations with respect to $\overset{h}{u}_3$, $\overset{h}{Q}_\alpha$ and $\overset{h}{\Phi}_1$ with arbitrary parameters γ_i. Choosing γ, we get refined theories in a wide sense [5], which correspond to plate bending equations for the elastic case.

Other systems of differential equations, appropriate for extension (compression) of a plate for piezoceramic bodies, are the following:

(i) when D_3 is given on $S_1^{\pm}$

$$b_{66}\,\Delta\bar{u}_+ + (P_{11} - b_{66})\text{ grad div }\bar{u}_+ = (2h)^{-1}\,[\int f_+ - (\overset{+}{g}_+-\overset{-}{g}_+) - \int\text{grad}\,(P_{11}\,\sigma_{33}-P_{14}\,D_3)];$$

(ii) when S is known on $S_2^{\pm}$, we have:

$$b_{66}\Delta\bar{u}_+ + (P'_{11}-b_{66}) \text{ grad div } \bar{u}_+ = (2h)^{-1}[\int f_+ \, dt - b_{13}b_{33}^{-1}\int \text{grad } \sigma_{33} - P'_{31} \text{ grad } (S^+ - S^-)].$$

Here

$$\bar{u}_\alpha = \frac{1}{2h}\int u_\alpha, \; u_+ = (u_1, u_2)^T.$$

Boundary conditions which follow from the data given on S must be in agreement with the order of the system of two-dimensional differential equations to avoid several paradoxes mentioned in the paper [5].

It should be noted that the coefficients involved in the above adduced equations are formed by means of the constants b, as well as the numbers e and $\ni$.

In order to obtain a complete formulation (compare e.g.,]7], Ch. III, §§13, 14), let us introduce a moment of the first order of $\Phi_2 = (t, S)$.

Then we get for the first case

$$(\ni_{11} + b_{44}^{-1} e_{15}^2)\Delta\Phi_2 + (3b_{44})^{-1} e_{15} h^2(1+2\gamma)(\overset{+}{g}_{\alpha,\alpha} - \overset{-}{g}_{\alpha,\alpha}) = R_{13}[\Phi_2; \gamma].$$

In the second case we have for Φ_2 te same expression:

$$(\ni_{11} + b_{44}^{-1} e_{15}^2)\Delta\Phi_2 + (3b_{44})^{-1} e_{15} h^2 (1+2\gamma)(\overset{+}{g}_{\alpha,\alpha} - \overset{-}{g}_{\alpha,\alpha}) + \overset{+}{D}_{3,3} - \overset{-}{D}_{3,3}] = R'_{13}[\Phi_2;\gamma],$$

where

$$\overset{\pm}{D}_{3,3} \approx (\ni_{11} + b_{44}^{-1} e_{15}^2)\Delta v_0^{\pm} + b_{44}^{-1} e_{15}^2 \overset{\pm}{g}_{\alpha,\alpha}\,.$$

References

[1] Vorovich, I. Some mathematical problems of the theory of plates and shells. *Proc. of 2nd All Union Cong. on Theor. and Appl. Mech. Review reports*, **3**, 116–136 (1965) (in Russian).

[2] Berdichevski, L. The approach to the dynamic theory of thin elastic plates. *Mechanics of Solid Bodies*, **6**, 99–109 (1973) (in Russian).

[3] Ciarlet, P. and Destuynder, P. A justification of two-dimensional linear plate model. *J. de Mécanique*, **18** (2), 315–344 (1979).

[4] Ciarlet, P. and Paumier, G.C. A justification of the Marguerre-von Karman equations. *Computational Mechanics*, **1**, 177-202 (1986).

[5] Vashakmadze, T. Theory of elastic plates. *Advances in Mechanics*, **11** (3), 43-76 (1988) (Russian).

[6] Vashakmadze, T. On the problem of constructing the mathematical theory of plates and shells. *Trends Appl. Pure Math. to Mech.*, (eds. J. Besseling, W. Eckhaus). Proc. 7-th Symp. (Wassenaar, 1987), Springer-Verlag, 273-279 (1988).

[7] Parton, V. and Kudriavtsev, B. *Electromagnetoelasticity of Piezoelectrics and Electrically Conductive Solids*, p. 472, Nauka, Moscow, 1988 (in Russian).

T.S. Vashakmadze
I.N. Vekua Institute of Applied Mathematics
Tbilisi University
University Str. 2
380043 Tbilisi, Georgia
USSR

R. ELLINGHAUS AND W. MUSCHIK

Liquids subjected to electromagnetic fields: evaluation of the dissipation inequality with computer algebraic methods

Abstract

We are able to apply Liu's lemma to quite big state spaces by doing the necessary algebra with the help of extensions of the REDUCE Language. Although the constitutive relations could not be presented in this short summary, and some of the resulting equations are difficult to interpret, we got them in full generality. The application of these equations will be treated elsewhere.

Introduction

Applying Liu's lemma [1] to a relativistic liquid in electromagnetic fields, almost all of the Lagrange multipliers appearing in this lemma can be uniquely determined, if suitable (i.e. physically induced) assumptions are made about special constitutive quantities. The only arbitrary multiplier is the one corresponding to the balance of moment of momentum, which can be chosen equal zero. The knowledge of these parameters is the first step in developing a constitutive theory. Because of lack of space we only sketch how to compute them. The calculations have been performed by use of extensions to the REDUCE package (software for algebraic computations), implemented in the underlying RLISP dialect.

Equations of balance

The equations of balance for materials without internal variables in special relativity read [2], [3]:

Balance of particle number: $$\partial_\alpha (n_0 u^\alpha) = 0 \tag{1}$$

Balance of energy-momentum: $$\partial_\beta \left(T^{\alpha\beta} + \underset{\mathbf{em}}{T}{}^{\alpha\beta} \right) = 0 \tag{2}$$

Balance of moment of momentum: $$T^{[\alpha\beta]} + \underset{\mathbf{em}}{T}{}^{[\alpha\beta]} = 0 \tag{3}$$

Maxwell's equations: $$\in^{\alpha\beta\gamma\mu} \partial_\mu F_{\beta\gamma} = 0 \tag{4}$$

$$\partial_\beta G^{\alpha\beta} = j^\alpha, \tag{5}$$

where, following Grot and Eringen [3], the electromagnetic energy-momentum tensor is given by:

$$\underset{\mathbf{em}}{T}^{\alpha\beta} = -F^{\alpha\nu} G_\nu{}^\beta + \frac{1}{4\mu_0} F^{\mu\nu} F_{\nu\mu} g^{\alpha\beta} \tag{6}$$

and the total energy-momentum tensor is decomposed according to

$$T^{\alpha\beta} + \underset{\mathbf{em}}{T}^{\alpha\beta} = \frac{1}{c^2}\left(n_0 e u^\alpha u^\beta + u^\alpha q^\beta\right) + p^\alpha u^\beta - t^{\alpha\beta}. \tag{7}$$

The dissipation inequality reads

$$n_0 D\eta + \Phi^\alpha{}_{,\alpha} \geq h. \tag{8}$$

The involved quantities are defined in the following manner: n_0 is the particle density in a local rest frame, $\mathbf{u}$ the 4-velocity, e the specific energy density, $\mathbf{q}$ the conductive energy flux, $\mathbf{p}$ the nonmechanical momentum, $\mathbf{t}$ the 4-stress tensor, $\mathbf{g}$ the metric, $\mathbf{F}$ and $\mathbf{G}$ are the electromagnetic field tensors, j the charge-current 4-vector, ϵ the total antisymmetric tensor, $D = u^\alpha \partial/\partial x^\alpha$ the covariant material time-derivative, η the specific density of entropy, $\boldsymbol{\Phi}$ the conductive flux of entropy, h the supply of entropy in a local rest frame and c is the velocity of light. We need furthermore the definitions:

$$P^{\alpha\beta} := g^{\alpha\beta} + \frac{u^\alpha u^\beta}{c^2}, \tag{9}$$

$$\overset{*}{\partial}_\alpha := P_\alpha{}^\beta \partial_\beta, \tag{10}$$

and

$$\overset{*}{e}^{\alpha\beta\gamma} := e^{\alpha\beta\gamma\nu} \frac{u_\nu}{c}. \tag{11}$$

State space and constitutive assumptions

An appropriate state space for a heat conducting viscous fluid in external electromagnetic fields is given by

$$\mathbf{X} := \left(n_0;\, u^\alpha;\, \partial_\beta u^\alpha;\, \Theta;\, \partial_\beta \Theta;\, F^{\alpha\beta}\right), \tag{12}$$

where Θ is the scalar temperature in a local rest frame [3]. We presuppose that the following relations hold:

- The tensor of differential heat conductivity is negative definite, i.e.

$$\forall \mathbf{x}\,(\mathbf{x} \neq \mathbf{0} \wedge x_\alpha u^\alpha = 0) \Rightarrow x_\beta \frac{\partial q^\beta}{\partial \partial_\alpha \Theta} x_\alpha < 0. \tag{13}$$

- The field tensor $\mathbf{G}$ is independent of the 4-gradient of temperature, i.e.:

$$\frac{\partial G^{\alpha\beta}}{\partial \partial_\gamma \Theta} = 0. \tag{14}$$

- The 3-tensor of differential dielectricity is regular in any local rest frame, which can be shown to be equivalent to

$$\operatorname{rank}\left(\frac{\partial G^{\alpha\mu}}{\partial F_{\gamma\nu}} u_\mu u_\nu\right) = 3. \tag{15}$$

Computing the Lagrange parameters

Introducing the variables $\mathbf{X}$ (12) into the equations of balance (1), (2), (4), and (5), applying the chain rule of differentiation, multiplying each equation by an appropriate Lagrange parameter, adding those objects to the inequality (8) and collecting all terms involving highest the derivatives, we get the following system of equations:

- Factors of $\overset{*}{\partial}_\beta n_0$:

$$-\underset{u}{\Lambda}_\alpha \frac{\partial t^{\alpha\beta}}{\partial n_0} + \underset{e}{\Lambda} \frac{\partial q^\beta}{\partial n_0} + \frac{\partial \Phi^\beta}{\partial n_0} + \underset{G}{\Lambda}_\alpha \frac{\partial G^{\alpha\nu}}{\partial n_0} P_\nu{}^\beta = 0. \tag{16}$$

- Factors of Dn_0:

$$\underset{u}{\Lambda}_\alpha \frac{\partial p^\alpha}{\partial n_0} + \underset{e}{\Lambda} n_0 \frac{\partial e}{\partial n_0} + \underset{n}{\Lambda} + n_0 \frac{\partial \eta}{\partial n_0} - \frac{1}{c^2} \underset{G}{\Lambda}_\alpha \frac{\partial G^{\alpha\nu}}{\partial n_0} u_\nu = 0. \tag{17}$$

- Factors of $\overset{*}{\partial}_\beta \partial_\gamma \Theta$:

$$-\underset{u}{\Lambda}_\alpha \frac{\partial t^{\alpha\beta}}{\partial \partial_\gamma \Theta} + \underset{e}{\Lambda} \frac{\partial q^\beta}{\partial \partial_\gamma \Theta} + \frac{\partial \Phi^\beta}{\partial \partial_\gamma \Theta} = 0. \tag{18}$$

- Factors of $D\partial_\gamma \Theta$:

$$\underset{u}{\Lambda}_\alpha \frac{\partial p^\alpha}{\partial \partial_\gamma \Theta} + \underset{e}{\Lambda}\, n_0 \frac{\partial e}{\partial \partial_\gamma \Theta} + n_0 \frac{\partial \eta}{\partial \partial_\gamma \Theta} = 0. \tag{19}$$

- Factors of $\overset{*}{\partial}_\beta \partial_\gamma u^\nu$:

$$-\underset{u}{\Lambda}_\alpha \frac{\partial t^{\alpha\beta}}{\partial \partial_\gamma u^\nu} + \underset{e}{\Lambda} \frac{\partial q^\beta}{\partial \partial_\gamma u^\nu} + \frac{\partial \Phi^\beta}{\partial \partial_\gamma u^\nu} + \underset{G}{\Lambda}_\alpha \frac{\partial G^{\alpha\mu}}{\partial \partial_\gamma u^\nu} P_\mu{}^\beta = 0. \tag{20}$$

- Factors of $D\partial_\gamma u^\nu$:

$$\underset{u}{\Lambda}_\alpha \frac{\partial p^\alpha}{\partial \partial_\gamma u^\nu} + \underset{e}{\Lambda}\, n_0 \frac{\partial e}{\partial \partial_\gamma u^\nu} + n_0 \frac{\partial \eta}{\partial \partial_\gamma u^\nu} - \frac{1}{c^2} \underset{G}{\Lambda}_\alpha \frac{\partial G^{\alpha\mu}}{\partial \partial_\gamma u^\nu} u_\mu = 0. \tag{21}$$

- Factors of $\overset{*}{\partial}_\beta F_{\gamma\nu}$:

$$-\underset{u}{\Lambda}_\alpha \frac{\partial t^{\alpha\beta}}{\partial F_{\gamma\nu}} + \underset{e}{\Lambda} \frac{\partial q^\beta}{\partial F_{\gamma\nu}} + \frac{\partial \Phi^\beta}{\partial F_{\gamma\nu}} + \underset{G}{\Lambda}_\alpha \frac{\partial G^{\alpha\mu}}{\partial F_{\gamma\nu}} P_\mu{}^\beta + \underset{F}{\Lambda}_\alpha e^{\alpha\mu\gamma\nu} P_\mu{}^\beta = 0. \tag{22}$$

- Factors of $DF_{\gamma\nu}$:

$$\underset{u}{\Lambda}_\alpha \frac{\partial p^\alpha}{\partial F_{\gamma\nu}} + \underset{e}{\Lambda}\, n_0 \frac{\partial e}{\partial F_{\gamma\nu}} + n_0 \frac{\partial \eta}{\partial F_{\gamma\nu}} - \frac{1}{c^2} \underset{G}{\Lambda}_\alpha \frac{\partial G^{\alpha\mu}}{\partial F_{\gamma\nu}} u_\mu - \frac{1}{c} \underset{F}{\Lambda}_\alpha \overset{*}{e}^{\alpha\gamma\nu} = 0. \tag{23}$$

These equations serve as determining relations for the Lagrange multipliers $\underset{n}{\Lambda}$, $\underset{e}{\Lambda}$, $\underset{u}{\Lambda}_\alpha$, $\underset{F}{\Lambda}_\alpha$ and $\underset{G}{\Lambda}_\alpha$. Their uniqueness is insured by (13), (14) and (15). The explicit form of the multipliers is too long to represent them here.

Tensorial operations in REDUCE

The REDUCE language allows at the current stage (Version 3.3, see [4]) no computing with tensorial objects, scalar or dyadic products, tensorial differentiation, etc. without going into component representation. Simplifying expressions in this representation is little help in formulae involving complicated tensorial objects. To overcome these restrictions we implemented some new operators, predefined quantities and simplification rules. Their use is sketched by the following source code fragments:

```
%--- predefined variables: -------------------
% U is the 4-velocity
% C is the velocity of light
% PROJ is the projection "P"
% ZEROV is the zero 4-vector
%--- declarations: --------------------------
% declare A, B and C3 to be tensors of order 1:
fourvector A,B, C3;
% declare X and Y to be tensors of order 2:
fourtensor X,Y;
% declar M, L1 and L2 to be tensors of order 4:
higherorder 4, M, L1, L2;
% declare X and M orthogonal to U in any
% contraction:
spacelike X, M;
% declare TET and U to be state space elements
variable TET, U;
% declare A and B to be constant, i.e. independent
% of state space elements
const X, M;
%--- operations ------------------------
U !@ M;      % denotes the dyadic product of U and M
U !. A;      % denotes the Lorentz Product of U and A
M !.!. L1;   % denotes the dual contraction between M and L1
M !.!.!. L1; % denotes the triple contraction between M and L1
%--- special functions --------------------------
TDF(M,L1); % yields the tensorial derivative of M by L1
TSIMP(XX);% yields the simplification of XX
%----------------------------------------------------
```

References

[1] I-S. Liu. Method of Lagrange multipliers for exploitation of the entropy principle. *Arch. Rat. Mech. Anal.*, **46**, 131-148, 1972.

[2] R. Ellinghaus and W. Muschik. Nichtgleichgewichtsthermodynamik mikropolarer Medien in elektromagnetischen Feldern. *ZAMM*, **68**, T171-T172, 1988.,

[3] R. A. Grot and A.C. Eringen. Relativisitic continuum mechanics part II - electromagnetic interations with matter. *Int. J. Engng. Sci*, **4**, 639-670, 1966.

[4] G. Rayna, *REDUCE Software for Algebraic Computation*, Springer Verlag, 1987.

R. Ellinghaus and W. Muschik
Institut für Theoretische Physik,
Sekretariat PN 7-1
Technische Universität Berlin,
Hardenbergstr. 36,
1000 Berlin 12
GERMANY

T. LEWIŃSKI AND J. J. TELEGA

Non-uniform homogenization and effective properties of a class of non-linear elastic shells

Abstract

By using the method of Γ-convergence the homogenization formula is given for a class of non-coercive functionals in the two-dimensional case. This result is next used for the determination of the homogenized elastic potential for the periodic elastic shells with moderately large rotations around tangents.

Introduction

In our previous paper [6] we performed homogenization of Koiter's linear shells and the geometrically nonlinear shallow shells (Mushtari-Marguerre model) provided that they exhibit periodic structure with respect to assumed curvilinear parametrization. On account of an influence of the geometry of the shell such averaging problems belong to non-uniform homogenization problems. By using the method of two-scale asymptotic expansions it was shown that for Koiter's shell model the local problems are coupled. Consequently, the constitutive equations describing the homogenized shell model are also coupled. Uncoupling is preserved in the case of shallow shells, disregarding linearity or nonlinearity of the model.

The objective of this contribution is to determine effective (homogenized) properties of the nonlinear shell undergoing moderately large rotations around tangents to the shell middle surface. Prior to this, we formulate new and general results concerning non-uniform homogenization of a class of non-coercive functionals in the two-dimensional case. Towards this end the method of Γ-convergence is very suitable and convenient.

1. Non-uniform homogenization of a class of non-coercive functionals

Having in mind determination of effective properties of a nonlinear shell model we shall first consider a more general problem. Namely we shall extend the results due to Braides [3], who performed homogenization in the n-dimensional case when the integrand $f = f(\xi, \xi/\varepsilon, u, \nabla u)$, $u \in W^{1,p}(\Omega)$, $\Omega \subset R^n$, $p \in [1, \infty]$. Below only two-dimensional problems are studied.

Let $\Omega \subset R^2$ be a bounded domain. By $Y \subset R^2$ we denote as usual a basic cell [1, 3, 6, 7]. $M^2(M^2_s)$ stands for the space of 2×2 (symmetric) matrices. An integrand

$f = f(\xi, y, \mathbf{u}, w, \nabla\mathbf{u}, \nabla w, \nabla^2 w)$ ($\mathbf{u} \in H^1(\Omega, R^2)$, $w \in H^2(\Omega)$) satisfies the following properties:

(i) f is measurable and Y-periodic in y, continuous in ξ, $\mathbf{u}$, w and ∇w and

$$f(\xi, y, \mathbf{c}, c, \cdot, \mathbf{d}, \cdot) : M^2 \times M_s^2 \to [0, \infty]$$

is convex for almost all y and for all ξ, $\mathbf{c}$, c and $\mathbf{d}$.

(ii) $0 \leq f(\xi, y, \mathbf{c}, c, \mathbf{H}_1, \mathbf{d}, \mathbf{H}_2) \leq \rho(\xi)\,[A(y) + |\mathbf{c}|^2 + |c|^2 + |\mathbf{H}_1|^2 + |\mathbf{d}|^4 + |\mathbf{H}_2|^2]$.

(iii) $|f(\xi, y, \mathbf{c}, c, \mathbf{H}_1, \mathbf{d}, \mathbf{H}_2) - f(\xi', y, \mathbf{c}', c', \mathbf{H}_1, \mathbf{d}', \mathbf{H}_2)|$

$$\leq \omega(|\xi - \xi'| + |\mathbf{c} - c'| + |\mathbf{c} - c'| + |\mathbf{d} - \mathbf{d}'|) \cdot (A(y) + f(\xi, y, \mathbf{c}, c, \mathbf{H}_1, \mathbf{d}, \mathbf{H}_2))$$

for $\xi, \xi' \in \Omega$, $y \in Y, \mathbf{c}, \mathbf{c}', \mathbf{d}, \mathbf{d}' \in R^2$; $c, c' \in R, \mathbf{H}_1 \in M^2, \mathbf{H}_2 \in M_s^2$.

Here $A \in L^1(\Omega)$ is a Y-periodic function, $\omega : R^+ \to R^+$ is an increasing function, continuous at 0 and such that $\omega(0) = 0$, while ρ is a continuous and nonnegative function.

It is worth noting that condition (ii) does not require coercivity of the integrand. Also, we observe that in this condition growth with respect to ∇w, or the gradient of w, is in the fourth power. However, due to the embedding $H^2(\Omega) \subset W^{1,4}(\Omega)$ we infer that $|\nabla w|^4 \in L^1(\Omega)$.

For a fixed $\varepsilon > 0$ we set

$$F^\varepsilon(\mathbf{u}, w) = \int_\Omega f(\xi, \xi/\varepsilon, \mathbf{u}, w, \nabla\mathbf{u}, \nabla w, \nabla^2 w)\, d\xi \tag{1.1}$$

where $\mathbf{u} \in H^1(\Omega, R^2)$ and $w \in H^2(\Omega)$.

Let us denote by τ the strong topology of the space $L^2(\Omega, R^2) \times H^1(\Omega)$. Let

$$W_{per} = H^1_{per}(Y, R^2) \times H^2_{per}(Y); H^\alpha_{per}(Y),\ \alpha = 1,2,$$

be the Sobolev spaces of Y-periodic functions, cf. [6].

It can be shown that the $\Gamma(\tau^-)$-limit of the sequence of functionals (1.1) is given by

$$F(\mathbf{u}, w) = \Gamma(\tau^-) \lim_{\varepsilon \to 0} F^\varepsilon(\mathbf{u}, w) = \int_\Omega f^h(\xi, \mathbf{u}, w, \nabla\mathbf{u}, \nabla w, \nabla^2 w) d\xi, \tag{1.2}$$

where

$$f^h(\xi, \mathbf{c}, c, \mathbf{H}_1, \mathbf{d}, \mathbf{H}_2)$$

$$= \inf_{(\mathbf{v}, v) \in W_{per}} \frac{1}{|Y|} \int_Y f(\xi, y, \mathbf{c}, c, \nabla_y \mathbf{v}(y) + \mathbf{H}_1, \mathbf{d}, \nabla_y^2 v(y) + \mathbf{H}_2) dy. \tag{1.3}$$

Here ∇_y denotes the gradient with respect to $y \in Y$.

Remarks

(1) To prove the homogenization formulae one can follow the arguments developed in [3], [4] for $f = f(\xi, y, u, \nabla u)$, $u \in W^{1,p}(\Omega)$. Detailed discussion of Γ-limits is provided by the book by Attouch [1].

(2) Let us assume that the integrand f can be split up as follows:

$$f(\xi, y, \mathbf{u}, w, \nabla \mathbf{u}, \nabla w, \nabla^2 w)$$

$$= f_1(\xi, y, \mathbf{u}, w, \nabla \mathbf{u}, \nabla w) + f_2(\xi, y, \mathbf{u}, w, \nabla w, \nabla^2 w). \tag{1.4}$$

Then the local problem (1.3) also splits up into two *independent* minimization problems over the spaces $H^1_{per}(Y, R^2)$ and $H^2_{per}(Y)$, respectively.

(3) For the general case considered the so called equi-coercivity condition [1] does not hold. Stronger assumption of coercivity in the condition (ii) implies equi-coercivity. In such a case we have convergence of infimium of F^ε to the infimum of the limit functional F.

2. Application: homogenization of elastic shells

Let $\Phi : \Omega \to S$ be a sufficiently regular mapping [2, 5, 8]. Here $S = \Phi(\Omega)$ denotes the middle surface of the undeformed shell. The plane R^2 containing Ω is referred to coordinates (ξ^α), whereas R^3 is referred to (x^i); $\alpha = 1,2$; $i = 1,2,3$. The covariant metric tensor of the middle surface is $a_{\alpha\beta} = \Phi_{,\alpha} \Phi_{,\beta}$, its determinant being denoted by a. The curvature tensor of the surface S is denoted by $\mathbf{b} = (b_{\alpha\beta})$. $\mathbf{U} = (\mathbf{u}, w)$ is the displacement vector, $\mathbf{u} = (u_\alpha)$ being its tangent component; w stands for the normal deflection. For the model of shells with moderately large rotations around tangents, strain-displacement relations are given by (cf. [2], [8]).

$$\gamma_{\alpha\beta}(\mathbf{u}, w) = \theta_{\alpha\beta}(\mathbf{u}, w) + \frac{1}{2} \varphi_\alpha(\mathbf{u}, w) \varphi_\beta(\mathbf{u}, w), \tag{2.1}$$

$$\kappa_{\alpha\beta}(\mathbf{u}, w) = w|_{\alpha\beta} + b_{\alpha}|_{\beta}\, \mathbf{u}_{\lambda} + b^{\lambda}_{\alpha}\, u_{\lambda}|_{\beta} + b^{\lambda}_{\beta}\, u_{\lambda}|_{\alpha} - b^{\lambda}_{\alpha}\, b_{\alpha\beta}\, w, \tag{2.2}$$

where

$$u_{\alpha}|_{\beta} = u_{\alpha,\beta} - \Gamma^{\lambda}_{\alpha\beta} u_{\lambda}, \quad \varphi_{\alpha}(\mathbf{u}, w) = w_{,\alpha} + b^{\lambda}_{\alpha}\, u_{\lambda}, \tag{2.3}$$

$$\theta_{\alpha\beta}(\mathbf{u}, w) = \frac{1}{2}(u_{\alpha}|_{\beta} + u_{\beta}|_{\alpha}) - b_{\alpha\beta} w. \tag{2.4}$$

$\Gamma^{\alpha}_{\beta\lambda}$ are Christoffel symbols of the undeformed shell middle surface S.

Let us observe that for Φ = identity we recover the von Kármán plate equations. The stored energy function W is quadratic and is expressed as follows:

$$W(\boldsymbol{\gamma}, \boldsymbol{\kappa}) = \frac{1}{2} A^{\alpha\beta\lambda\mu}\, \gamma_{\alpha\beta}\, \gamma_{\lambda\mu} + \frac{1}{2} D^{\alpha\beta\lambda\mu}\, \kappa_{\alpha\beta}\, \kappa_{\lambda\mu} \tag{2.5}$$

under the usual assumptions on stiffness tensors $\mathbf{A}$ and $\mathbf{D}$ [2, 5, 8]. For shells these tensors depend on $\xi \in \Omega$ (even in the case of homogeneity).

Let us consider now an εY-periodically inhomogeneous shell, cf. [6]. The stiffnesses are assumed as: $A^{\alpha\beta\lambda\mu}_{\varepsilon} = A^{\alpha\beta\lambda\mu}(\xi, y)$, $D^{\alpha\beta\lambda\mu}_{\varepsilon} = D^{\alpha\beta\lambda\mu}(\xi, y)$, $y = \xi/\varepsilon$, where $\varepsilon > 0$, $y \in Y$. The functions $A^{\alpha\beta\lambda\mu}(\xi, \ )$ and $D^{\alpha\beta\lambda\mu}(\xi, \ )$ are Y-periodic.

For a fixed ε the functional $J^{\varepsilon}(\mathbf{u}, w)$ of the total potential energy of the clamped shell is a sum of functionals $F^{\varepsilon}(\mathbf{u}, w)$ and $G(\mathbf{u}, w)$. The latter, representing the virtual work of the external surface loading, plays the role of the continuous perturbation and is not influenced by the Γ-limit process [1]. The former is given by

$$F^{\varepsilon}(\mathbf{u}, w) = \frac{1}{2}\int_{\Omega} [A^{\alpha\beta\lambda\mu}_{\varepsilon}\, \gamma_{\alpha\beta}(\mathbf{u}, w)\gamma_{\lambda\mu}(\mathbf{u}, w) + D^{\alpha\beta\lambda\mu}_{\varepsilon}\, \kappa_{\alpha\beta}(\mathbf{u}, w)\kappa_{\lambda\mu}(\mathbf{u}, w)]\sqrt{a}\, d\xi, \tag{2.6}$$

where

$$\mathbf{u} \in H^{1}_{0}(\Omega, R^{2}), \quad w \in H^{2}_{0}(\Omega).$$

By using the results of the previous section we deduce that the macroscopic potential W^{h} of the homogenized shell can be expressed as

$$W^{h}(\xi, \boldsymbol{\gamma}^{h}, \boldsymbol{\kappa}^{h}) = \frac{\sqrt{a(\xi)}}{2|Y|} \inf_{(\mathbf{v}, v) \in W_{per}} \int_{Y} \{A^{\alpha\beta\lambda\mu}(\xi, y)\, [e_{y\,\alpha\beta}(\mathbf{v})$$

$$+\gamma^h_{\alpha\beta}(\xi)][e_{y\,\lambda\mu}(\mathbf{v})+\gamma^h_{\lambda\mu}(\xi)]+D^{\alpha\beta\lambda\mu}(\xi,y)\Big[\frac{\partial^2 v}{\partial y_\alpha\,\partial y_\beta}$$

$$+2b^\rho_\alpha(\xi)\frac{\partial v_\rho}{\partial y_\beta}+\kappa^h_{\alpha\beta}(\xi)\Big]\Big[\frac{\partial^2 v}{\partial y_\lambda\,\partial y_\mu}+2b^\delta_\lambda(\xi)\frac{\partial v_\delta}{\partial y_\lambda}+\kappa^h_{\lambda\mu}(\xi)\Big]\Big\}\,dy$$

$$=f^h(\xi,\mathbf{c},c,\ \mathbf{H}_1,\mathbf{d},\mathbf{H}_2)$$

$$=\frac{\sqrt{a(\xi)}}{2|Y|}\inf_{(\mathbf{v},v)\in W_{per}}\int_Y\Big\{A^{\alpha\beta\lambda\mu}(\xi,y)\,[e_{y\,\alpha\beta}(\mathbf{v})+H_{1\alpha\beta}-\Gamma^\rho_{\alpha\beta}(\xi)c_\rho$$

$$-b_{\alpha\beta}(\xi)c+\frac{1}{2}(d_\alpha+b^\rho_\alpha(\xi)c_\rho)(d_\beta+b^\delta_\beta(\xi)c_\delta)]\,[e_{y\lambda\mu}(\mathbf{v})+H_{1\lambda\mu}$$

$$-\Gamma^\omega_{\lambda\mu}(\xi)c_\omega-b_{\lambda\mu}(\xi)c+\frac{1}{2}(d_\lambda+b^\omega_\lambda(\xi)c_\omega)(d_\mu+b^\varphi_\mu(\xi)c_\varphi)]$$

$$+D^{\alpha\beta\lambda\mu}(\xi,y)\Big[\frac{\partial^2 v}{\partial y_\alpha\,\partial y_\beta}+H_{2\alpha\beta}-\Gamma^\rho_{\alpha\beta}(\xi)\,d_\rho+b^\rho_{\alpha|\beta}(\xi)c_\rho$$

$$+b^\rho_\alpha(\xi)\frac{\partial v_\rho}{\partial y_\beta}-b^\rho_\alpha(\xi)\Gamma^\delta_{\rho\beta}(\xi)c_\delta+b^\rho_\beta(\xi)\frac{\partial v_\rho}{\partial y_\beta}-b^\rho_\beta(\xi)\,\Gamma^\delta_{\rho\alpha}(\xi)c_\delta$$

$$-b^\rho_\alpha(\xi)\,b_{\rho\beta}(\xi)c\Big]\Big[\frac{\partial^2 v}{\partial y_\lambda\,\partial y_\mu}+H_{2\lambda\mu}-\Gamma^\varphi_{\lambda\mu}(\xi)d_\varphi+b^\varphi_{\lambda|\mu}(\xi)c_\varphi$$

$$+b^\varphi_\lambda(\xi)\frac{\partial v_\varphi}{\partial y_\mu}-b^\varphi_\lambda(\xi)\,\Gamma^\psi_{\varphi\mu}(\xi)c_\psi+b^\varphi_\mu(\xi)\frac{\partial v_\varphi}{\partial y_\lambda}$$

$$-b^\varphi_\mu(\xi)\Gamma^\psi_{\varphi\lambda}(\xi)c_\psi-b^\varphi_\lambda(\xi)b_{\varphi\mu}(\xi)c\Big]\Big\}\,dy, \tag{2.7}$$

where

$$e_{y\alpha\beta}(\mathbf{v})=\Big(\frac{\partial v_\alpha}{\partial y_\beta}+\frac{\partial v_\beta}{\partial y_\alpha}\Big)/2.$$

Here $\boldsymbol{\gamma}^h(\xi) \in M_s^2$ and $\boldsymbol{\kappa}^h(\xi) \in M_s^2$ $(\xi \in \Omega)$ are macroscopic strain measures. The last part of the relation (2.7) suggests that these measures are calculated according to Eqs. (2.1) and (2.2), provided that $(\mathbf{u}, w)$ is a displacement field of the homogenized shell.

The averaged membrane forces and moments are given by the following macroscopic constitutive relationships: $N_h^{\alpha\beta} = \partial W^h/\partial \gamma_{\alpha\beta}^h$, $M_h^{\alpha\beta} = \partial W^h/\partial \kappa_{\alpha\beta}^h$, respectively. Similarly as in the case of a linear Koiter's shell [6] the above macroscopic constitutive relations are coupled, though for every $\varepsilon > 0$ they are uncoupled. Uncoupling always occurs for shallow shells and, more generally, when the term $b^\lambda{}_{(\alpha}u_{\lambda|\beta)}$ appearing in Eq. (2.2) may be dropped.

References

[1] H. Attouch, *Variational Convergence for Functions and Operators*, Pitman, London 1984.

[2] W.R. Bielski and J.J. Telega, ZAMM, **68**, 1988, T155–T157.

[3] A. Braides, *Ric. di Mat.*, **32**, 1983, 347–368.

[4] G. Buttazzo and G. Dal Maso, *J. Analyse Math.*, **37**, 1980, 145–185.

[5] Ph. Destuynder, *Acta Appl. Math.*, **4**, 1985, 16–63.

[6] T. Lewiński and J.J. Telega, *Arch. Mech.*, **40**, 1988, 705–723.

[7] J.J. Telega, Variational methods and convex analysis in contact problems and homogenization, IFTR Reports **38**, 1990, Warsaw (in Polish).

[8] J.J. Telega and W.R. Bielski. In: *Selected Problems of Modern Continuum Theory*, pp. 177–193 (ed. W. Kosinsky, T. Manacorda, A. Morro, T. Ruggeri), Pitagora Editrice, Bologna 1987.

T. Lewiński
Warsaw Technical University,
Institute of Structural Mechanics,
Armii Ludowej 16,
00–637 Warsaw,
POLAND

J.J. Telega
Polish Academy of Sciences,
Institute of Fundamental Technological Research, ZTK,
Świetokrzyska 21,
00–049 Warsaw,
POLAND

D. CIORANESCU AND J. SAINT JEAN PAULIN

Asymptotic techniques to study tall structures

Dedicated to Professor E. Kröner on his 70th birthday

Introduction

We study the elasticity problem for periodic structures consisting of very thin bars. They are tower-like structures. The case we consider here is that of four vertical bars linked with horizontal bars which are periodically distributed along the height. The height L is fixed and all the other dimensions are small. More precisely the period is ε, the horizontal bars are of length e and of cross section $e\delta/2 \times \varepsilon\delta$ and the vertical bars have the cross section $e\delta/2 \times e\delta/2$ (see Fig. 1). We consider the case when the three parameters e, ε, δ are small and our aim is to give the asymptotic behaviour of the displacement. For simplicity we suppose that the material is homogeneous and isotropic. A natural guess is that we get two fourth-order systems and one second-order system to characterize the displacements and that the coefficients are expressed in terms of Young's modulus E and Poisson's ratio ν.

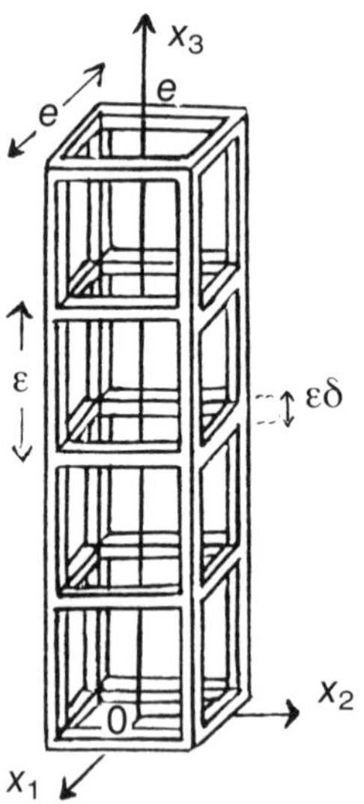

Figure 1

In the present paper we prove the above statement and give the overall global coefficients. It is important to notice that these limit coefficients are very simple as was already the case for honeycomb and reinforced structures [1] and for networks [2]. In our asymptotic study we make first $e \to 0$ and use beam techniques. The second step $\varepsilon \to 0$ is a homogenization process. Finally $\delta \to 0$ comes easily from an explicit calculation of coefficients.

Let us detail the system and the domain in which we work. The displacements $u^{e\varepsilon\delta}$ satisfy the system of linearized elasticity on the domain $\Omega_{e\varepsilon\delta}$ occupied by the material

$$\begin{cases} -\dfrac{\partial}{\partial x_i}\left(a_{ijkh}\dfrac{\partial u_k^{e\varepsilon\delta}}{\partial x_h}\right) = F_i^e \quad \text{in } \Omega_{e\varepsilon\delta} \\ u^{e\varepsilon\delta} = 0 \text{ on the bottom of the crane} \\ \left(a_{ijkh}\dfrac{\partial u_k^{e\varepsilon\delta}}{\partial x_h}\right) n_j = 0 \text{ on the top of the crane} \\ \left(a_{ijkh}\dfrac{\partial u_k^{e\varepsilon\delta}}{\partial x_h}\right) n_j = G_i^{e\varepsilon} \quad \text{on } \partial T_{e\varepsilon\delta} \text{ (boundary of the crane except the two bases)} \end{cases}$$

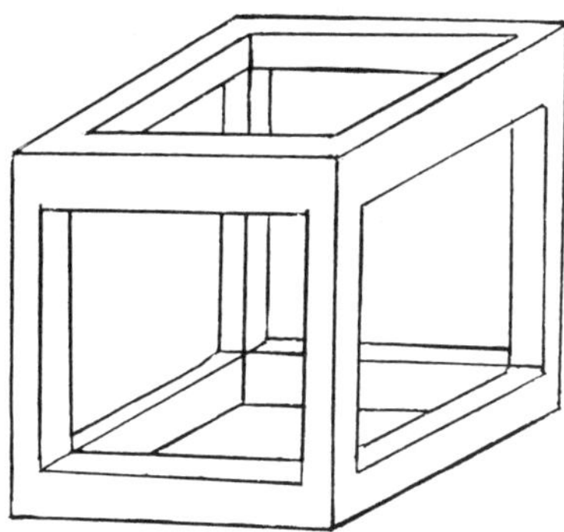

Figure 2. Representation period corresponding to Figure 1.

We make the following assumptions:

(i) $F^e \in L^2((-e/2, +e/2) \times (-e/2, +e/2) \times (0, L))$, $\quad G^{e\,\varepsilon} \in L^2(\partial T_{e\varepsilon\delta})$

(ii) the a_{ijkh} are Lamé constants: $a_{ijkh} = \lambda\delta_{ij}\delta_{kh} + \mu(\delta_{ik}\delta_{jh} + \delta_{ih}\delta_{jk})$.

1. Limit for $e \to 0$ with ε and δ fixed

In order to study the dependence on u, we work on a fixed domain by making the change of variables and of functions

$$\begin{cases} z_\alpha = x_\alpha/e, \ \alpha = 1, 2; \quad z_3 = x_3 \\ u_\alpha^{\varepsilon\delta}(z_1,z_2,z_3) = e^{-1} u_\alpha^{e\varepsilon\delta}(ez_1, ez_2, z_3)\ (\alpha = 1,2),\ u_3^{\varepsilon\delta}(z_1, z_2, z_3) = e^{-2} u_e^{e\varepsilon\delta}(ez_1,ez_2,z_3). \end{cases}$$

The domain $\Omega_{e\varepsilon\delta}$ is changed into $\Omega_{1\varepsilon\delta} \subset Q = (-1/2, +1/2) \times (-1/2, +1/2) \times (0, L)$. We denote by $S_{e\varepsilon\delta}(z_3)$ the cross section of the structure at the height z_3. By the above transformation, it is changed into $S_{1\varepsilon\delta}(z_3)$. We prove the following theorem:

Theorem 1.1: *Suppose that we have the following convergences*:

$$\int_{S_{e\varepsilon\delta}(z_3)} e^{-2} F_\alpha^e(x_1, x_2, x_3)\, dx_1\, dx_2 + \int_{\partial S_{e\varepsilon\delta}(z_3)} e^{-2} G_\alpha^{e\varepsilon}(\sigma, z_3)\, d\sigma \to \mathcal{F}_\alpha^{\varepsilon\delta}(z_3)$$

$$\int_{S_{e\varepsilon\delta}(z_3)} e^{-2} x_\alpha F_3^e(x_1, x_2, x_3)\, dx_1\, dx_2 + \int_{\partial S_{e\varepsilon\delta}(z_3)} e^{-2} x_\alpha G_\alpha^{e\varepsilon}(\sigma, z_3)\, d\sigma \to \mathcal{G}_\alpha^{\varepsilon\delta}(z_3)$$

$$\int_{S_{e\varepsilon\delta}(z_3)} e^{-1} F_\alpha^e(x_1, x_2, x_3)\, dx_1\, dx_2 \to \mathcal{F}_3^{\varepsilon\delta}(z_3).$$

Then: $e^3 u^{\varepsilon\delta} \to u^{*\varepsilon\delta}$ in $H^1(\Omega_{1\varepsilon\delta})$ *weakly with*

$$\begin{cases} u_\alpha^{*\varepsilon\delta}(z_1, z_2, z_3) = V_\alpha(z_3)\ (\alpha = 1, 2) \text{ in } \Omega_{1\varepsilon\delta} \\ u_3^{*\varepsilon\delta}(z_1, z_2, z_3) = -z_1 \dfrac{\partial V_1}{\partial z_3}(z_3) - z_2 \dfrac{\partial V_2}{\partial z_3}(z_3) + V_3(z_3) \text{ in } \Omega_{1\varepsilon\delta}. \end{cases} \tag{1.1}$$

The functions $V_i^{\varepsilon\sigma}$ *satisfy respectively*

$$\begin{cases} V_\alpha \in H^2(0,L),\ V_\alpha(0) = \dfrac{\partial V_\alpha}{\partial z_3}(0) = 0,\ \dfrac{\partial^3 V_\alpha}{\partial z_3^3}(L) = \dfrac{\partial^2 V_\alpha}{\partial z_3^2}(L) = 0 \\ E\dfrac{\partial^2}{\partial z_3^2}\left[\left(\displaystyle\int_{S_{1\varepsilon\delta}(z_e)} z_1^2\, dz_1\, dz_2\right)\dfrac{\partial^2 V_\alpha}{\partial z_3^2}\right] = -\mathcal{F}_\alpha + \dfrac{\partial \mathcal{G}_\alpha}{\partial z_3} \text{ in } (0,L) \end{cases}$$

$$\begin{cases} V_3 \in H^1(0,L),\ V_3(0) = 0,\ \dfrac{\partial V_3}{\partial z_3}(L) = 0 \\ -E\dfrac{\partial}{\partial z_3}\left[\operatorname{meas} S_{1\varepsilon\delta}(z_3)\,\dfrac{\partial V_3}{\partial z_3}\right] = \mathcal{F}_3 \text{ in } (0,L) \end{cases} \tag{1.2}$$

Remarks

(i) Though the initial problem is set on a perforated 3-dimensional domain with big holes, the limit systems are given on a non-perforated 1-dimensional domain. The holes interfere via the integrals over the cross section $S_{1\varepsilon\delta}(z_3)$.

(ii) The limit systems obtained in this theorem also hold for any structure such that

$$\begin{cases} \displaystyle\int_{S_{e\varepsilon\delta}(z_3)} z_1^2\, dz_1\, dz_2 = \int_{S_{1\varepsilon\delta}(z_3)} z_2^2\, dz_1\, dz_2 \\ \displaystyle\int_{S_{e\varepsilon\delta}(z_3)} z_1\, dz_1\, dz_2 = \int_{S_{e\varepsilon\delta}(z_3)} z_2\, dz_1\, dz_2 = \int_{S_{e\varepsilon\delta}(z_3)} z_1 z_2\, dz_1\, dz_2 = 0. \end{cases}$$

Proof: Using the coercivity of the tensor (a_{ijkh}) and Korn's inequality, we obtain the *a priori* estimates

$$\begin{cases} \|\gamma_{\alpha\beta}(u^{\varepsilon\delta})\|_{L^2(\Omega_{1\varepsilon\delta})} \le Ce^{-1}; & \|\gamma_{\alpha 3}(u^{\varepsilon\delta})\|_{L^2(\Omega_{1\varepsilon\delta})} \le Ce^{-2} \\ \|\gamma_{33}(u^{\varepsilon\delta})\|_{L^2(\Omega_{1\varepsilon\delta})} \le Ce^{-3}; & \|u^{\varepsilon\delta}\|_{H^1(\Omega_{1\varepsilon\delta})} \le Ce^{-3} \end{cases}$$

hence: $e^3 u^{\varepsilon\delta} \to u^{*\varepsilon\delta}$ in $H^1(\Omega_{1\varepsilon\delta})$ weakly with $u^{*\varepsilon\delta} = 0$ on the bottom basis and $\gamma_{\alpha j}(u^{*\varepsilon\delta}) = 0$, $\alpha = 1,2$. These last relations imply formula (1.1). The limit equations (1.2) are derived by using estimates on the first-order moments.

2. LIMIT FOR $\varepsilon \to 0$, WITH δ FIXED

To pass to the limit in system (1.2), we notice that $S_{1\varepsilon\delta}(z_3)$ is ε-periodic and has the form: $S_{1\varepsilon\delta}(z_3) = S_{11\delta}(z_3/\varepsilon) = \Sigma_\delta(z_3/\varepsilon)$. We use homogenization techniques and establish the following result:

Theorem 2.1: *Let* $\varepsilon \to 0$. *Then*

$$V_\alpha^{\varepsilon\delta} \to V_\alpha^\delta \text{ in } H^2(0,L) \text{ weak and } V_3^{\varepsilon\delta} \to V_3^\delta \text{ in } H^1(0,L) \text{ weak.}$$

The V_i^δ *are characterized by the homogenized equations:*

$$\left\{\begin{array}{l} E\, q_\delta^* \dfrac{\partial^4 V_\alpha^\delta}{\partial z_3^4} = -\mathcal{F}_\alpha^\delta + \dfrac{\partial}{\partial z_3} \mathcal{G}_\alpha^\delta,\ (\alpha = 1,2) \text{ in } (0,L) \\ V_\alpha^\delta(0) = \dfrac{\partial V_\alpha^\delta}{\partial z_3}(0) = 0,\ \dfrac{\partial^3 V_\alpha^\delta}{\partial z_3^3}(L) = \dfrac{\partial^2 V_\alpha^\delta}{\partial z_3^2}(L) = 0 \end{array}\right. \tag{2.1}$$

$$\left\{\begin{array}{l} -E\, \tilde{q}_\delta \dfrac{\partial^2 V_3^\delta}{\partial z_3^2} = \mathcal{F}_3^\delta \text{ in } (0,L) \\ V_3^\delta(0) = 0,\ \dfrac{\partial V_3^\delta}{\partial z_3}(L) = 0, \end{array}\right.$$

where $\mathcal{F}_j^\delta = \lim \mathcal{F}_j^{\varepsilon\delta}$ *and* $\mathcal{G}_\alpha^\delta = \lim \mathcal{G}_\alpha^{\varepsilon\delta}$ ($\alpha = 1,2$). *The coefficients* q_δ^* and $\tilde{q}_\delta$ *are given by*

$$q_\delta^* = \int_0^1 \left[\int_{\Sigma_\delta(y)} y_1^2\, dy_1\, dy_2 \right] \frac{\partial^2 w_\delta}{\partial y^2}\, dy \quad \text{and} \quad [\tilde{q}_\delta]^{-1} = \int_0^1 \frac{dy}{\text{meas}(\Sigma_\delta(y))}.$$

The function w_δ *is defined by*

$$\begin{cases} \dfrac{\partial^2}{\partial y^2}\left[\left(\displaystyle\int_{\Sigma_\delta(y)} y_1^2 \, dy_1 \, dy_2\right) \dfrac{\partial^2 \mathcal{W}_\delta}{\partial y^2}\right] = 0 \; in \; (0,1) \\ \mathcal{W}_\delta - \dfrac{1}{2} y^2 \; periodic \; in \; (0,1) \; and \; \displaystyle\int_0^1 (\mathcal{W}_\delta - \dfrac{1}{2} y^2) \, dy = 0. \end{cases}$$

A consequence of Theorem 2.2 is that, if $\underset{\sim}{u}_\alpha^{\varepsilon\delta}$ denotes the extension of $u_\alpha^{\varepsilon\delta}$ by zero, we have

Corollary 2.2. *Let* $\varepsilon \to 0$. *Then*: $\underset{\sim}{u}_\alpha^{\varepsilon\delta} \to u_\alpha^{*\delta}$ *in* $H^2(Q)$, $\underset{\sim}{u}_3^{\varepsilon\delta} \to u_3^{*\delta}$ *in* $H^1(Q)$, *with*

$$u_\alpha^{*\delta} = V_\alpha^\delta \, (\alpha = 1,2) \; and \; u_3^{*\delta} = - z_1 \frac{\partial V_1^\delta}{\partial z_3} - z_2 \frac{\partial V_2^\delta}{\partial z_3} + V_3^\delta,$$

where V^δ *is given by (2.1).*

3. Limit when $\delta \to 0$

A simple calculation in (2.1) gives: $[\tilde{q}_\delta]^{-1} = (2 - 3\delta + 2\delta^2)/(\delta^2(2 - \delta))$, hence $\delta^{-2} \tilde{q}_\delta \to 1$. Moreover we also have $[q_\delta^*]^{-1} = \int_0^1 (1/a(y)) \, dy$, with

$$a(y) = \int_{\Sigma_\delta(y)} y_1^2 \, dy_1 \, dy_2 = \begin{cases} \dfrac{\delta^2}{4}(1 - \delta + \dfrac{\delta^2}{3}) \text{ if } \delta/2 < y < 1 - (\delta/2) \\ \delta\,(1 - \dfrac{3\delta}{2} + \delta^2 - \dfrac{\delta^3}{4}) \text{ if } 0 < y < \delta/2 \text{ or } 1 - (\delta/2) < y < 1. \end{cases}$$

It follows that $\delta^{-2} q_\delta \to 1/4$. To summarize, we have proved

Theorem 3.1: *Under regularity assumptions on the data, we have, when* $\delta \to 0$,

$$V_\alpha^\delta \to V_\alpha^* \; in \; H^2(0,L) \; weakly \; and \; V_3^\delta \to V_3^* \; in \; H^1(0,L),$$

where V^* *is characterized by the equations*:

$$\begin{cases} q^* E \dfrac{\partial^4 V^*_\alpha}{\partial z_3^4} = - \mathcal{F}^*_\alpha + \dfrac{\partial}{z_3} \mathcal{G}^*_\alpha \, (\alpha = 1,2) \text{ in } (0, L) \\ V^*_\alpha(0) = \dfrac{\partial V^*_\alpha}{\partial z_3}(0) = 0, \quad \dfrac{\partial^3 V^*_\alpha}{\partial z_3^3}(L) = \dfrac{\partial^2 V^*_\alpha}{\partial z_3^2}(L) = 0 \end{cases}$$

$$(3.1)$$

$$\begin{cases} - \tilde{q} E \dfrac{\partial^2 V^*_3}{\partial z_3^2} = \mathcal{F}^*_3 \text{ in } (0, L) \\ V^*_3(0) = 0, \quad \dfrac{\partial V^*_3}{\partial z_3}(L) = 0, \end{cases}$$

with: $\mathcal{F}^*_j = \lim \delta^{-2} \mathcal{F}^\delta_j$, $\mathcal{G}^*_\alpha = \lim \delta^{-2} \mathcal{G}^\delta_\alpha$, $(\alpha = 1,2)$ *and*

$$q^* = 1/4, \quad \tilde{q} = 1.$$

Corollary 3.2: *Under the hypotheses of Theorem 3.1,* $u^{*\delta}_\alpha \to u^*_\alpha$ *in* $H^2(Q)$ *weakly* $u^{*\delta}_3 \to u^*_3$ *in* $H^1(Q)$ *weakly with*

$$u^*_\alpha = V^*_\alpha, \; (\alpha = 1,2) \quad u^*_3 = - z_1 \frac{\partial V^*_1}{\partial z_3} - z_2 \frac{\partial V^*_2}{\partial z_3} + V^*_3 .$$

4. Applications to other cases

4.1. Towers with oblique bars (see Fig. 3)

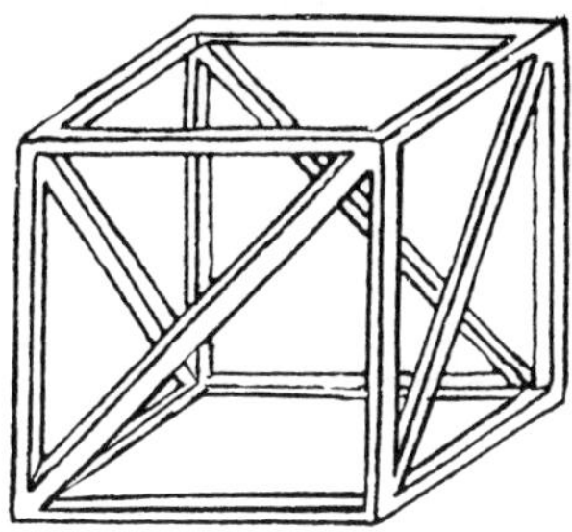

Figure 3. Tower with oblique bars, a representative period

We have the same limit problem as (3.1) with

$$\tilde{q} = 1 + 2\sqrt{2} \quad \text{and} \quad q^* = \frac{\sqrt{2}}{\sqrt{2 - \sqrt{2}}\ \arctan\sqrt{2 - \sqrt{2}}}$$

4.2. Towers where the material is concentrated along layers on the faces of the period (see Fig. 4).

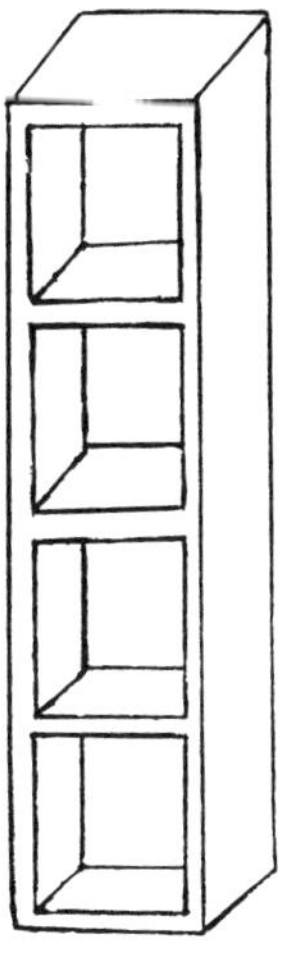

Figure 4.

The limit problem (3.1) still holds with

$$\tilde{q} = 2 \quad \text{and} \quad q^* = 1.$$

References

[1] D. Cioranescu and J. Saint Jean Paulin, Reinforced and honeycomb structures, *J. Math. Pures et Appl.* **65** (1986), 403–422.

[2] D. Cioranescu and J. Saint Jean Paulin, Global behaviour of very thin cellular structures - *Applications to networks, Trends in Applications of Mathematics to Mechanics, Proc. 7th Symposium*, (ed. J. Besseling and W. Eckhauss), Springer Verlag (1988), 26–34.

D. Cioranescu
CNRS-Lab.d'Analyse Numérique
4 Place Jussieu
75252 Paris Cedex 05,
FRANCE

J. Saint Jean Paulin
Département de Mathématiques
Ille de Saulcy
57045 Metz Cedex 01,
FRANCE

P. MAZILU

Hereditary constitutive laws and the principle of inertia

1. Definition of inertial causal dependences

Let us consider the relation between two scalar or vector functions $Y = Y(t)$ and $X = X(t)$, $t \in (-\infty, \infty)$. Assume that $Y(t)$ depends linearly on the past history of X; i.e. it is a linear functional of $\overset{t}{\underset{-\infty}{X}} = \{X(\tau) \mid \tau \leq t\}$

$$Y(t) = \mathcal{F}\Big(\overset{t}{\underset{-\infty}{X}}\Big). \tag{1.1}$$

One says that (1.1) defines an *inertial causal dependence* if $Y(t_0) = 0$ and $X(t) = 0$ for any $t \geq t_0$ implies $Y(t) = 0$ for all $t > t_0$. Some simple examples of inertial or non-inertial causal dependences are given in [2].

A particular class of linear functional dependences is given by Volterra integrals

$$Y(t) = \int_{-\infty}^{t} K(t - \tau)\, X(\tau) \mathrm{d}\tau. \tag{1.2}$$

For scalar functions $X(\cdot)$ and $Y(\cdot)$ a main theorem proved in 1973 (see [1]) states that, provided some conditions of regularity hold, (1.2) forms an inertial causal dependence precisely if $K(\tau)$ is of the form $Ce^{\alpha\tau}$ with α and C real constants. This theorem has been successively refined. In the last years (see [3]) causal dependences of more general form

$$\mathbf{Y}(t) = \mathbf{M}\dot{\mathbf{X}}(t) + \mathbf{N}\mathbf{X}(t) + \int_{-\infty}^{t} \mathbf{L}(t - \tau)\mathbf{X}(\tau)\mathrm{d}r \tag{1.3}$$

have been considered, where one denotes

$$\mathbf{X} = (x_1,\ldots,x_m),\ \ \mathbf{Y} = (y_1,\ldots,y_n),$$

$$\mathbf{M} = (M_{ij}),\ \ \mathbf{N} = (N_{ij}),\ \ \mathbf{L}(t) = (L_{ij}(t)),\ \ t \in (0, \infty),\ \ i = 1,2,\ldots,n, j = 1,2,\ldots,m.$$

For continuous and bounded $L_{ij}(\cdot)$ a theorem states that (1.3) form inertial causal dependences only if

$$\mathbf{L}(t) = e^{-\mathbf{A}t}\mathbf{C}, \tag{1.4}$$

where $\mathbf{A}$ and $\mathbf{C}$ denote $n \times n$ and $n \times m$ constant matrices.

In the following applications of this theorem in particle dynamics and in viscoelasticity will be briefly presented.

2. Particle dynamics

Let us denote by $m, \mathbf{v}$ and $\mathbf{p}$ the mass, velocity and impulse $\mathbf{p} = m\mathbf{v}$ of a particle. According to the principle of causality the impulse $\mathbf{p}(t)$ of the particle at the moment t depends on the past history of the acting force $\mathbf{F}$

$$\mathbf{p}(t) = \mathcal{F}\Big(\overset{t}{\underset{-\infty}{\mathbf{F}}} \Big). \tag{2.1}$$

By $\mathcal{F}$ one denotes a functional which is assumed to be linear and smooth enough to allow the application of the Riesz–Fréchet representation theorem.

The principle of inertia discovered by Galilei states. 'A body remains at rest or in motion with constant velocity if it is not subjected to the influence of forces'. The first part of this principle implies that (2.1) must be an inertial causal dependence in the sense of the above definition. Provided the above conditions of linearity and smoothness hold and under the assumption of isotropy this dependence must have necessarily the form

$$\mathbf{p}(t) = C \int_{-\infty}^{t} e^{\alpha(t-\tau)}\mathbf{F}(\tau)d\tau \tag{2.2}$$

with α and C real constants. In the form of a differential dependence (2.2) reads

$$\frac{d\mathbf{p}}{dt} = C\mathbf{F} + \alpha\mathbf{p}, \ \ \mathbf{p}(-\infty) = \mathbf{0}. \tag{2.3}$$

The second part of the principle of inertia holds only if (1) $\alpha = 0$ or (2) the product $\alpha t = 0$ is negligible.

If $\alpha = 0$ (2.3) reduces to

$$\frac{d\mathbf{p}}{dt} = C\mathbf{F}. \tag{2.4}$$

By a suitable definition of mass one can take $C = 1$. Then (2.4) can be recasted into

$$\frac{\mathrm{d}(m\mathbf{v})}{\mathrm{d}t} = \mathbf{F}. \tag{2.5}$$

Under the supplementary condition that m = const., (2.5) reduces to Newton's second law of mechanics. The general case of variable mass requires particular consideration.

Let us denote by $\mathcal{L}$ the mechanical work defined by

$$\frac{\mathrm{d}\mathcal{L}}{\mathrm{d}t} = \mathbf{F}\mathbf{v}. \tag{2.6}$$

According to the Lenard hypothesis [4] this mechanical work is transformed into mass according to the law

$$\Delta m = \frac{\Delta\mathcal{L}}{c^2}, \tag{2.7}$$

where by c one denotes the velocity of light.

Remark. The idea of energy–mass conversion was first formulated by Newton [5]. The physical and mathematical aspects of this conversion have been investigated by Hasenöhrl [6] and Einstein [7].

Using (2.5) and (2.7) one obtains the following differential equation for the mass determination

$$\frac{\mathrm{d}m}{\mathrm{d}t} = \frac{1}{c^2}\,\mathbf{v}\,\frac{\mathrm{d}(m\mathbf{v})}{\mathrm{d}t}. \tag{2.8}$$

By integrating (2.8) one finds

$$m = \frac{m_0}{\sqrt{1 - v^2/c^2}} \tag{2.9}$$

with m_0 the mass corresponding to $v = 0$. The second law of dynamics reads

$$\frac{\mathrm{d}}{\mathrm{d}t}\left(\frac{m_0\mathbf{v}}{\sqrt{1 - v^2/c^2}}\right) = \mathbf{F}. \tag{2.10}$$

In this way we have derived the second law of dynamics as a consequence of the principles of causality and inertia and without any hypothesis about the time-space structure. In particular the equation (2.10) can be derived by using the classical concept of uniform time as well as the concept of non-uniform (relativisitc) time. This fact will be very useful for performing comparative evaluations between Poincaré's and Einstein's models of relativity.

One remarks that the equation of motion (2.10) has the same form in all inertial frames of reference only if the chrono-geometries of any two reference systems having a constant relative velocity $\mathbf{v}$ are related by the Lorentz transformations. In accordance with the type of time utilized one regains Poincaré's or Einstein's model of relativity. As is known, both of these models use the same equations but differ by the interpretations of the two different times involved in Lorentz's transformations. Actually by using an absolute and a local time Poincaré's model deviates from the principle of relativity because the frame in which these two times coincide has a particular position. We have exhibited a situation in which the axiom of the existence of the uniform time and its negation do not lead to contradictory models. This logical inconsistency disappears if one considers the second case of compatibility with Galilei principle of inertia - the case when the product αt is negligible.

Let us turn back to the general equation (2.5). Using again the above-mentioned Lenard hypothesis one finds the mass equation

$$m = \frac{m_0}{\sqrt{1 - v^2/c^2}}\, e^{-\alpha \int_{-\infty}^{t} \frac{v^2/c^2}{1 - v^2/c^2} \mathrm{d}t} \tag{2.11}$$

The following lemma can be easily proved:

Lemma. *The mass given by* (2.11) *is constant if only if*

$$\mathbf{v} = \mathbf{v}_0\, \mathrm{e}^{\alpha t}. \tag{2.12}$$

Using this lemma in Eq. (2.5) one finds that the velocity (2.12) corresponds to null external force $F = 0$. This means that for $\alpha \neq 0$ the inertial motion implies an exponential increasing or decreasing velocity. The natural question arises of whether behaviour is observed in our universe.

Let us denote by $\mathbf{r}$ the distance between two particles; i.e. the solution of the differential equation

$$\frac{\mathrm{d}\mathbf{r}}{\mathrm{d}t} = \mathbf{v}, \tag{2.13}$$

where by $\mathbf{v}$ one denotes the relative velocity between the particles and t the uniform time. A particular solution of (2.12) reads

$$\mathbf{r} = \frac{\mathbf{v}_0}{\alpha} e^{\alpha t}, \quad \mathbf{v} = \mathbf{v}_0 \, e^{\alpha t}. \tag{2.14}$$

It is easily verified that between the velocity $\mathbf{v}$ and the distance $\mathbf{r}$ the following relation of proportionality holds:

$$\mathbf{v} = \alpha \mathbf{r}. \tag{2.15}$$

One regains in a pure deductive way the famous law of galaxy recessions discovered empirically by Huble in 1929. We have a numerical evaluation of $\alpha : \alpha \approx 10^{-18} s^{-1}$. This means that for t small enough in comparison with 10^{18} s the product αt is actually negligible. By this we have obtained an important result which states that the concept of the uniform time and the principle of inertia are consistent with a universe in expansion.

The relation (2.15) associates to each point in the universe a velocity. This means the requirements of the homogeneity of the universe and those of relativity coincide. Assuming the existence of uniform time, then the homogeneity and hence relativity are ensured if one constructs at each point of the expanding universe a Poincaré model of relativity.

3. Materials with memory

The underlying assumption is that the state of material is completely determined by a finite set of state variables. The principle of inertia states that any given equilibrium, charactrized by equilibrium values of the state parameters, is preserved if these states of parameters remain constant.

For the class of Boltzmann–Volterra materials considered it has been proven (see [1], [2]) that only for Maxwell materials do the stress components form a complete system of state parameters. For other materials the agreement with the principle of inertia can be restored if one introduces, in addition to the stress components, suitable internal parameters. Then the most general form of the Boltzmann–Volterra constitutive law turns out to be (see [3])

$$\begin{pmatrix} \sigma_{11} \\ \sigma_{22} \\ \sigma_{33} \\ \sigma_{12} \\ \sigma_{13} \\ \sigma_{23} \end{pmatrix}(t) = \mathbf{G} \int_{-\infty}^{t} e^{-\mathbf{A}(t-r)}\mathbf{C} \begin{pmatrix} \dot{\varepsilon}_{11} \\ \dot{\varepsilon}_{22} \\ \dot{\varepsilon}_{33} \\ \dot{\varepsilon}_{12} \\ \dot{\varepsilon}_{13} \\ \dot{\varepsilon}_{23} \end{pmatrix}(r)\mathrm{d}r \tag{3.1}$$

where $\mathbf{G}, \mathbf{A}$ and $\mathbf{C}$ are $6 \times n$, $n \times n$ and $n \times 6$ constant matrices respectively; n denotes the number of state parameters (i.e. the number of integral parameters + the number of stress components). For viscoelastic materials of Kelvin-Voigt type the general form of the stress-strain relation in agreement with the principle of inertia reads

$$\dot{\varepsilon}(t) = \mathbf{GM}\sigma(t) + \mathbf{GN}\sigma(t) + \mathbf{G} \int_{-\infty}^{t} e^{-A(t-\tau)}\, \mathbf{C}\sigma(\tau)\mathrm{d}\tau. \tag{3.2}$$

By $\boldsymbol{\sigma}$ and $\dot{\boldsymbol{\varepsilon}}$ one denotes the stress and strain rate written in the same vector form as in Eq. (3.1) and $\mathbf{G}, \mathbf{M}, \mathbf{N}, \mathbf{A}, \mathbf{C}$ are $6 \times n$, $n \times n$ and $n \times 6$ matrices. For Boltzmann-Volterra materials the identification procedure of the matrices $\mathbf{G}, \mathbf{A}$ and $\mathbf{C}$ is presented in [8].

References

[1] Mazilu P. (1973), Sur les lois constitutive de Boltzmann-Volterra. *Comptes Rendus Acad. Sci. Paris*, Serie A, 951-952.

[2] Mazilu P. (1985), Die Onsagerschen Reziprozitätsbeziehungen in der Thermodynamik der Boltzmann-Volterra-Materialien, *ZAMM*, **65**, S.137-149.

[3] *Identifizierung der Zustandsparameter für viskoelastische Materialien*, DFG-Zwieschenbericht, Institut für Mechanik, TH-Darmstadt, January 1985.

[4] Lenard P. (1944), *Deutsche Physik*, Erster Band, J.F. Lehmanns Verlag, München, Berlin.

[5] Newton I. (1719), *Optice: Sive de Reflexibus, Refractionibus, Inflexionibus & Coloribus Lucis*, Libri tres Impensis Gul. & Joh. Innys, Londini, 1719.

[6] Hasenöhrl F. (1904), Theorie der Strahlung in Bewegten Körpern, *Annalen der Physik* **15**, S.344-370.

[7] Einstein A. (1906), Das Prinzip von der Erhaltung der Schwerpunktsbewegung und die Trägheit der Energie, *Annalen der Physik* **20** , S.627-633.

[8] Buggisch H., Mazilu P. and Weber H (1988) Parameter identification for viscoelastic materials, *Rheologica Acta* **27**, S.363-368.

P. Mazilu
Technische Hochschule Darmstadt
Petersenstraße 30
BD-6100 Darmstadt
GERMANY

Part V
Free Section: Significant Interactions of Mathematics and Mechanics

D.R. AXELRAD

Stochastic analysis of the fluid flow in a fully saturated porous medium

1. Introduction

Most formulations of transport phenomena in porous media have been based on the principles of continuum mechanics with the acceptance of empirical relations such as Darcy and the generalized Fick laws. A common feature of these theories is the adoption of an 'appropriate averaging procedure' in order to achieve the transition from the description at the 'pore level', i.e. the microscopic one to the required macroscopic relations [1], [2]. Such a transition is usually carried out by taking mean values over an elementary volume considered to be representative of the material structure. The transition can also be achieved by probabilistic arguments, which are more in line with the experimental observations.

The class of materials considered in this paper represents a two-phase structure in which the solid α-phase (particles) are embedded in the fluid or β-phase. Thus the space between the solid particles or clusters of them is entirely filled by the fluid. This type of structure, as observed by electron microscopy, is schematically shown in Fig. 1.

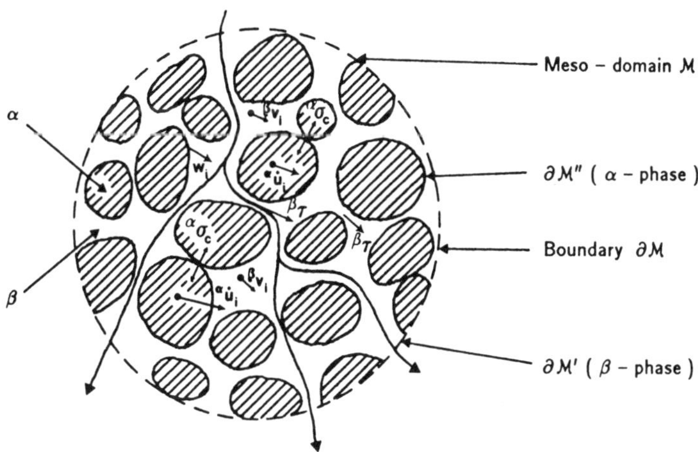

Dispersion of flow structure (deformable α - phase)

$w_i \in W$ — relative velocity of α , β - phases

$\underset{\sim}{\sigma} = \underset{\sim}{\sigma}({}^{\alpha}\underset{\sim}{\sigma}_c , {}^{\beta}\underset{\sim}{T}) : {}^{\alpha}\underset{\sim}{\sigma}_c$ = surface stress, ${}^{\beta}\underset{\sim}{T}$ = viscous stress

${}^{\alpha}\dot{u}_i$ — velocity of α - phase

${}^{\beta}v_i$ — velocity of β - phase

Figure 1. Meso-domain structure and flow characteristics of the β-phase.

There are in general during the flow of the β-phase through a porous medium two sources of microscopic effects. The first is due to chemical reactions and diffusion, particularly if the β-phase consists of several molecular species and the second, which is always present, is caused by the randomness of the 'pore structure'. The latter effect is studied here, which generally induces a non-uniform constraint flow of the β-phase.

2. Stochastic analysis of the β-phase flow

In the formulation of stochastic transport phenomena in general, the basic principles and definitions of probabilistic mechanics [3] can be used. These principles are based on the mathematical theory of probability and the axiomatics of measure theory (see also [3–7]). One of the main concepts of the probabilistic mechanics theory is the introduction of three measuring scales [7] in which the material body as a whole is partioned into:

(i) micro-elements of the structure;
(ii) a statistical ensemble of such elements forming a 'meso-domain' in the material body;
(iii) a denumerable number of non-intersecting meso-domains representing the macroscopic material body.

Hence in the present case concerned with the transport of the β-phase, a micro-element (molecule) of this phase $\beta \in \mathcal{B}$ is defined as a set of mathematical manifolds β representing each element of the fluid structure. A 'configuration' is then the image of β at time t, or

$$\mathbf{x}(\beta, t) \in C \text{ or briefly } {}^{\beta}\mathbf{x}_t \in C \tag{1}$$

where C is the configuration space as a subspace of a probabilistic function space $\mathcal{X}$, which can also be regarded as a state-space [3]. The motion of an element β is defined by

$$\{{}^{\beta}\mathbf{x}(t)\,; \quad -\infty < t < \infty\} \tag{2}$$

in the discrete and time-continuous case, where $\mathbf{x}$ refers to the current position vector of β with respect to the Euclidean frame. In probabilisitc mechanics these motions are in general regarded as 'stochastic processes' designated by $\{{}^{\beta}\mathbf{x}(t)\}$ or ${}^{\beta}\mathbf{x}_t$ for each element β. The state-space, denoted by Z, can be identified with the more general probabilisitic function space $\mathcal{X}$ [3]. It is formed by the state vectors ${}^{\beta}\mathbf{z} \in Z$ defined by a set of r-parameters representing the 'states $i = 1,...,r$' of an element β of the structure. From the probability theory point of view ${}^{\beta}\mathbf{z}$ is an 'outcome or elementary event E' in $\mathcal{X}$ as a result of the 'statistical experiment β'. However, such an event can only be defined by the set of state vectors within a certain range $\Delta^{\beta}\mathbf{z}$ due to experimental constraints and the accuracy with which relevant observations can be carried out.

The events are subsets $E_n \subset \mathcal{X}$ defined by:

$$E_n \stackrel{\Delta}{=} \left\{ \mathbf{z}_n^{(i)} < {}^{\beta}\mathbf{z}_n^{(i)} < \mathbf{z}_n^{(i)} + \Delta\mathbf{z}_n^{(i)} \right\}; \; \bigcup_n E_n = \mathcal{Z} \equiv \mathcal{X} \tag{3}$$

$$E_n \cap E_k = \emptyset, \; (k \in \mathbb{R}^1; \; i = 1,\dots,r)$$

yielding a σ-algebra $\mathcal{F}_n$ as a subalgebra of the σ-algebra $\mathcal{F}$ on $\mathcal{X} = \mathbb{R}^n$. The elements E_n of $\mathcal{F}_n$ are Borel sets which togegher with $\mathcal{F}$ define a measurable space $[\mathcal{X}, \mathcal{F}]$ (see also [3], [4]). An appropriate measure on these subsets is then given by

$$0 \le \mathcal{P}\{E_n\} \le 1; \; \mathcal{P}\{E_n\} = 0, \text{ if } E_n = \emptyset \text{ and } \mathcal{P}\{\mathcal{X}\} = 1. \tag{4}$$

Hence one can designate a probability space $[\mathcal{X}, \mathcal{F}, \mathcal{P}]$ and characterize the flow behaviour by a set of state vectors belonging to $\mathcal{X}$ or in terms of the 'abstract dynamical system $[\mathcal{X}, \mathcal{F}, \mathcal{P}]$'.

In continuum models of the flow of the β-phase, the overall diffusive flux appearing in the mass balance equations is considered to be composed of a molecular diffusion and a mechanical dispersion. The former is usually represented by Fick's law, whilst the mechanical dispersion is often characterized by assuming that its magnitude increases proportional to the Darcy velocity. This involves then the use of a 'dispersivity tensor' in the mass balance relations. In the stochastic formulation an equivalent result can be obtained by the use of the theory of Markov processes. In particular, the constraint flow indicated schematically in Fig. 1 can be represented by a Markov diffusion process or equivalently by a Langevin type equation as shown subsequently. From a molecular fluid dynamics point of view one may assume that the β-phase behaves in a Newtonian manner and that it consists of a collection of undistinguishable particles (i.e. molecules $\beta_1, \beta_2, \dots$).

Thus, considering the unconstraint motion of single particles $\beta_1, \beta_2, \dots$ or the collective motion of an ensemble of such elements within a given meso-domain $\mathcal{M}$, i.e. $\{\beta\} = \mathcal{M}$ as a random endomorphism [3], it will be represented by a set of functions $\{\mathbf{x}(\beta, t)\} = \mathcal{M} \subset \mathcal{C}$ in the configuration space contained in $\mathcal{X}$.

Thus a set of position vectors to the C.M. of each β at time t characerizing the collective mode of motion will also be a one-parameter family of configurational meso-domains such that:

$$\mathcal{M}_t = \{\mathbf{x}(\beta, t) \equiv {}^{\beta}\mathbf{x}_t\}_1^N \subset \mathcal{C} \text{ or } \mathbb{R}^3. \tag{5}$$

In this sense fluid motion (unconstraint) is a mapping:

$$\mathcal{M}_t = G_t\, \mathcal{M}_0; \quad \mathcal{M}_0 = \mathcal{M}_t \text{ at } t = 0. \tag{6}$$

The sets $\mathcal{M}_0$, $\mathcal{M}_t$ are bounded point sets in $\mathbb{R}^3$ and correspond to domains that are occupied by the same fluid elements at time t in $\mathbb{R}^3$. For further details of this transformation see also [3], [8].

However, the fluid motion can be characterized by the velocity field $\mathbf{v}$ as a function of $\mathbf{x}(\beta, t)$. Hence in the hydrodynamic theory one can use the Lagrangian representation in which $\mathbf{x}(\beta, t)$ is the solution of the deterministic differential equation:

$$\frac{\partial^{\beta}\mathbf{x}(t)}{\partial t} = \mathbf{v}(^{\beta}\mathbf{x}, t) \tag{7}$$

with the initial condition: $^{\beta}\mathbf{x}_0$ in $\mathcal{M}_0$ and the velocity vector $\mathbf{v} \in \mathcal{C}^{\infty}(G_t\, \mathcal{M}_0)$. By using the triple $\{\mathcal{V}; \mathcal{F}^{v}, \mathcal{P}^{v}]$ where $\mathcal{V}$ is the velocity space as a subspace of $\mathcal{X}$, and $\mathcal{P}^{v}$ a corresponding measure, the velocity $\mathbf{v} \in \mathcal{V}$ will be a $\mathcal{P}^{v}$-regular measurable function in $\mathcal{V}$. Its expected value is then

$$E\{\mathbf{v}\} = \langle {}^{\beta}\mathbf{v} \rangle = \int_{\mathcal{V} \subset \mathcal{X}} {}^{\beta}\mathbf{v}\, d\mathcal{P}^{v} \tag{8}$$

in which the integral is understood in the Lebesgue-Stieltjes sense. The standard deviation of $\mathbf{v}$ is given by

$$D^{v} = \left\{ \int \left[{}^{\beta}\mathbf{v} - \langle {}^{\beta}\mathbf{v} \rangle \right]^2 d\mathcal{P}^{v} \right\}^{\frac{1}{2}}, \tag{9}$$

which makes the velocity space $\mathcal{V}$ a Hilbert space with the inner product $(^{\beta_1}\mathbf{v}, {}^{\beta_1}\mathbf{v}) = E\{^{\beta_1}\mathbf{v}, {}^{\beta_2}\mathbf{v}\}$. It has been shown in [9] that the simplest models of random behaviour for the flow can be obtained from the theory of Markov processes. In the present case concerned with the 'constraint' flow of the β-phase, the deterministic differential equation (7) will have to be replaced by a set of non-linear stochastic differential equations to account for the dispersion and possible small (molecular) diffusion of the elements of the fluid. This type of flow within an arbitrary meso-domain of the structure is also indicated in Fig. 1.

It has been shown in [10] that for the convenience of the analysis a 'system functional' can be introduced that contains in its argument some of the controlling factors affecting the random flow phenomenon. Hence by characterizing the random flow field by such a vector functional leads to the hydrodynamic description of the flow. In this manner the unobservable velocity field can be related to some of the observable or measurable quantities. It can be expressed formally by

$$\mathbf{f}(\mathbf{v}) \Rightarrow \mathbf{f}\,[{}^{\beta}\mathbf{x}(t), t\,] \Rightarrow \mathcal{F}_{\varepsilon}^{t}\{\theta_{\alpha}, \theta_{\beta}, \boldsymbol{\varepsilon}, \dot{\boldsymbol{\varepsilon}}, \mathbf{q}, \dot{\mathbf{q}}, ...\} \tag{10}$$

in which $\theta_{\alpha}, \theta_{\beta}$ denote the volume fraction of the α– and β–phase, respectively, as elements of a 'control space Θ' contained in $\mathcal{X}$, $\boldsymbol{\varepsilon}, \dot{\boldsymbol{\varepsilon}}$ the volumetric strain and strain rate and $\mathbf{q}$, $\dot{\mathbf{q}}$ the relative discharge and discharge rate. Obviously $\boldsymbol{\varepsilon}, \mathbf{q}$ or their rates are observable quantities belonging to an 'observation space $\mathcal{E}$' (see also [10]).

Hence by using the above vector functional, for the description of the transport of the β–phase and a set of 'control parameters $\boldsymbol{\theta}_t$' relevant to the occurring microscopic effects during the flow, leads to the following non-linear stochastic differential equation:

$$\mathrm{d}^{\beta}\mathbf{x}(t) = \mathbf{f}\,[{}^{\beta}\mathbf{x}(t), t, \boldsymbol{\theta}_t]\mathrm{d}t\,;\ t \geq t_0,\ (t, t_0) \in T. \tag{11}$$

The set of parameters $\boldsymbol{\theta}_t$ implicitly stated above can be recognized as disturbances on the constraint flow of the β–phase and expressed by a function $\underline{g}\,(\mathbf{x}, t)$. It is evident that if the microscopic effect, including small diffusion of the elements, are small and of the order of white noise, one can write for eqn. (11) a Langevin type equation of the form:

$$\frac{\mathrm{d}^{\beta}\mathbf{x}(t)}{\mathrm{d}t} = \mathbf{f}\,[{}^{\beta}\mathbf{x}(t), t] + \underline{g}\,[{}^{\beta}\mathbf{x}(t), t\,]\,\boldsymbol{\theta}_t;\ t \geq t_0, (,t_0) \in T, \tag{12}$$

where $\mathbf{f}$ is the velocity function in (10) characteristic for the constraint flow and $\underline{g}$ an $n \times n$–matrix function with the initial condition $\mathbf{x}(0) = \mathbf{x}_{t_0}$, $t = 0 = t_0$, assumed to be independent of the noise process $\{\boldsymbol{\theta}_t,\ t \geq t_0\}$. However, $\boldsymbol{\theta}_t$ is a delta-correlated random process (see for instance [11]) and not integrable in the mean square sense. For this reason, the formal derivative of $\boldsymbol{\theta}_t$ is usually considered to express a vector process of independent Brownian motion or a Wiener process so that (12) can be written as

$$\mathrm{d}^{\beta}\mathbf{x}(t) = \mathbf{f}\,[{}^{\beta}\mathbf{x}(t), t\,]\,\mathrm{d}t + \underline{g}\,[{}^{\beta}\mathbf{x}(t), t\,]\,\mathrm{d}\mathrm{w}(t), \tag{13}$$

which is an Itô stochastic differential equation. The existence, uniqueness and solution of this equation is extensively treated in the literature (see for instance [12]–[16]). On the assumption that the initial condition for the process ${}^{\beta}\mathbf{x}(t)$ is independent of the Wiener process, the solution of ${}^{\beta}\mathbf{x}(t)$ is a strong Markov or diffusion process in the velocity space $\mathcal{V}$ (see also [10]).

Thus regarding the constraint flow of the β–phase as a Markov diffusion process and recognizing that the corresponding velocity field $\mathbf{v} \in \mathcal{V}$ can be represented by the configuration vector:

$$\mathbf{x} = [\,{}^{\beta}x_1, {}^{\beta}x_2, \ldots, {}^{\beta}x_n]^{\mathrm{T}},$$

the velocity vector:

$$\mathbf{v} = [v_1(x,t),\ v_2(x,t), \ldots, v_n(x,t)]^{\mathrm{T}},$$

one obtains the following Itô vector differential equation:

$$\mathrm{d}\mathbf{x}(t) = \mathbf{v}(\mathbf{x},t)\mathrm{d}t + \underline{\sigma}(\mathbf{x},t)\mathrm{d}\mathbf{w}(t) \tag{14}$$

in which $\mathbf{w}(t)$ is the vector-valued Wiener process and the diffusion matrix $\underline{\sigma}$ or $\{\sigma_{jk}(t\}_{jk\,\leq n}$ corresponds to the previous disturbance function $\underline{g}$ (Eq. 13). By using an operational formalism and the veleocity functional $\mathbf{f}\,[\,{}^{\beta}\mathbf{x}(t), t\,]$, the Kolomogorov equations (see also [10]) can be expressed by:

$$L_s\,\mathbf{f} = \frac{\partial \mathbf{f}}{\partial s} + A_s\,\mathbf{f}\,;\ A_s = \{a\nabla_x + \frac{1}{2}\sum_{j,k=1}^{n} a_{jk}\,\frac{\partial^2}{\partial x_j \partial x_k}\},\ a_{jk} = (\sigma\sigma^{\mathrm{T}})_{jk} \tag{15}$$

and

$$L_t\,\mathbf{f} = \frac{\partial \mathbf{f}}{\partial t} + A_t\,\mathbf{f}\,;\ A_t = \{\nabla_y(a\cdot\mathbf{f}) - \frac{1}{2}\sum_{j,k=1}^{n} a_{jk}\,\frac{\partial^2}{\partial y_j \partial y_k}\} \tag{16}$$

where (15) represents the backward Kolmogorov and (16) the forward or Fokker–Planck equation, respectively (see also [10]). The operators L_s, L_t are also called diffusion operators.

The boundary $\partial\mathcal{M}$ of an arbitrary meso-domain $\mathcal{M}$ (Fig. 1) is given by

$$\partial\mathcal{M} = \partial\mathcal{M}' \cup \partial\mathcal{M}'' : \partial\mathcal{M}' = \bigcup_{k=1}^{m} \partial\mathcal{M}'_k\,;\ \partial\mathcal{M}'' = \bigcup_{j=1}^{p} \partial\mathcal{M}''_j \tag{17}$$

where $\partial\mathcal{M}'$ is that part of the boundary which permits a 'through flow' of the β–phase to any neighbouring domains, whilst at $\partial\mathcal{M}''$ the fluid can only be 'absorbed or reflected'. Denoting the boundary value functions of the Kolomogorov equations by $\mathbf{u}(\mathbf{x}, s)$ and $\mathbf{u}(\mathbf{y}, t)$ respectively, it can be stated that:

(i) Absorption or reflection of the β–phase with small diffusion will occur at $\partial_{\mathcal{M}}''$ if

$$\lim_{s \uparrow t} \mathbf{u}(\mathbf{x}, s) = 0; \ s < t \text{ on } \partial\mathcal{M}'' \cap \partial\mathcal{M}$$

(ii) Through flow of the β-phase and the condition of a subclass of boundary value problems of the parabolic type will occur when

$$\lim_{t \downarrow s} \mathbf{u}(\mathbf{y}, t) = 1; \ t > s \text{ on } \partial\mathcal{M}' \cap \partial\mathcal{M}$$

For a perfectly absorbing boundary at $\partial\mathcal{M}''$ or $\cup\, \partial\mathcal{M}_j''$ of $\partial\mathcal{M}$, one can write the following probability:

$$P\{\mathbf{x}(t) = \mathbf{x}(s) \mid \mathbf{x}(0) = \mathbf{x}_0 \in \mathcal{M}\} \text{ for all } \tau,$$

where τ is referred to as 'first exit time of $\mathbf{x}(t)$' from the domain $\mathcal{M}$ (see also [14]-[16]). Hence for no transition from any point $\mathbf{x} \in \partial\mathcal{M}''$ to any point or configuration $\mathbf{y}$ inside the domain to occur, the probability density must be zero, i.e.:

$$p\{\mathbf{x}, s; \mathbf{y}, t\} = 0 \text{ for all } \mathbf{x} \in \partial\mathcal{M}'', \ \mathbf{y} \in \mathcal{M}, \tag{18}$$

so that the flow is absorbed. The reflec;tion at the boundary $\partial\mathcal{M}''$ is more complex and involves the reflection of the process $\mathbf{x}(t)$ into the domain $\mathbb{R}^n - \mathcal{M}$. The exit problem and that of reflecting boundaries are considered in some detail in [17]-[18].

3. Experimental observations

To illustrate the above stochastic analysis, some of the experimental observations concerning the flow through the medium will be briefly discussed here. It is evident that such observations involve the microstructure as a whole, i.e. both the α and β-phases.

Thus in the simplest case, if a material specimen is subjected to a uni-axial confined compression test, the main observable quantities are the volumetric strain ε_v and the discharge q as elements of $\mathcal{E} \subset \mathcal{X}$. Evidently a more complete description of the material response requires consideration of the motion of the α-phase as well. This necessitates then the use of a relative velocity $\mathbf{w} = {}^{\beta}\mathbf{v} - {}^{\alpha}\dot{\mathbf{u}}$, $\mathbf{w} \in \mathcal{W}$, or relative velocity space contained in $\mathcal{X}$. Furthermore, the load application causes changes in the volumetric fractions $\theta_\alpha, \theta_\beta$ and induces an internal variable σ, which is a function of ${}^{\alpha}\sigma_c$ (surface contact stress) between elements of the α-phase and a viscous stress ${}^{\beta}\tau$ in the fluid phase. This is indicated in Fig. 1. The topological aspects of these subspaces of $\mathcal{X}$ and the corresponding mappings between them have been discussed in [10]. In order to reduce the three-dimensional analysis to the simple case of uni-axial loading one may consider the mean velocity defined in (8), and a simplified system functional (10) in terms of one control parameter only or $\mathcal{F}^t_\varepsilon\{\theta_\alpha\}$.

Since the velocity space $\mathcal{V}$ can be regarded as a Hilbert space the following inner product can be written:

$$< \mathbf{f}(\mathbf{v}),\ \mathcal{P}^v > = \int_{\mathcal{V}\subset\mathcal{X}} \mathbf{f}(\mathbf{v})\, d\mathcal{P}^v \tag{19}$$

in which $\mathcal{P}^v \in \mathcal{M}(\mathbb{R}^n)$ is a set of measures, $\mathbf{f}\ (\mathbf{v}) \in \mathcal{B}_0(\mathbb{R}^n)$ a class of bounded functions on $\mathcal{B}_0$ or the space of all real-valued bounded Borel measurable functions on $\mathbb{R}^n$. Thus letting $\mathbf{v} \Rightarrow v_t \in \mathcal{V}$ the one-dimensional process for the β-phase will be characterized by the probability density:

$$p(v, t) = \int_{t_0}^{t_1} p(v, t)\, dt, \tag{20}$$

where the upper limit of the integral refers to the instant of time at which no measurable flow occurs. At this point the α-phase under the effect of the external load and constraint boundary conditions reaches a maximum of compaction so that only a minimum discharge or no noticeable flow of the β-phase can take place. It has been shown in [10]-[19] that the evolution of the probability density in (20) is of an equivalent form as the diffusion equations, which for a single control parameter or its average becomes

$$\frac{d}{dt} p\,\{\bar{\theta}_\alpha\} = L_\theta\, p\{\bar{\theta}_\alpha\} \tag{21}$$

in which L_θ represents the earlier diffusion operators L_s, L_t, respectively. It is evident that for a constant applied stress the conditions of: $\bar{\sigma} = T_\theta\, \bar{\theta}_\alpha$ and $\bar{\varepsilon}_v = \varepsilon_v(\theta_\alpha)$ will hold, where T_θ is a composition operator discussed in [10]. Since experimentally the evolution with time of the volumetric strain ε_v and the distributions of θ_α can be obtained, the evaluation of the latter can be specified by

$$\theta_\alpha(t) = \int_{t_0}^{t_1} K(t-\tau)\, \{\dot{\theta}_\alpha\}\, d\tau \tag{22}$$

in which the Kernel function can be obtained from

$$\theta_\alpha(t) = \theta^0_\alpha + A\, e^{-Bt}, \tag{23}$$

where the constants θ_α^0 and A, B are evaluated from the initial condition and the rate of change of the experimental values of θ_α immediately after the initial condition. The constant θ_α^0 can be determined from consideration of the 'curve θ_α' to approach the steady-state condition after the time t_1 has been reached.

For the determination of the distributions of the control parameter, scanning-electron microscopy with the subsequent application of quantitative stereology to the obtained micrographs of the cross-section of material specimen have been used. It has been found convenient for the purpose of these tests to employ an aluminium-resin composite consisting of ethylene-glycol-ether as the β-phase in which aluminium particles of $\sim 5\,\mu m$ diameter were inserted as the α-phase.

The numerical evaluation of Eq. (23) shows good agreement with the experimentally determined results (circles) in Fig. 2. For an applied stress of $\sigma = 140$ kPa, the upper time limit t_1 (i.e. no flow condition) was reached at 2.2 minutes.

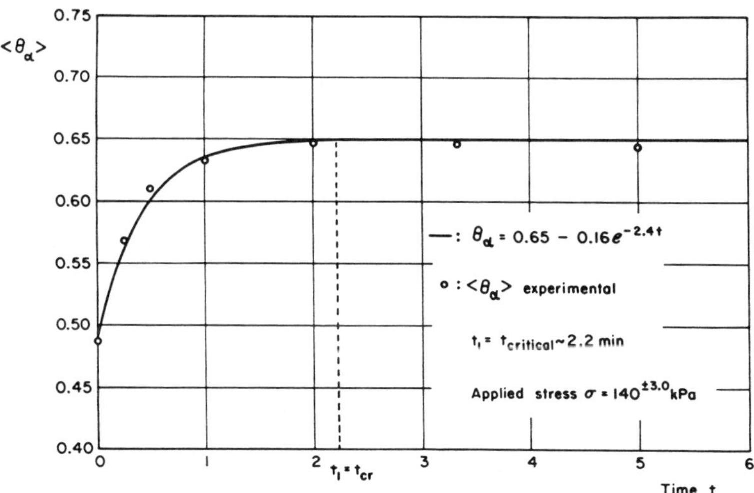

Figure 2. Mean value of θ_α vs. time.

References

[1] J. Baer. *Dynamics of Fluids in Porous Media*. Elsevier, New York (1972).

[2] C.M. Marle. On macroscopic equations governing multi-phase flow with diffusion and chemical reactions in porous media. *Int. J. Eng. Sci.*, **20** (5) (1982).

[3] D.R. Axelrad. *Foundations of the Probabilistic Mechanics of Discrete Media*. Foundations and Philosophy of Science and Technology Series, Pergamon

Press, Oxford (1984).

[4] P.R. Halmos. *Measure Theory*, Van Nostrand, New York (1950).

[5] A.N. Kolmogorov. *Foundations of the Theory of Probability*, Chelsea Publications, New York (1950).

[6] A. Renyi. *Foundations of Probability*, Holden-Day, San Francisco (1970).

[7] D.R. Axelrad. *Random Theory of Deformation of Structured Media*, Lecture Notes **71**, Centre for Mechanical Sciences, Udine, Italy (1971), Springer-Verlag, Vienna, New York (1972).

[8] R.E. Meyer. *Introduction to Mathematical Fluid-Dynamics*, Wiley-Interscience, New York (1971).

[9] D.R. Axelrad. Markov theory in the mechanics of discrete fluids, *Int. J. Eng. Sci*. **20** (2) (1982).

[10] D.R. Axelrad. Stochastic transport theory for porous media, *J. Appl. Math. and Physics*, ZAMP, Birkhauser Verlag, Basel, Vol. 41 March 1990.

[11] W. Horsthemke and R. Lefever. *Noise Induced Transitions: Theory and Application in Physics, Chemistry and Biology*, Springer-Verlag, Berlin (1984).

[12] I.I. Gihman and A.V. Skorohod. *Stochastic Differential Equations*, Springer-Verlag, Berlin (1972).

[13] R.H. Elliott. *Stochastic Calculus and Applications*, Springer-Verlag, Berlin (1982).

[14] E.B. Dynkin. *Markov Processes*, Vol. I, II. Springer-Verlag, New York (1965).

[15] Yu. V. Prohorov and Yu. A. Rozanov. *Probability Theory*. Springer-Verlag, New York (1969).

[16] I.I. Gihman and A.V. Skorohod. *The Theory of Stochastic Processes,* Vol. I, II. Springer-Verlag, Berlin (1975).

[17] R.F. Anderson and S. Orey. Small random perturbations of dynamical systems with reflecting boundary, *Nagoya Math. J.* **60** (1976).

[18] Z. Schuss and B. Matkovsky. The exit problem: a new approach to diffusion across potential barriers, *SIAM J. Appl. Math*. **36** (1979).

[19] D.R. Axelrad. Stochastic analysis of structural changes in Solids. In *Constitutive Laws and Microstructure* Proc. Seminar: Inst. for Advanced Study (Berlin), Springer Verlag (1988).

D.R. Axelrad
Micromechanics Research Laboratory
Department of Mechanical Engineering
McGill University
817 Sherbrooke Street West
Montreal, QC, H3A 2K6
CANADA

A. DI CARLO AND A. TIERO

The geometry of linear heat conduction

> No geometric structure is imposed on the body manifold *a priori*; rather, that structure is *determined by precise mathematical process* in terms of the constitutive assumptions laid down. [1]

0. Introduction

The so-called *heat equation*

$$\dot{\theta} = \Delta\theta + s \tag{1}$$

is introduced on a Riemannian manifold for the sake of studying the geometrical properties of the associated Laplace-Beltrami operator Δ (see, for instance, the recent introductory survey by Chavel [2]). A meagre physical motivation is usually adjoined (see again [2], pp. 134-5), which runs more or less as follows. You are presented with a smooth manifold M, already endowed with some God-given Riemannian structure. Now, you consider that manifold as a *homogeneous* and *isotropic* conducting medium, and define $\theta(x, t)$ and $s(x, t)$, respectively, as the temperature and the heating supply of $x \in M$ at time t. Then, from the postulation of a balance principle, you get - *modulo* some further (implied) hypothesis relating internal energy to temperature - the evolution equation (1).

We dislike this lack of balance between the wealth and depth of analytical and geometrical results gained through the study of the heat equation on Riemannian manifolds, and its dim and unduly special physical foundation. What is most distasteful to us is that, seemingly, a Riemannian structure should be imposed on the manifold even *before* constitutive assumptions could be stated: you are not entitled to stipulate that the medium M is homogeneous without a connection on M, nor that it is isotropic at a point x without an inner product on the tangent space $T_x M$. Moreover, what about *in*homogeneous and/or *an*isotropic conductors?

In the following sections, we show how to remove the limitations in clarity and scope of the customary presentation of the heat equation. To this end, we strive to state the assumptions behind the standard linear theory of heat conduction (and drift) with non-standard explicitness in an uncommonly weak geometrical setting. At this stage (beginning of Section 2), we view the medium as a mere differentiable manifold, and describe the phenomenon through a number of differential forms, and pointwise linear mappings between such forms. We conclude Section 2 by showing that the constitutive assumptions laid down do determine a Riemannian metric.

Within Section 3, we obtain a generalized heat operator, whose principal part is just the Laplacian associated with the intrinsic Riemannian sturcture previously defined. Some further hypotheses - which are usually taken for granted - finally boil down our generalized heat equation to the standard form (1).

For the sake of giving an orderly presentation, in Section 1 we assemble a compact array of purely algebraic implements, to rely upon in the following sections. We would nevertheless advise the reader who wants motivation to skip over this section, and come back to pay it a visit once in a while. Being short of space, we do not give any detailed proof, but only a few hints. It turns out, however, that the tough part of this game is to imagine the right conjectures; then, to prove them is altogether elementary, if not trivial. The careful reader should also draw - for his, or her, benefit - the diagrams we omit.

1. Algebraic preliminaries

Let V be a real, finite dimensional vector space, and let $n := \dim V$. We shall denote by A^r and W_r, respectively, the spaces of alternating covariant and contravariant tensors of order r on V (r-forms and r-vectors, for short). The space of r-forms is (identified with) the dual of the space of r-vectors: we shall write $<\cdot\,,\cdot>$ for any member of the corresponding hierarchy of duality pairings (of course, W_1 and A^1 are but serialized bynames for V itself and its dual).

We shall also consider the following spaces of linear maps:

$$\mathrm{lin}_r := \mathrm{Hom}\,(A^r, W_r), \quad \mathrm{LIN}_r := \mathrm{Hom}\,(A^r, A^{n-r})\,, \tag{2}$$

and presume lin_r to be identified with $W_r \otimes W_r$ (the space of bilinear forms on A^r) through

$$(\forall \alpha, \beta \in A^r)\ l\,(\alpha, \beta) = <\alpha, l\,\beta>. \tag{3}$$

Proposition 1. *The spaces A^{n-r} and W_r are naturally isomorphic: to each volume element (i.e. nonzero n-form) μ there corresponds the Weyl isomorphism $\iota_\mu : A^{n-r} \to W_r$, defined by*

$$(\forall \alpha \in A^r)\,(\forall \beta \in A^{n-r})\ \ <\alpha, \iota_\mu\,\beta>\mu = \alpha \wedge \beta\,. \tag{4}$$

Moreover, for any nonzero scalar ρ, $\iota_{\rho\mu} = \rho^{-1}\,\iota_\mu$.

Remark. The Weyl isomorphism is to the Hodge star isomorphism as the duality pairing is to the inner product. Nonetheless, it is far less mentioned in the literature: after rediscovering it, we came across the concept and its attribution to Weyl only in the broad textbook by Maurin [3].

Proposition 2. *The spaces* LIN_r *and* lin_r *are naturally isomorphic: the function that maps* $L \in \mathrm{LIN}_r$ *to its* μ-*correspondent* $l_\mu := \iota_\mu L \in \mathrm{lin}_r$ *is an isomorphism. Moreover, for any nonzero scalar* ρ, $l_{\rho\mu} = \rho^{-1} l_\mu$.

Remark. Through the identification (3) and the isomorphism just defined, each bilinear form on A^r is related to a unique element of LIN_r, and vice versa. In particular, *nondegenerate* bilinear forms are related to *bijective* maps.

Proposition 3. *For each* $L \in \mathrm{LIN}_r$, *there exists a unique* $\hat{L} \in \mathrm{LIN}_r$ - *which we name its wedge-adjoint - such that*

$$(\forall \alpha, \beta \in A^r) \quad \alpha \wedge L\beta = \beta \wedge \hat{L}\alpha . \tag{5}$$

Moreover, if you let $\mathrm{SYM}_r := \{L \in \mathrm{LIN}_r \mid \hat{L} = L\}$, $\mathrm{SKW}_r := \{L \in \mathrm{LIN}_r \mid \hat{L} = -L\}$, *then the following decomposition holds:*

$$\mathrm{LIN}_r = \mathrm{SYM}_r \oplus \mathrm{SKW}_r . \tag{6}$$

Hint. The wedge-adjoint of L is the μ-correspondent of the dual of the μ-correspondent of L. Note, however, that the map $L \mapsto \hat{L}$ and the decomposition (6) are independent of the choice of any particular volume element.

Proposition 4. *Given a volume element* μ, *each bijection* $L \in \mathrm{LIN}_r$ *induces on* W_r *the nondegenerate bilinear form*

$$g := (\iota_\mu L)^{-1}, \tag{7}$$

which is symmetric if, and only if, $L \in \mathrm{SYM}_r$. We say that L *is* μ-*positive definite when its* μ-*correspondent is positive definite: in this case,* g *is a Euclidean inner product on* W_r.

Remark. The most important case is $r = 1$, for then the inner product structure is induced on the underlying vector space V. This result - and the next, which is peculiar to $r = 1$ - will play a crucial role in the following treatment.

Proposition 5. *The subspace* SKW_1 *is canonically isomorphic to* A^{n-2}: *each* $(n - 2)$-*form* ψ *may be identified with the unique homomorphism* $L_\psi \in \mathrm{SKW}_1$ *such that*

$$(\forall \alpha \in A^1) \;\; L_\psi \alpha = - \alpha \wedge \psi . \tag{8}$$

2. Balance principle and constitutive assumptions

We envisage the conducting medium as an orientable smooth manifold M of dimension n ($< \infty$), devoid - for the moment - of any further structure. The space of differential r-forms on M will be denoted by Λ^r, while A_x^r will denote the space of external r-forms built on $T_x M$: if $\alpha \in \Lambda^r$, then $\alpha(x) \in A_x^r$, for any $x \in M$. Under standard smoothness assumptions, the balance of energy admits the following localization (a superposed dot denoting time rates):

$$\dot{\varepsilon} = \sigma + \mathrm{d}\varphi , \tag{9}$$

where ε and σ are the n-forms that, integrated on any finite domain $P \subset M$, give respectively the *internal energy* and the *body heating* of P, while φ is the $(n - 1)$-form that gives the *contact heating* of P, when integrated on its boundary ∂P.

We admit that the state of the medium is adequately described by the 0-form θ, giving the difference between the actual and a (fixed) reference *temperature* field, and characterize its behaviour through the following constitutive assumptions.

Assumption 1. *Let the present values of* ε, σ *and* φ *depend only on the present value of* θ.

Assumption 2. *Let the external n-forms* $\varepsilon(x)$, $\sigma(x)$ *at a point* $x \in M$ *depend only on* $\theta(x)$, *and the point* x *itself*: $\varepsilon(x) = \hat{\varepsilon}(\theta(x); x)$, $\sigma(x) = \hat{\sigma}(\theta(x); x)$.

Assumption 3. *Let the external* $(n - 1)$*-form* $\varphi(x)$ *at a point* $x \in M$ *depend only on the values of* θ *and its differential* $\mathrm{d}\theta$ *at* x, *and* x *itself*: $\varphi(x) = \hat{\varphi}(\theta(x), \mathrm{d}\theta(x); x)$.

Assumption 4. *Let the constitutive functions* $\hat{\varepsilon}(\cdot\, ; x)$, $\hat{\sigma}(\cdot\, ; x)$, $\hat{\varphi}(\cdot\, , \cdot; x)$ *be affine, for all* $x \in M$.

Remark. Without loss of generality, we may assume the functions $\hat{\varepsilon}(\cdot\, ; x)$, $\hat{\sigma}(\cdot\, ; x)$, to be linear: were it not so, we could subtract from them their values at $\theta = 0$, while adding $-\dot{\hat{\varepsilon}}(0; \cdot) + \mathrm{d}\hat{\varphi}(0, 0; \cdot)$ to $\sigma_0 := \hat{\sigma}(0; \cdot)$, without affecting the balance equation (9). Hence, we arrive at the following

Representation theorem. *The above constitutive assumptions* (1)-(4) *are satisfied if and only if*

$$\varepsilon = \theta \wedge \mu , \quad \sigma = \sigma_0 + \theta \wedge \tau , \quad \varphi = \theta \wedge \omega + L\, \mathrm{d}\theta , \tag{10}$$

where L *acts as a pointwise linear map from* 1*-forms to* $(n-1)$*-forms*:

$$(\forall \alpha \in \Lambda^1)\,(\forall x \in M)\ (L\alpha)(x) := L(x)\,\alpha(x),\ L(x) \in \mathrm{Hom}\,(A^1_x, A^{n-1}_x). \qquad (11)$$

Remark. The first and second terms on the right-hand side of (10_3) are labelled as *convective* and *conductive*, respectively: we call ω the *drift* form, and L the *conductivity* field.

By substituting (10) into (9), the following evolution equation for θ is obtained:

$$(\theta \wedge \mu)^{\cdot} = \sigma_0 + \theta \wedge \tau + \mathrm{d}(\theta \wedge (\omega + \mathrm{d}\psi) + L_+\,\mathrm{d}\theta), \qquad (12)$$

after L has been decomposed into its *symmetric* and *skew* parts (see Proposition 3 in Section 1):

$$(\forall x \in M)\ L_+(x) := \mathrm{sym}\, L(x) \equiv \tfrac{1}{2}(L(x) + \hat{L}(x)),\ L_-(x) := \mathrm{skw}\, L(x) \equiv \tfrac{1}{2}(L(x) - \hat{L}(x)), \qquad (13)$$

and the skew part L_- has been identified with a *stream-function* ψ (by virtue of Proposition 5 in Section 1). It is of paramount importance that, because of this, L_- yields no net contribution to the principal part of the right-hand side of (12), viewed as a differential operator acting on θ. Secondly, the structure of the left-hand side of equation (12) strongly invites the next assumption, both on analytical and physical grounds.

Assumption 5. *Let the* n*-form* μ *in* (10_1) *be a volume form:* $\mu(x) \neq 0$, *for all* $x \in M$.

Remark. We call μ the *energy volume form.* Then, recalling the definition of μ-definiteness given in Section 1 (Proposition 4), we state the following :

Assumption 6. *Let the symmetric part of the conductivity* L_+ *be positive definite with respect to the energy volume form* μ .

In conclusion, we endow the medium M with the intrinsic Riemannian metric

$$g := (\iota_\mu L_+)^{-1}, \qquad (14)$$

where ι_μ is the Weyl isomorphism corresponding to the energy volume form. We call g the *conductivity metric*, and *conductivity volume form* the g-volume μ_g that belongs to the same orientation as μ ; the (positive) ratio between the energy and conductivity volume forms will be denoted by ρ :

$$\mu = \rho\, \mu_g \,, \tag{15}$$

while ι_g (instead of the cumbersome ι_{μ_g}) will denote the Weyl isomorphism corresponding to the conductivity volume form: because of (15), one has (recall Proposition 1 in Section 1)

$$\iota_g = \rho\, \iota_\mu \,. \tag{16}$$

3. The generalized heat equation

We define the μ-*divergence* of a vector field X on M through:

$$(\mathrm{div}_\mu X)\, \mu = \mathrm{d}(\iota_\mu^{-1} X), \tag{17}$$

and call simply *g-divergence* the μ_g-divergence of X. It turns out that

$$\rho\, \mathrm{div}_\mu X = \mathrm{div}_g(\rho X) = \rho\, \mathrm{div}_g X + g(\mathrm{grad}\, \rho\, , X), \tag{18}$$

where the *gradient* operator is defined as grad := $g^{-1}\mathrm{d}$. Equation (12) may hence be rewritten in the form:

$$\dot{\theta} = \Delta_g \theta + \rho^{-1}\, [\, g\, (\mathrm{grad}\, \rho\, , \mathrm{grad}\, \theta) + \mathrm{div}_g(\theta W)\,] + s + \theta t\, , \tag{19}$$

which we call the *generalized heat equation*, where $\Delta_g := \mathrm{div}_g\, \mathrm{grad}$ is the *Laplacian* associated with the conductivity metric g, while the *intrinsic driving vector field* W is defined as $W := \iota_g(\omega + \mathrm{d}\varphi)$, and the *intrinsic heating supply* $\theta t + s$ comes from the positions: $\sigma_0 = s\mu$, $\tau - \dot{\mu} = t\, \mu$ (we shall call the n-form $t\, \mu$ the *net specific body heating*). Let us now state a few simplifying hypotheses, restricting the scope of the constitutive assumptions considered so far.

Hypothesis 1. *Suppose the intrinsic driving vector field W to vanish identically.*

Remark. Compare with 'Casimir's reciprocal relation', as presented and discussed by Truesdell [4] (p. 373 ff.). Under this hypothesis, the operator on the right-hand side of (19) is self-adjoint.

Hypothesis 2. *Suppose the net specific body heating $t\, \mu$ to be identically zero.*

Hypothesis 3. *Suppose the medium to be intrinsically homogeneous, that is the ratio ρ between the energy and the conductivity volume forms to be constant over M.*

Concluding remark. Note that the last hypothesis is less restrictive that the standard assumption of homogeneity with respect to a given foreign metric. Of course, under hypotheses 1-3, the generalized heat equation (19) reduces to the standard heat equation (1).

References

[1] C. Truesdell, Preface to *Continuum Theory of Inhomogeneities in Simple Bodies*, a reprint of six memoirs by W. Noll, R.A. Toupin, and C.-C. Wang. Springer-Verlag, Berlin, 1968.

[2] I. Chavel, *Eigenvalues in Riemannian Geometry*, Academic Press, Orlando, Florida, 1984.

[3] K. Maurin, *Analysis* (revised and enlarged English edition), Part II. PWN-Polish Scientific Publishers, Warszawa, 1980.

[4] C. Truesdell, *Rational Thermodynamics* (second edition, corrected and enlarged), Lecture 7: The Onsager Relations. Springer-Verlag, New York, 1984.

A. Di Carlo and A. Tiero
Università di Roma "La Sapienza"
Dipartimento di Ingegneria Strutturale
Via Eudossiana, 18
I-00184 Roma
ITALY

T.M. FISCHER

Theoretical and numerical investigation of boundary-layer instabilities

1. Introduction

Different types of instabilities are observed in boundary-layer flows. These instabilities are produced by two and three-dimensional disturbances in a two-dimensional boundary layer as, for instance, in Blasius flow over a flat plate, whereas a three-dimensional boundary layer is governed by stationary as well as non-stationary disturbances. However, in both cases, we can distinguish between primary and secondary disturbances, where the primary disturbances develop from a linear instability of the basic flow to a nonlinear saturation state, which then becomes unstable against another type of disturbances, the so-called secondary disturbances (cf. [5] [8]). Our aim is to put this conception on a mathematical basis. In particular, we have to develop an appropriate nonlinear model, which describes the saturation state of certain disturbances. Details will appear in [4]. Furthermore, for solving the various stability problems, we need an efficient and reliable numerical method.

In the present paper, we shall concentrate on the mathematical analysis of the fundamental linear eigenvalue problem, i.e. the *Orr-Sommerfeld equation*:

$$\mathcal{L}(\omega)\varphi := \frac{1}{\mathrm{Re}}\left(\frac{\mathrm{d}^4\varphi}{\mathrm{d}z^4} - 2\alpha^2\frac{\mathrm{d}^2\varphi}{\mathrm{d}z^2} + \alpha^4\varphi\right) - \mathrm{i}\alpha\left[U\left(\frac{\mathrm{d}^2\varphi}{\mathrm{d}z^2} - \alpha^2\varphi\right) - \frac{\mathrm{d}^2U}{\mathrm{d}z^2}\varphi\right]$$

$$+\,\mathrm{i}\omega\left(\frac{\mathrm{d}^2\varphi}{\mathrm{d}z^2} - \alpha^2\varphi\right) = 0 \text{ in } 0 < z < \infty, \tag{1.1}$$

with the *boundary conditions*,

$$\varphi(0) = \frac{\mathrm{d}\varphi}{\mathrm{d}z}(0) = 0 \text{ and } \varphi(z),\ \frac{\mathrm{d}\varphi}{\mathrm{d}z}(z) \to 0 \text{ as } z \to \infty. \tag{1.2}$$

Here the parameters Re, $\alpha > 0$, denote the Reynolds number and the disturbance wave number, respectively. The boundary-layer profile U is a given, smooth function of the wall-normal coordinate z. The complex circular frequency ω and the amplitude function φ of the disturbance have to be determined as an eigenvalue and a corresponding eigenfunction, respectively.

The Orr-Sommerfeld eigenvalue problem, which is defined on a finite interval, has been investigated in [2] (see also the references therein). There it has been proved that the spectrum is discrete and that the eigenfunctions are complete in a certain Hilbert space. The proof is based on a compactness theorem, which is not available in the semi-infinite case. However, the Orr-Sommerfeld operator, which is defined by the eigenvalue problem (1.1), (1.2), can be decomposed into a main part and a compact remainder, and we shall show that the spectrum is discrete outside a certain one-dimensional subset of the complex plane. Moreover, using a theorem by Rannacher [12] on the stability of isolated eigenvalues, we shall be able to refine some previous convergence results [3] concerning the Galerkin approximation of the problem (1.1), (1.2).

The set of eigenvalues of the Orr-Sommerfeld eigenvalue problem (1.1), (1.2) has been studied recently in [10], [11]. Whereas the characterization of the spectrum which is given there partially exceeds that of the present paper, our approach is less tricky and has the advantage that it can be applied to a reasonable approximation scheme.

2. The linear eigenvalue problem

Let L^2 be the Hilbert space of all measurable, square-integrable functions on the semi-infinite interval $(0, \infty)$. Let H^2 be the Sobolev space of all functions in L^2, for which the first and second weak derivatives exist and are in L^2 [1]. Furthermore, let C_0^∞ be the space of all functions, which are differentiable of infinite order and have compact supports in $(0, \infty)$. Let H_0^2 be the closure of C_0^∞ in H^2. Then we consider the following weak formulation of the Orr-Sommerfeld eigenvalue problem:

Find $\omega \in \mathbb{C}$ *and* $\varphi \in H_0^2$, $\varphi \neq 0$, *such that*

$$(\mathcal{L}(\omega)\varphi, \psi)_{L^2} = 0 \quad (\psi \in H_0^2), \tag{2.1}$$

where

$$(\mathcal{L}(\omega)\varphi, \psi)_{L^2} = \frac{1}{\mathrm{Re}} \int_0^\infty \left(\frac{d^2\varphi}{dz^2} \frac{d^2\overline{\psi}}{dz^2} + 2\alpha^2 \frac{d\varphi}{dz} \frac{d\overline{\psi}}{dz} + \alpha^4 \varphi \overline{\psi} \right) dz$$

$$+ i\alpha \left[\int_0^\infty U \left(\frac{d\varphi}{dz} \frac{d\overline{\psi}}{dz} + \alpha^2 \varphi \overline{\psi} \right) dz - \int_0^\infty \frac{dU}{dz} \varphi \frac{d\overline{\psi}}{dz} dz \right]$$

$$- i\omega \int_0^\infty \left(\frac{d\varphi}{dz} \frac{d\overline{\psi}}{dz} + \alpha^2 \varphi \overline{\psi} \right) dz \tag{2.2}$$

is obtained formally from (1.1), (1.2) by using integration by parts (cf. [3]). With regard to the boundary-layer profile U, we assume that there exist real numbers U_∞ and $q \geq 2$ such that

$$(U - U_\infty)^{q/2}, \frac{dU}{dz}, \frac{d^2U}{dz^2} \in L^2. \tag{2.3}$$

According to the Sobolev imbedding theorem [1], condition (2.3) implies, in particular, that $\lim_{z \to \infty} U(z) = U_\infty$ and

$$\sup_{0 \leq z < \infty} |U(z)| < \infty, \quad \mu := \frac{1}{2} \sup_{0 \leq z < \infty} \left| \frac{dU}{dz}(z) \right| < \infty \tag{2.4}$$

hold.

The bilinear form defined by (2.2) is continuous on $H_0^2 \times H_0^2$ for every $\omega \in \mathbb{C}$. Consequently, there exists a uniquely determined continuous, linear operator $L(\omega)$: $H_0^2 \to H_0^2$, with

$$(L(\omega)\varphi, \psi)_{H^2} = (\mathcal{L}(\omega)\varphi, \psi)_{L^2} \quad (\varphi, \psi \in H_0^2), \tag{2.5}$$

and the weak Orr–Sommerfeld eigenvalue problem (2.1) is equivalent to the following problem:

Find $\omega \in \mathbb{C}$ *and* $\varphi \in H_0^2$, $\varphi \neq 0$, *such that*

$$L(\omega)\varphi = 0. \tag{2.6}$$

Every $\omega \in \mathbb{C}$, for which $\text{imag}(\omega) \geq \mu$ with μ from (2.4), belongs to the resolvent set of L [3]. Moreover, the operator L admits the representation

$$L(\omega) = L_0(\omega) + K, \quad L_0(\omega) := A_0 - i\omega B_0 \quad (\omega \in \mathbb{C}), \tag{2.7}$$

where $A_0, B_0, K : H_0^2 \to H_0^2$ are continuous, linear operators defined by

$$(A_0\varphi, \psi)_{H^2} = \frac{1}{\mathrm{Re}} \int_0^\infty \left(\frac{d^2\varphi}{dz^2} \frac{d^2\bar{\psi}}{dz^2} + 2\alpha^2 \frac{d\varphi}{dz} \frac{d\bar{\psi}}{dz} + \alpha^4 \varphi \bar{\psi} \right) dz$$

$$+ i\alpha\, U_\infty \int_0^\infty \left(\frac{d\varphi}{dz} \frac{d\bar{\psi}}{dz} + \alpha^2 \varphi \bar{\psi} \right) dz \quad (\varphi, \psi \in H_0^2), \tag{2.8}$$

$$(B_0\varphi, \psi)_{H^2} = \int_0^\infty \left(\frac{d\varphi}{dz} \frac{d\bar{\psi}}{dz} + \alpha^2 \varphi \bar{\psi} \right) dz \quad (\varphi, \psi \in H_0^2), \tag{2.9}$$

$$(K\varphi, \psi)_{H^2} = i\alpha \left[\int_0^\infty (U - U_\infty) \left(\frac{d\varphi}{dz} \frac{d\bar{\psi}}{dz} + \alpha^2 \varphi\psi \right) dz - \int_0^\infty \frac{dU}{dz} \varphi \frac{d\bar{\psi}}{dz} dz \right]$$

$$(\varphi, \psi \in H_0^2). \tag{2.10}$$

Thus, we have decomposed L into a main part L_0 and a remainder K. By using (2.3), the operator K can be shown to be compact [4]. In addition, the inequality

$$(L_0(\omega)\varphi, \varphi)_{H^2} \neq 0 \quad (\varphi \in H_0^2,\ \varphi \neq 0) \tag{2.11}$$

holds and $L_0(\omega)$ has an inverse whenever $\mathrm{real}(\omega) \neq \alpha U_\infty$ or $\mathrm{imag}(\omega) > -\alpha^2/\mathrm{Re}$. We then obtain the following result:

Theorem 1. *Let* $\sigma(L)$ *denote the spectrum of* L *and let*

$$\theta_0 := \left\{ \omega \in \mathbb{C} \mid \mathrm{real}(\omega) = \alpha U_\infty \text{ and } \mathrm{imag}(\omega) \leq -\frac{\alpha^2}{\mathrm{Re}} \right\}. \tag{2.12}$$

Then the intersection of $\sigma(L)$ *and* $\mathbb{C}\backslash\theta_0$ *consists entirely of isolated eigenvalues with finite algebraic multiplicities. Furthermore, for every* $\omega \in \sigma(L)$, *we have*

$$|\,\mathrm{real}(\omega) - \alpha U_\infty| \leq \alpha \sup_{0 \leq z < \infty} |\,U(z) - U_\infty| + \mu\,,$$

$$\mathrm{imag}(\omega) \leq -\frac{\alpha^2}{\mathrm{Re}} + \mu\,, \tag{2.13}$$

with μ *from* (2.4).

Similarly, the set of eigenvalues of the Orr–Sommerfeld eigenvalue problem (1.1), (1.2) has been described in [11] (see also [10]). There the analysis is based on an integro–differential operator formulation of the linearized disturbance differential equations. Instead of (2.3), the assumptions $U - U_\infty$, $dU/dz \in L^2$ are used. Moreover, it has been proved [10] that, if the boundary–layer profile approaches a constant exponentially at infinity, then the number of eigenvalues is finite. This is what is observed in the numerical calculations (cf. [9]). The exception set θ_0 is also part of the spectrum [11]. A certain characterization of θ_0 can be found in [7].

3. The Galerkin approximation

Let $\varphi^{(n)}$, $n = 1,2,...,$ be a complete set of trial functions in H_0^2. Let N be a positive integer and let H_N denote the finite–dimensional space which is spanned by $\varphi^{(1)},...,\varphi^{(N)}$. – In our numerical computations, we use Jacobi polynomials, which are combined with an exponential mapping of the semi–infinite interval onto a finite interval, as basic components of the trial functions (see [3] and the references therein). Furthermore, let Q_N be the orthogonal projection $H_0^2 \to H_N$. Then the Galerkin approximation of the eigenvalue problem (2.1) is given by the following discrete problem:

Find $\omega_N \in \mathbb{C}$ *and* $\varphi_N \in H_N$, $\varphi_N \neq 0$, *such that*

$$(\mathcal{L}(\omega_N)\varphi_N,\ \psi_N)_{L^2} = 0 \quad (\psi_N \in H_N). \tag{3.1}$$

Let $L_N(\omega_N) : H_N \to H_N$ denote the continuous, linear operator, which is uniquely determined by the bilinear form in (3.1). We then have the representation

$$L_N(\omega_N) = Q_N L(\omega_N) \mid_{H_N}, \tag{3.2}$$

and there is a decomposition of this operator analogous to that of L in (2.7). Accordingly, the general convergence results of [12] (see also [13]) concerning the stability of isolated eigenvalues can be applied, and we obtain:

Theorem 2. *Let* θ_0 *be defined as in Theorem* 1.

(i) *If* $\{\omega_N\} \subset \mathbb{C}$ *is a sequence of eigenvalues of the Galerkin Eqs.* (3.1), $N = 1,2,...,$ *for which a subsequence converges to* $\omega \in \mathbb{C}\backslash\theta_0$, *then* ω *is an eigenvalue of the weak Orr-Sommerfeld eigenvalue problem* (2.1).

(ii) *If* $\omega \in \mathbb{C}\backslash\theta_0$ *is an eigenvalue of the equation* (2.1), *then there exists a sequence* $\{\omega_N\} \subset \mathbb{C}$ *of eigenvalues of the discrete eigenvalue problems* (3.1), $N = 1,2,\ldots,$ *which converges to* ω.

We remark that, in addition, the eigenprojections on the algebraic eigenspaces for ω_N, $N = 1,2,\ldots,$ can be shown to be discrete convergent [12].

Furthermore, as a direct consequence of these convergence results, we have [6]:

Theorem 3. *Let* $\omega \in \mathbb{C}\backslash\theta_0$ *be an eigenvalue of* L, *let* V *be a basis of the algebraic eigenspace for* ω, *and let* p *denote the order of the pole in the Laurent expansion of* $L^{-1}(\cdot)B_0$ at ω. *Assume that* $\{\omega_N\} \subset \mathbb{C}$ *is a sequence of eigenvalues of the operators* L_N, $N = 1,2,\ldots,$ *which converges to* ω. *Then the following estimate holds:*

$$|\omega - \omega_N|^p = O(\max_{\varphi \in V} \|\varphi - Q_N\varphi\|_{H^2}). \tag{3.3}$$

In particular, if ω is a simple eigenvalue of L, then $p = 1$, and the accuracy of the Galerkin approximation is determined by the order of the truncation error.

Acknowledgement

The author is indebted to Prof. Dr K. Kirchgässner and Prof. Dr W.L. Wendland for their constant encouragement.

References

[1] Adams, R.A. *Sobolev Spaces*, New York, Academic Press, 1975.

[2] Di Prima, R.C. and Habetler, G.J. A completeness theorem for non selfadjoint eigenvalue problems in hydrodynamic stability. *Arch. Rat. Mech. Anal* **34** (1969), 218–227.

[3] Fischer, T.M. A Galerkin approximation for linear eigenvalue problems in two and three-dimensional boundary-layer flows. In: *The Navier-Stokes Equations: Theory and Numerical Methods* (J.G. Heywood et al. eds.), pp. 100–108, Berlin, Springer-Verlag, 1990.

[4] Fischer, T.M. Ein mathematisch-physikalisches Modell zur Beschreibung transitioneller Grenzschichtströmungen. Report DLR-FB 90-24 (1990) (with corrigendum).

[5] Fischer, T.M. and Dallmann, U. Primary and secondary stability analysis applied to the DFVLR-transition swept-plate experiment, *AGARD Conf. Proc. No. 438* (1989), Ref. 15.

[6] Grigorieff, R.D. Diskrete Approximation von Eigenwertproblemen. II. Konvergenzordnung, *Numer. Math.* **24** (1975), 415–433.

[7] Grosch, C.E. and Salwen, H. The continuous spectrum of the Orr–

Sommerfeld equation. Part 1. The spectrum and the eigenfunctions. *J. Fluid Mech.* **87** (1978), 33-54.

[8] Herbert, T. Secondary instability of boundary layers, *Ann Rev. Fluid Mech.* **20** (1988), 487-526.

[9] Mack, L.M. A numerical study of the temporal eigenvalue spectrum of the Blasius boundary layer, *J. Fluid Mech.* **73** (1976), 497-520.

[10] Miklavcic, M. Eigenvalues of the Orr-Sommerfeld equation in an unbounded domain, *Arch. Rat. Mech. Anal.* **83** (1983), 221-228.

[11] Miklavcic, M. and Williams, M. Stability of mean flows over an infinite flat plate, *Arch. Rat. Mech. Anal.* **80** (1982), 57-69.

[12] Rannacher, R. Zur asymptotischen Störungstheorie für Eigenwertaufgaben mit diskreten Teilspektren, *Math. Z.* **141** (1975), 219-233.

[13] Stummel, F. Diskrete Konvergenz linearer Operatoren. I. *Math. Ann.* **190** (1970), 45-92; II. *Math. Z.* **120** (1971), 231-264.

T.M. Fischer
Deutsche Forschungsanstalt für Luft- und
Raumfahrt (DLR)
Institut für Theoretische Strömungsmechanik
Bunsenstraße 10
D-3400 Göttingen
GERMANY

J. BRILLA

Functional analytic approach to linear viscoelastodynamics

1. Introduction

A characteristic feature of the dynamics of viscoelastic structures is damping. In fact it is a natural feature, because in reality free vibration does not exist.

When dealing with analysis of equations of linear viscoelastodynamics we find out that there are two classes of materials with different damping properties.

When considering initial-boundary value problems of linear viscoelasticity we arrive at the following class of equations

$$\mathrm{Au} = \sum_{\alpha=0}^{s} \rho a_\alpha D_t^{s+2} u + \sum_{\alpha=0}^{r} A_\alpha D_t^\alpha u$$
$$= \sum_{\alpha=0}^{s} a_\alpha D_t^\alpha f, \quad x \in \Omega, \ t \in (0, \infty), \tag{1}$$

where Ω is a bounded domain in R^2 with a sufficiently smooth boundary S, A_k are strongly elliptic operators of order $2m$, m is 1 in the case of a viscoelastic membrane, 2 in the case of a viscoelastic plate and 4 in the case of a viscoelastic shell. Then u is a transverse displacement, a stress function or a shell function. $D_t^\alpha = \partial^\alpha / \partial t^\alpha$, f is a transverse loading and ρ is the mass density per unit area. For real materials $r = s$ or $r = s + 1$.

Operators A_k can assume also a tensor form

$$A_\alpha(\cdot) \equiv - C_{ijkl}^{\alpha}(\cdot)_{k,jl}, \tag{2}$$

where $i, j, k, l = 1{,}2$ for $\Omega \in R^2$ and $i, j, k, l = 1, 2, 3$ for $\Omega \in R^3$. Here we apply the summation rule with respect to double indices and a comma followed by indices denotes the differentiation with respect to space variables. C_{ijkl}^{α} are tensors of elastic and viscoelastic moduli and $\mathbf{u}$, $\mathbf{f}$ are vectors of displacement and volume forces, respectively.

We shall consider Dirichlet boundary conditions and nonhomogeneous initial conditions

$$D_\nu^j u(s, t) = g_j(s, t), \ 0 \leq j \leq m - 1, \tag{3}$$

$$D_t^j u(x, 0) = u_j(x), \; 0 \leq j \leq s + 1. \tag{4}$$

Multiplying (1) by the inverse of the operator $\Sigma_{\alpha=0}^{s} a_\alpha D_t^\alpha$ and considering $u_j(x0) = 0$ for $j \geq 2$ we arrive at

$$\rho D_t^2 u + \int_0^t G_r(t - \tau) \, D_\tau u(x, \tau) \, d\tau = f, \tag{5}$$

for $r = s$ and at

$$\rho D_t^2 + A_r D_t u + \int_0^t G_{r-1}(t - \tau) D_\tau u(x, \tau) \, d\tau = f, \tag{6}$$

where $G_r(t)$ and $G_{r-1}(t)$ are exponential kernels. Thus we have arrived at two classes of equations, solutions of which exhibit different damping properties. However, for their analysis, i.e. for spectral analysis, it is more convenient to consider their differential form (1).

2. Spectral analysis

For an analysis of dynamics of viscoelastic bodies and structures it is necessary to analyse their spectral properties.

Assuming that f and $g_j(x, t)$ belong to the weighted Hilbert space of H^m-valued functions on R^+ with the norm

$$\|f\|^2_{L_2(R^+, H^m, \sigma)} = \int_0^\infty \|f\|^2_{H^m(\Omega)} e^{-2\sigma t} \, dt, \tag{7}$$

and applying Laplace transform to (1) we arrive at

$$\tilde{A}\,\tilde{u} = \sum_{\alpha=0}^{s} \rho a_\alpha p^{\alpha+2}\,\tilde{u} + \sum_{\alpha=0}^{r} p^\alpha A_\alpha\,\tilde{u} = \sum_{\alpha=0}^{s} a_\alpha p^\alpha \tilde{f} + \tilde{f}_i = \tilde{F}, \tag{8}$$

where tildes denote Laplace transforms and $\tilde{f}_i$ is the Laplace transform of initial conditions.

The corresponding linear eigenvalue equation assumes the form

$$\tilde{A}\,\tilde{X}_n(p) = \Big(\sum_{\alpha=0}^{s} \rho a_\alpha \, p^{\alpha+2} + \sum_{\alpha=0}^{r} p^\alpha A_\alpha \Big) \tilde{X}_n(p) = \lambda_n(p)\,\tilde{X}_n(p), \tag{9}$$

where $\lambda_n(p)$ are linear eigenvalues and $\tilde{X}_n(p)$ are eigenfunctions. In [1], [2] we have proved that non-self-adjoint operators of linear viscoelasticity are semisimple and have a denumerable set of eigenvalues and a complete biorthonormal set of eigen-functions. Thus the solutions of dynamic problems of viscoelasticity can be expressed in the form of eigenfunction expansions.

For analysis of dynamic properties we have to decompose $1/\lambda_n(p)$ into partial fractions and thus to find non-linear eigenvalues of the problem, which are the roots of the equation

$$\lambda_n(p) = 0. \tag{10}$$

In order to find characteristic features of free vibrations of viscoelastic structures we analyse asymptotic properties of nonlinear eigenvalues of equations of the second order.

In the case of $r = s = 0$, it is in the case of elastic structures with damping, we arrive at the equation

$$\rho D_t^2 u + kD_t u + A_0 u = f. \tag{11}$$

Its nonlinear eigenvalues are roots of the equation

$$\rho p^2 + kp + \lambda_{0n} = 0, \tag{12}$$

where λ_{0n} are eigenvalues of the operator A_0.

Hence

$$p_{n1,2} = -\frac{k}{2\rho}\left[1 \pm (1 - 4\frac{\rho\lambda_{0n}}{k^2})^{1/2}\right] \tag{13}$$

Thus for higher values of n the nonlinear eigenvalues of (12) are complex conjugate. Their imaginary parts tend to infinity as n tends to infinity and are of order m.

For $r = 1$ and $s = 0$ we arrive at the equation

$$\rho D_t^2 u + A_1 D_t u + A_0 u = f. \tag{14}$$

For the sake of simplicity we assume that A_1 and A_2 are spectrally similar operators with eigenvalues λ_{1n} and λ_{2n} and with eigenfunctions X_n. Then nonlinear eigenvalues of (15) are given by the equation

$$\rho p^2 + \lambda_{1n} p + \lambda_{0n} = 0, \tag{15}$$

roots of which are

$$p_{n1,2} = -\frac{\lambda_{1n}}{2\rho}\left[1 \pm (1 - 4\rho\frac{\lambda_{0n}}{\lambda_{1n}^2})^{1/2}\right]. \tag{16}$$

Hence

$$\lim_{n\to\infty} p_{n1} = -\lim_{n\to\infty} \lambda_{1n}/\rho = -\infty, \tag{17}$$

$$\lim_{n\to\infty} p_{n2} = -\lim_{n\to\infty} 2\frac{\lambda_{0n}}{\lambda_{1n}} > -\infty. \tag{18}$$

Thus one group of nonlinear eigenvalues has the limit of order $2m$ equal to $-\infty$ and the second group has a finite limit of order 0.

Applying relations between roots and coefficients of polynomial equations we can prove that for $r = s$,

$$\lambda_n^r(p) = (p - p_{1n})(p - \bar{p}_{1n}) \prod_{k=0}^{r} (p - c_{0kn}), \tag{19}$$

where p_{1n} are of order m and c_{0kn} are of order 0.

Similarly for $r = s + 1$ we can prove that

$$\lambda_n^r(p) = (p - p_{1n})(p - p_{2n}) \prod_{k=0}^{r-1} (p - c_{0kn}), \tag{20}$$

where p_{1n} are of order $2m$ and p_{2n} has for $n \to \infty$ a finite limit and thus is of order 0. p_{1n} and p_{2n} can be complex conjugate for smaller values of n.

In order to show the meaning of nonlinear eigenvalues in analysis of dynamic properties of viscoelatic structures we consider eigenfunction expansions of solutions of the equation (1).

In the case $r = s + 1$ we consider the equation (1) for $r = 1$, $f = 0$, $u_0 \neq 0$, $u_1 = 0$.

Then we have

$$u = \frac{1}{\rho}\sum_{n=1}^{\infty} \frac{1}{p_{1n} - p_{2n}}\left[(\rho p_{1n} + \lambda_{1n})\, e^{p_{1n}t} - (\rho p_{2n} + \lambda_{1n})\, e^{p_{2n}t}\right](u_0, X_n)X_n. \tag{21}$$

As according (17) p_{1n}, p_{2n} can be for smaller n complex conjugate with negative real parts, a vibration for these frequencies has undercritical damping and for higher frequencies it has critical damping.

In the case $r = s$ we consider the equation (1) for $r = 1$, $f = 0$, $u_0 \neq 0$, $u_1 = u_2 = 0$.

Then we arrive at

$$u = \frac{1}{\rho}\Big(\sum_{n=1}^{\infty} \frac{\rho p_{1n}^2 + a_0 \rho p_{1n} + \lambda_{1n}}{(p_{1n} - \bar{p}_{1n})(p_{1n} - c_{0n})} e^{p_{1n}t} + \frac{\rho \bar{p}_{1n}^2 + \alpha_0 \rho \bar{p}_{1n} + \lambda_{1n}}{(\bar{p}_{1n} - p_{1n})(\bar{p}_{1n} - c_{0n})} e^{\bar{p}_{1n}t}$$

$$+ \frac{\rho c_{0n}^2 + \alpha_0 \rho c_{0n} + \lambda_{1n}}{(c_{0n} - p_{1n})(c_{0n} - \bar{p}_{1n})} e^{c_{0n}t}\Big)\,(u_0, X_n)X_n. \tag{22}$$

As real parts of p_{1n} and $\bar{p}_{1n}$ are negative and c_{0n} are real negative, the solution u consists of terms with an undercritical damping: $\exp(p_{1n}t)$, $\exp(\bar{p}_{1n}t)$ and of terms with a critical damping: $\exp(c_{0n}t)$.

3. Analysis in weighted anisotropic spaces with dominant mixed derivatives

Now we shall deal with analysis of dynamic problems modelled by the class of equations with $r = s$, dynamic properties of which appear more real.

For analysis of these equations we have introduced [3] weighted anisotropic functional spaces with dominant mixed derivatives $H^{mk,k}(\Omega, R^+, \sigma)$ endowed with the norm

$$\|f\|^2_{H^{mk,k}} = \int_0^\infty \int_\Omega (|f| + |D^{mk}\, D_t^k f|)^2\, e^{-2\sigma t}\, d\Omega\, dt, \tag{23}$$

where $mk = \alpha$ is the multi-index.

For the trace space at $t = 0$ we have derived the following inequality:

$$\| D_t^j f\|^2_{H^\mu(\Omega)} \le c^2 \int_0^\infty \int_\Omega (|f| + |D^{mk}\, D_t^k f|)^2\, e^{-2\sigma t}\, d\Omega\, dt, \tag{24}$$

where

$$\mu = mk - j\,. \tag{25}$$

For trace properties at $x \in S$ we receive

$$\| D_u^j f \|_{H^{\mu_j, \rho_j}(S, R^+, \sigma)} \leq c^2 \| f \|_{H^{mk,k}(\omega, R^+, \sigma)}, \tag{26}$$

where

$$\mu_j = mk - j - 1/2, \; \rho_j = k - (j - 1/2)/m \tag{27}$$

for $j < mk - 1/2$. Here D_ν^j is the jth derivative in the direction of an exterior normal to S.

Then we can prove

Theorem 1. *Let* $f \in H^{0,0}(\Omega, R^+, \sigma)$, $u_0 \in H^{3m}(\Omega)$, $u_1 \in H^{2m}(\Omega)$, $u_2 \in H^m(\Omega)$ and $g_j \in H^{3m-j-1/2}(S, R^+, \sigma)$, $0 \leq j \leq m-1$. *Then there exists a solution of* (1) *for* $r = s = 1$ *such that* $u \in H^{3m,3}(\Omega, R^+, \sigma)$.

These spaces are convenient also for analysis of hyperbolic equations. When applying these spaces for analysis of hyperbolic equations we arrive at compatibility of initial conditions and trace spaces, which do not occur in analysis of J.L. Lions and E. Magenes [4], who for analysis of hyperbolic equations have applied anisotropic functional spaces with dominant ordinary derivatives.

References

[1] Brilla, J. On spectral analysis of non-selfadjoint elliptic operators, *Equadiff 6 - Proc. of the Int. Conf. on Diff. Eqs,* Brno 1985, (ed. Vosmansky J. and Zlamal M.), Springer, Berlin, Heidelberg, 1986.

[2] Brilla, J. On spectral analysis of non-selfadjoint operators in mechanics, *Trends in Applications of Pure Mathematics to Mechanics*, (ed. Kroener E. and Kirchgaessner K.), Lecture Notes in Physics., Vol. 249, Springer, Berlin, Heidelberg, New York, Tokia, 1986.

[3] Brilla, J. On analysis of quasihyperbolic equations of linear viscoelasticity, *Equadiff 7 - Proc. of the Int. Conf. on Diff. Eqs.,* Prague 1989, Teubner, Leipzig, in print.

[4] Lions, J.L. and Magenes, E. *Non-Homogeneous Boundary Value Problems and Applications*, Vol. II. Springer, Berlin, Heidelberg, New York, 1972.

J. Brilla
Institute of Applied Mathematics and Computing Technique
Comenius University
Mlynska dolina
842-15 Bratislava
CZECHOSLOVAKIA

J. VERHAS

Liquid crystal membranes with morphogenetic relevance

Abstract

The equilibrium shapes of a closed liquid crystal membrane have morphogenetic relevance. A model for cell proliferation and gastrulation is given. Some forms resembling segmentation are also found. The mathematical model has only one relevant adjustable parameter.

The membrane model

The closed liquid crystal membrane is a rather simple mechanical structure with curvature energy ([1]-[3]) given as

$$E_c = \int [\, K_1(c_1 + c_2 - a)^2 + K_2 c_1 c_2]\, d\Omega \tag{1}$$

Here c_1 and c_2 are the principal curvatures K_1 K_2 and a are material constants; the latter is referred to as natural curvature.

This structure is simple enough for spontaneous formation yet able to model cell division and gastrulation. Figures 1 and 2 show the equilibrium shapes. The equilibrium is meant in the mechanical sense only. The structure can also perform

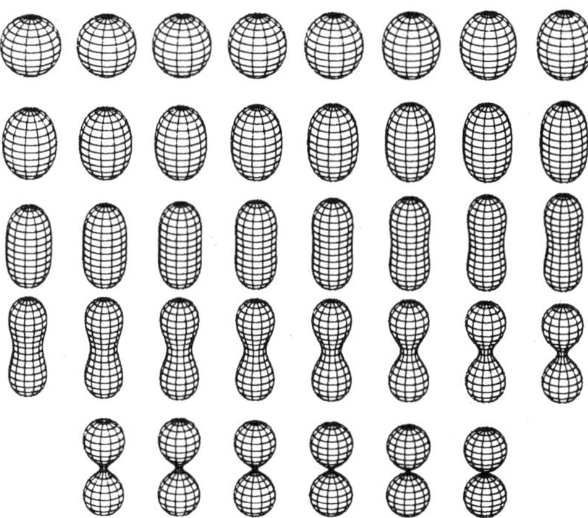

Figure 1. The forms of the proliferating cell.

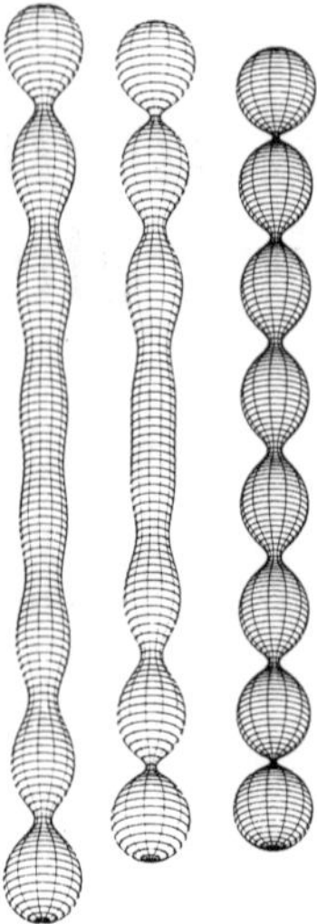

Figure 2. The forms of gastrulation.

some segmented forms (Fig. 3). The mathematical model is rather plausible but complicated and the differential equations are non-linear and have a number of removable singularities. They can be treated by numerical methods and computers. The minimization of the curvature energy with constant volume and surface area leads to the system of differential equations, which, in non-dimensional form, for surfaces of revolution, reads

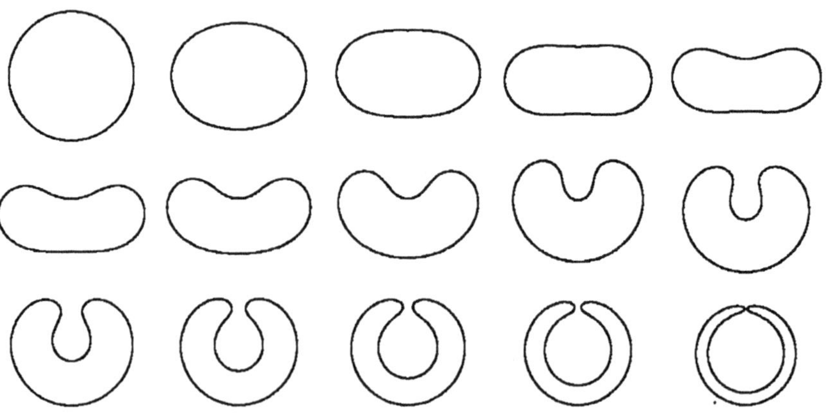

Figure 3. Forms resembling segmentation.

$$\frac{d^2\alpha}{ds^2} = \left(\frac{\sin\alpha}{x} - \frac{d\alpha}{ds}\right)\frac{\cos\alpha}{x} + \frac{\sin\alpha}{2\cos\alpha}\frac{\alpha}{\alpha}\left[\left(\frac{\sin\alpha}{x} - \mathcal{A}\right)^2 - \left(\frac{d\alpha}{ds}\right)^2 - \mathcal{K} + \mathcal{K}\frac{x}{\sin\alpha}\right]$$

(where $\mathcal{K}$ is the non-dimensional underpressure) and

$$\frac{dx}{ds} = \cos\alpha.$$

The notation is explained in Fig. 4. The shapes obtained by computation speak for themselves.

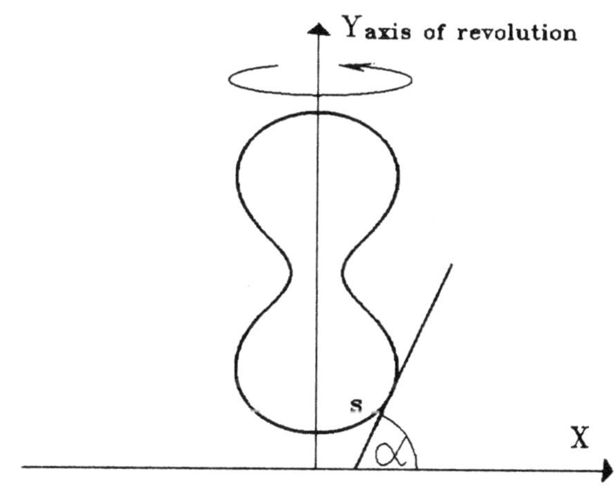

Figure 4. Explanation of the notations

The parameters are given in Figs. 5 and 6, respectively. The derivation of the equations has been published [4] and the details can be found there, together with further references.

A new computational method has been elaborated that starts the computation at the bottom. This way the computations were able to perform the form of the gastrulation, too. The break in symmetry is simply a loss of stability. The sequence of shapes is determined by the non-dimensional natural curvature $\mathcal{A}$. If it is below $-39/23$ the gastrulated forms are exhibited. The elongated shapes are preferred otherwise. If it is over a number slightly smaller than 1.8, determined numerically, the path to proliferation is denoted.

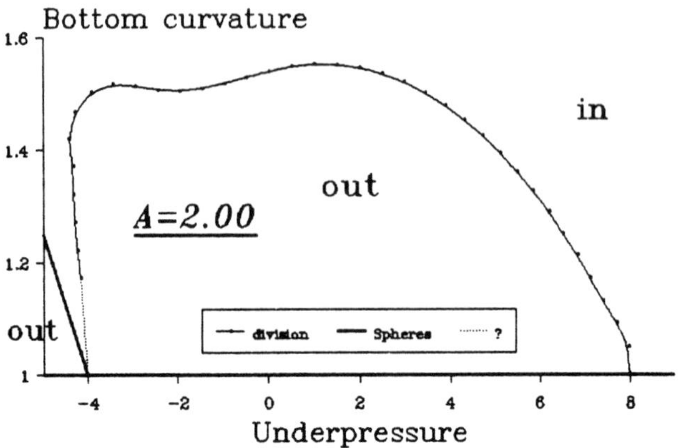

Figure 5. Cell division – bifurcational diagram

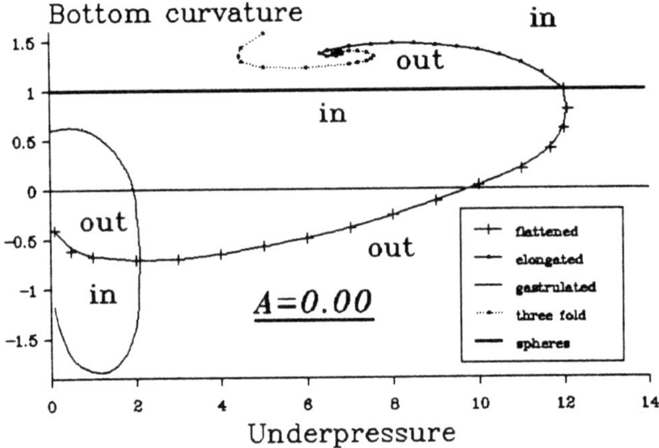

Figure 6. Gastrulation – bifurcational diagram

The forms of gastrulation presented here outline also a new idea on how the first spheres were formed in prebiotic circumstnaces. The sequences of shapes starting from a sphere have not been discovered completely for the values of $\mathcal{A}$ between $-39/23$ and 1.8.

References

[1] F.C. Frank, *Discuss. Faraday Soc.* **25**, 19 (1958).

[2] W. Helfrich, *Phys. Lett.* A. **43**, 409 (1973); *Z. Naturf.* (c), **28**, 693 (1973); *Phys. Lett.* A **50**, 115 (1974); *Z. Naturf.* (c), **29**, 510 (1974).

[3] H.J. Deuling and W. Helfirch, *J. Phys. Paris* **37**, 1335 (1976)' *Biophys. J.* **16/8**, 861 (1976).

[4] J. Verhas, *Liquid Crystals* **3**, 1183 (1988).

J. Verhas
Technical University Budapest
H-1521 Budapest
HUNGARY

P. CHADWICK

Elastic surface waves polarized in a plane of material symmetry

1. Introduction

In the theory of steady plane motions of an anisotropic elastic body a precise distinction can be made between motions which are subsonic and those which are supersonic in relation to the slowest body wave [1, sect. 1]. A surface wave, travelling along the traction-free boundary of a semi-infinite body, is a particular kind of steady plane motion and normally it is subsonic [2, p. 306]. During the last 20 years, however, examples have been found of supersonic surface waves [3, sect. V; 4, p. 412; 5, sect. 8; 6, sect. 12]. They occur in media with tetragonal, hexagonal (or transversely isotropic) and cubic symmetries, and in all these cases the reference plane of the surface wave (that is the plane containing the direction of propagation and the normal to the boundary) is a plane of material symmetry.

This observation suggests that a broad framework for a study of supersonic surface-wave propagation can be defined by (a) considering an elastic medium for which the symmetry group admits a single plane of reflectional symmetry and is otherwise unrestricted, and (b) supposing that the reference plane of the wave coincides with the symmetry plane of the medium. It turns out that a complete theory of symmetric surface-wave propagation can be developed under these simple assumptions and the main results are outlined in the sequel. A full account of the investigation has been giveen elsewhere [7, 8].

We are concerned with an elastic material which is anisotropic in relation to a natural reference configuration N. The linear elasticity tensor $\mathbf{B}$ in N, with components B_{ijkl} relative to an arbitrary orthonormal basis, is taken to be symmetric, so that $B_{jikl} = B_{klij} = B_{ijkl}$, and positive definite. The acoustical and associated tensors $\mathbf{Q}(\mathbf{n})$ and $\mathbf{R}(\mathbf{m}, \mathbf{n})$ have components

$$Q_{ij}(\mathbf{n}) = B_{pirj} n_p n_r, \quad R_{ij}(\mathbf{m}, \mathbf{n}) = B_{pirj}\, m_p n_r, \tag{1.1}$$

$\mathbf{m}$ and $\mathbf{n}$ being arbitrary orthonormal vectors with components m_i, n_i, and summation applying on repeated suffixes. The stipulated properties of $\mathbf{B}$ imply that $\mathbf{Q}(\mathbf{n})$ is symmetric and positive definite. The eigenvalues are $\rho c_a^2(\mathbf{n})$, $\rho c_b^2(\mathbf{n})$, $\rho c_c^2(\mathbf{n})$, where ρ is the density in N and $c_a(\mathbf{n})$, $c_b(\mathbf{n})$, $c_c(\mathbf{n})$ are the speeds of the homogeneous plane waves W_a, W_b, W_c which can propagate in the direction of $\mathbf{n}$ in an unbounded homogeneous body. Associated orthonormal eigenvectors $\mathbf{p}_a(\mathbf{n})$, $\mathbf{p}_b(\mathbf{n})$, $\mathbf{p}_c(\mathbf{n})$ are the polarizations of W_a, W_b, W_c. The slowness surface $\mathcal{S}$ of the material is the union of the sheets $\mathcal{S}_a, \mathcal{S}_b, \mathcal{S}_c$ given by

$$\mathbf{s} = \{c_k(\mathbf{n})\}^{-1}\mathbf{n} \ \forall\ \mathbf{n} \in \mathcal{U}, \quad k = a, b, c, \tag{1.2}$$

where $\mathbf{s}$ is the slowness and $\mathcal{U}$ denotes the set of all unit vectors.

2. Properties of plane waves propagating in a plane of material symmetry

When an elastic material possesses a plane of symmetry Π the elasticities B_{ijkl} satisfy a linear algebraic relation involving the components of $\mathbf{e}$, a unit vector normal to Π [9, sect. 2]. With the use of $(1.1)_1$ it can be deduced from this relation that, for all $\mathbf{n} \in \mathcal{U} \cap \Pi$,

$$\mathbf{Q}(\mathbf{n}) = \mathcal{Q}_{cc}(\mathbf{n})\mathbf{c} \otimes \mathbf{c} + \mathcal{Q}_{dd}(\mathbf{n})\mathbf{d} \otimes \mathbf{d} + \mathcal{Q}_{cd}(\mathbf{n})(\mathbf{c} \otimes \mathbf{d} + \mathbf{d} \otimes \mathbf{c}) + \mathcal{Q}_{ee}(\mathbf{n})\mathbf{e} \otimes \mathbf{e}, \tag{2.1}$$

$\mathbf{c}$ and $\mathbf{d}$ being fixed unit vectors in Π forming with $\mathbf{e}$ an orthonormal basis. When $\mathbf{m}, \mathbf{n} \in \Pi$, the associated tensor $\mathbf{R}(\mathbf{m}, \mathbf{n})$ decomposes similarly into the sum of *in-plane* and *ex-plane* parts with ranges Π and the orthogonal complement of Π respectively.

It is seen from (2.1) that $\mathbf{e}$ is an eigenvector of $\mathbf{Q}(\mathbf{n})$ and hence that one of the plane waves which can travel in the direction of $\mathbf{n}$ is a *transverse wave, polarized normally to* Π [10, sect. 7]. Identifying this wave with W_b, we gather from (2.1) that

$$\rho c_b^2(\mathbf{n}) = \mathcal{Q}_{ee}(\mathbf{n}) \tag{2.2}$$

and that W_a and W_c, the other plane waves with wave normal $\mathbf{n}$, are polarized in Π.

Let $S = \mathcal{S} \cap \Pi$ be the slowness section in Π, with branches $S_a = \mathcal{S}_a \cap \Pi$ etc. From (1.2), (2.2) and $(1.1)_1$ the equation of S_b is

$$\mathcal{Q}_{ee}(s_c\mathbf{c} + s_d\mathbf{d}) = \rho, \tag{2.3}$$

where $s_c = \mathbf{c}\cdot\mathbf{s}$ and $s_d = \mathbf{d}\cdot\mathbf{s}$ are the in-plane slowness coordinates. The left-hand side of (2.3) is a homogeneous quadratic form in s_c and s_d, so S_b *is an ellipse.* The other branches of S are nested and we choose S_a to enclose S_c.

In relation to orthonormal vectors $\mathbf{e}_1$ and $\mathbf{e}_2$ in Π, the transonic states on S are the points of contact of the tangents parallel to $\mathbf{e}_2$ on which $\mathbf{e}_1\cdot\mathbf{s} > 0$ [11, sect. 1]. There are at most 7 such states (cf. [12, sect. 8]). The value of $(\mathbf{e}_1\cdot\mathbf{s})^{-1}$ on the tangent is the limiting speed of a transonic state and the first transonic state and the first transonic state on S_a are, in turn, the state with the smallest limiting speed and the state involving contact with S_a which has the smallest limiting speed. Denoting these speeds by $\hat{v}$ and v_a respectively, and by v_b the limiting speed of the unique transonic state involving contact with S_b, we have $\hat{v} = \min(v_a, v_b)$.

The transonic state with limiting speed v_b falls exclusively on S_b when it is of type 1 [2, sect. VI; 11, sect.1]. The limiting wave, represented by the state, is then W_b which, regardless of the orientation of $\mathbf{c}, \mathbf{d}$ relative to $\mathbf{e}_1, \mathbf{e}_2$, leaves free of traction the planes with normal $\mathbf{e}_2$. This state is accordingly *exceptional* [11, sect. 1]. The first transonic state on S_a is exceptional only if it is of type 1 and the polarization of the limiting wave, now W_a, is in the ray direction, normal to S_a. It can be shown, by exploiting the property that $S_a \cup S_c$ and S_b are determined by disjoint subsets of the elastic moduli of the material, that these requirements are never met [8, sect. 2]. Necessarily, therefore, the first transonic state on S_a is *normal* whenever it is of type 1.

3. The pure modal character of surface waves

A surface wave in a homogeneous body of semi-infinite extent may be regarded as a linear combination of inhomogeneous plane waves, each giving rise to a displacement of the form

$$\exp\{i\kappa(\mathbf{e}_1\cdot\mathbf{x} + p\mathbf{e}_2\cdot\mathbf{x} - vt)\}\mathbf{a}. \tag{3.1}$$

Here $\mathbf{x}$ denotes spatial position and t time, κ is the wave number, v the speed of propagation and the orthonormal vectors $\mathbf{e}_1$ and $\mathbf{e}_2$ define the direction of propagation and the outward normal to the boundary of the body respectively. Substitution of (3.1) into the equations of motion leads to a *fundamental eigenvalue problem* yielding 6 values of the scalar p and corresponding values of the displacement amplitude $\mathbf{a}$. The p's form 3 pairs of complex conjugates when $0 \le v < \hat{v}$, $\hat{v}$ being the limiting speed of the first transonic state relative to $\mathbf{e}_2$ in the reference plane R spanned by $\mathbf{e}_1$ and $\mathbf{e}_2$ [1, sect. 1]. (The notation is consistent with that introduced in Section 2, but R has not yet been assumed to be a plane of material symmetry.)

In a surface-wave motion the displacement tends to zero as $-\mathbf{e}_2\cdot\mathbf{x} \to \infty$ and the boundary of the transmitting body is traction free. The decay requirement is met by (3.1) only if $\operatorname{Im} p < 0$. The surface-wave displacement is consequently of the form

$$\mathbf{u}(\mathbf{x}, t) = \sum_{\alpha=1}^{3} \gamma_\alpha \exp\{i\kappa(\mathbf{e}_1\cdot\mathbf{x} + p_\alpha\mathbf{e}_2\cdot\mathbf{x} - v_s t)\}\mathbf{a}_\alpha, \tag{3.2}$$

where the pairs $(p_\alpha, \mathbf{a}_\alpha)$, with $\operatorname{Im} p_\alpha < 0$, satisfy the fundamental problem. The condition at the boundary $\mathbf{e}_2\cdot\mathbf{x} = 0$ imposes on the traction amplitudes

$$\mathbf{l}_\alpha = \{p_\alpha \mathbf{Q}(\mathbf{e}_2) + \mathbf{R}(\mathbf{e}_2, \mathbf{e}_1)\}\mathbf{a}_\alpha \tag{3.3}$$

the linear dependence

$$\sum_{\alpha=1}^{3} \gamma_\alpha \mathbf{l}_\alpha = \mathbf{0}, \tag{3.4}$$

implying the relation

$$[\mathbf{l}_1, \mathbf{l}_2, \mathbf{l}_3] = 0. \tag{3.5}$$

This is the secular equation, determining the speed of propagation v_s, and (3.4) then specifies, to within a common complex multiplier, the coefficients γ_α. The surface wave is subsonic when $v_s < \hat{v}$ and supersonic when $v_s > \hat{v}$.

When the reference plane R coincides with a plane of material symmetry Π, two of the eigenvectors of the fundamental problem are ex-plane (that is, scalar multiples of $\mathbf{e}$) and the others are in-plane (orthogonal to $\mathbf{e}$). In view of (2.1) and the analogous representation of $\mathbf{R}(\mathbf{m}, \mathbf{n})$, these properties of the displacement amplitudes are transferred by (3.3) to the traction amplitudes. One of $\mathbf{a}_\alpha$, say $\mathbf{a}_3$, is ex-plane, as is $\mathbf{l}_3$, and (3.4) dictates that

$$\gamma_1 \mathbf{l}_1 + \gamma_2 \mathbf{l}_2 = \mathbf{0}, \quad \gamma_3 = 0.$$

The $\alpha = 3$ contribution is thus removed from the displacement field (3.2) and the terms which remain are in-plane. This means that *a surface wave having as its reference plane a plane of material symmetry* $\Pi : \mathbf{e}\cdot\mathbf{x} = 0$ *is a pure mode, with* $\mathbf{e}\cdot\mathbf{u} = 0$.

4. Existence-uniqueness considerations

From this point onwards the development relates to the Barnett-Lothe theory of surface waves and specifically to the key tensors $\mathbf{S}_1(v)$, $\mathbf{S}_2(v)$, $\mathbf{S}_3(v)$, the behaviour of which must be characterized when R coincides with Π. Only a brief sketch of the conclusions can be given: a short survey of the essential background may be found in [1, sects. 2, 3].

The critical finding is that the symmetric tensors $\mathbf{S}_2(v)$ and $\mathbf{S}_3(v)$ break down into in-plane and ex-plane parts in the same way as $\mathbf{Q}(\mathbf{n})$ does in (2.1), while $\mathbf{S}_1(v)$ *is entirely in-plane*. By interpreting each tensor as an integral over the branches of the slowness section S it is found that the in-plane parts are wholly determined by $S_a \cup S_c$ and the ex-plane parts by S_b alone. Concurrently, the domains of the in-plane and ex-plane constituents are $[0, v_a)$ and $[0, v_b)$ respectively, where v_a and

v_b are the limiting speeds defined in Section 2.

The indifference of $\mathbf{S}_1(v)$ to the presence of the elliptical branch S_b entails the possibility of supersonic surface-wave propagation in the following way. The complex secular equation (3.5) has the real equivalent

$$F(v) := 1 + \tfrac{1}{2}\operatorname{tr}\mathbf{S}_1^2(v) = 0$$

[1, sect. 3], and $F(v)$, like $\mathbf{S}_1(v)$, has domain $[0, v_a)$. When $v_a > v_b$, that is when the first transonic state is on $S_b, F(v)$ can therefore have a zero greater than v_b and when this happens a supersonic surface wave exists.

The basic questions of the existence and uniqueness of surface waves can be resolved by studying the eigenvalues of $\mathbf{S}_3(v)$ (cf. [1, sect. 2]). By virtue of the above-mentioned decomposition, two of these eigenvalues, $\lambda_1(v)$ and $\lambda_2(v)$ say, have domain $[0, v_a)$ and the third eigenvalues, $\lambda_3(v)$, has domain $[0, v_b)$. Each eigenvalue is monotonic increasing on its domain, with a negative value at $v = 0$, and $\lambda_1(v)$ and $\lambda_2(v)$ vanish together at a zero of $F(v) : \lambda_3(v)$ vanishes at v_b and terminates there in a cusp. An asymptotic analysis of $\lambda_1(v)$ and $\lambda_2(v)$ near v_a (as in [2, sect. IX]) shows that the larger of these eigenvalues tends to positive infinity as $v \to v_a$.

The patterns of variation of the eigenvalues of $\mathbf{S}_3(v)$ with v permitted by these properties give rise to the following existence-uniqueness theorem, valid.

> *When R is a plane of material symmetry a unique surface wave exists with displacement polarized in R. If the first transonic state is on S_a, or jointly on S_a and S_b, this wave is subsonic. If the first transonic state is on S_b, the surface wave is subsonic or supersonic according as $F(v_b)$ is negative or positive.*

References

[1] Chadwick, P. Recent developments in the theory of elastic surface and interfacial waves. In: *Elastic Wave Propagation* (McCarthy, M.F. and Hayes, M.A., eds.), pp. 3-16, North-Holland, Amsterdam, 1989.

[2] Chadwick, P. and Smith, G.D. Foundations of the theory of surface waves in anisotropic elastic materials. In: *Advances in Applied Mechanics*, Vol. 17 (Yih, C-S., ed.), pp. 303-376, Academic Press, New York, 1977.

[3] Farnell, G.D. Properties of elastic surface waves. In: *Physical Acoustics*, Vol. VI (Mason, W.P. and Thurston, R.N., eds.), pp. 109-166, Academic Press, New York, 1970.

[4] Barnett, D.M., Gavazza, S.D. and Lothe, J. Slip waves along the interface

between two anisotropic elastic half-spaces in sliding contact, *Proc. R. Soc. Lond.* A **415**, 389–419, 1988.

[5] Chadwick, P. Aspects of the behaviour of surface waves in transversely isotropic elastic media. In: *Recent Developments in Surface Acoustic Waves* (Parker, D.F. and Maugin, G.A., eds.), pp. 80–90, Springer–Verlag, London, 1988.

[6] Chadwick, P. Wave propagation in transversely isotropic elastic media II. Surface waves, *Proc. R. Soc. Lond.* A **422**, 67–101, 1989.

[7] Chadwick, P. The behaviour of elastic surface waves polarized in a plane of material symmetry I. General analysis, *Proc. R. Soc. Lond.* A **430**, 213–240, 1990.

[8] Barnett, D.M., Chadwick, P. and Lothe, J. The behaviour of elastic surface waves polarized in a plane of material symmetry. Addendum to part I, *Proc. R. Soc. Lond.*, to appear.

[9] Cowin, S.C. and Mehrabadi, M.M. On the identification of material symmetry for anisotropic elastic materials, *Q. Jl Mech. Appl. Math.* **40**, 451–476, 1987.

[10] Cowin, S.C. Properties of the anisotropic elasticity tensor, *Q. Jl Mech. Appl. Math.* **42**, 249–266, 1989.

[11] Chadwick, P. A general analysis of transonic states in an anisotropic elastic body, *Proc. R. Soc. Lond.* A **401**, 203–223, 1985.

[12] Chadwick, P. Wave propagation in transversely isotropic elastic media I. Homogeneous plane waves, *Proc. R. Soc. Lond.* A **422**, 23–66, 1989.

P. Chadwick
School of Mathematics
University of East Anglia
Norwich, NR4 7TJ,
U.K.

A. TATONE AND N. RIZZI

A one-dimensional model for thin-walled beams

Abstract

A one-dimensional continuum model endowed with local structure, whose kinematical states are characterized by a four-dimensional differentiable manifold, is presented. It is shown that this model allows a suitable description of the relevant aspects of the mechanical behaviour of thin-walled beams. Further it is shown that, by introducing a suitable characterization of the material, the linearized model coincides with Vlasov's model.

1. Introduction

In the first decades of this century there was a large increase in the use of open-section beams with thinner walls, essentially due to the needs of the aeronautical industry. The growing interest in the structural behaviour of such members resulted in efforts to give a general theory to be adopted as a proper design tool.

In a pioneer work H. Wagner [1] pointed out as distinctive features of the behaviour of a thin-walled beam: (a) the considerable increase in its torsional rigidity when warping is prevented in some section; (b) the arising of flexural-torsional buckling under axial load long before the reaching of the Eulerian buckling load. Then, starting from the classical Saint-Venant results for torsion and introducing the so called *unitary warping* concept, he proposed a *linear one-dimensional theory* able to describe the aforementioned phenomena.

That theory, later refined by R. Kappus [2] in order to attain a better agreement with experimental results, was finally recast in a more systematic way by Vlasov [3] and so accepted as the standard, technical, one-dimensional *linear* theory for thin-walled beams. Furthermore, progresses in nonlinear mechanics stimulated researches also in the field of thin-walled beams.

Looking at these works it seems to the writers that they share a basic idea: one-dimensional models do have a meaning only if obtained by a two- or three-dimensional one, via more or less consistent projection methods (see e.g. [4]). A different point of view - as far as the authors know - can be found only in a work by M. Epstein [5], who resorts to the *direct* theory of one-dimensional structured continua in order to introduce a model that, conceived for beams with particular cross sections, turns out to be very complicated.

The aim of this work is to present in a direct way a one-dimensional continuum model with a local structure conceived to be the *minimal* one able to describe at least the phenomena remarked by Wagner. From the kinematical point of view the configuration space for the local structure is a four-dimensional differential manifold obtained by adding one parameter to the descriptor of the rotations of the beam section. The additional parameter is a coarse descriptor of the warping of the section.

By introducing a standard expression for the mechanical power, *contact* and *body* dynamical actions are defined as quantities algebraically dual to those describing the velocity field. An expression for the mechanical power in terms of the sole contact actions is obtained.

A characterization of the material is given by introducing an internal constraint between warping and torsion and by defining a suitable class of constitutive functions. In particular it is assumed that the torsional component of the contact couple depends on the axial and flexural components of the strain measures.

It is finally shown that Vlasov's model can be recovered by the proposed model in the context of a linearized theory.

2. One-dimensional continuum model

We regard a beam as a differentiable manifold $\mathcal{B}$, and its motion as a function

$$\chi : \mathcal{B} \times I \to \mathcal{E} \times \mathcal{F} \tag{2.1}$$

where I is an interval in the field of real numbers $\mathcal{R}$, $\mathcal{E}$ is a three-dimensional Euclidean space, whose translation space is $\mathcal{V}$, $\mathcal{F}$ is the four-dimensional product manifold $SO(\mathcal{V}) \times \mathcal{R}$.

Let the motion of $\mathcal{B}$ be such that for any placement $\chi_t := \chi(\cdot, t)$ the natural projection $C_t := p(\chi_t(\mathcal{B}))$ of $\chi_t(\mathcal{B})$ into $\mathcal{E}$ is a one-dimensional differentiable manifold.

Let us consider a reference placement

$$\kappa : \mathcal{B} \to \mathcal{E} \times \mathcal{F}, \tag{2.2}$$

such that $\kappa : \mathcal{B} \to \kappa(\mathcal{B})$ is a diffeomorphism, and a parametrization $\mathbf{X}(s)$ of $C_\kappa := p(\kappa(\mathcal{B}))$, being $s \in [0, L]$ a curve-length parameter. With respect to the reference shape $\kappa(\mathcal{B})$, any placement χ_t can be described by a function $\mathbf{x}$ which maps C_κ to C_t and, for each point of C_κ, by a proper rotation $\mathbf{R}$ and a scalar quantity α. By using the parametrization of C_κ we will regard $\mathbf{x}, \mathbf{R}$ and α as functions on $[0, L] \times I$ and will assume them to be sufficiently smooth.

In our interpretation, $\mathbf{R}(s, t)$ and $\alpha(s, t)$ describe the rotation and the *warping*, with respect to the reference shape, of the beam cross-section corresponding to the point $\mathbf{x}(s, t)$ of the beam axis C_t.

The velocity is defined by

$$\mathbf{v} := \dot{\mathbf{x}},$$

$$\mathbf{W} := \dot{\mathbf{R}}\,\mathbf{R}^{\mathrm{T}}, \tag{2.3}$$

$$a := \dot{\alpha},$$

where a dot denotes differentiation with respect to time, while a suitable definition of strain measures, suggested by a similar definition in [6], is given as follows

$$\mathbf{e} := \mathbf{R}^{\mathrm{T}}\,\mathbf{x}' - \mathbf{X}',$$

$$\mathbf{E} := \mathbf{R}^{\mathrm{T}}\mathbf{R}', \tag{2.4}$$

$$\beta := \alpha',$$

where a prime denotes differentiation with respect to s.

Corresponding to each point $\mathbf{x}(s, t)$, the contact actions are characterized by a vector $\mathbf{t}$, a skew-symmetric tensor $\mathbf{T}$ and a scalar quantity ϖ, which in this context will be given the meaning of *bimoment*, while the body actions are characterized by a vector $\mathbf{b}$, a skew-symmetric tensor $\mathbf{B}$ and a scalar quantity η. Such a characterization of the actions results from the following expression of the power[1]

$$\mathcal{W} := \int_0^L (\mathbf{b}\cdot\mathbf{v} + \mathbf{B}\cdot\mathbf{W} + \eta a)\mathrm{d}S + [\,\mathbf{t}\cdot\mathbf{v} + \mathbf{T}\cdot\mathbf{W} + \varpi\, a\,]_0^L\,. \tag{2.5}$$

The conditions for the power to be objective are expressed by the following balance equations

$$\mathbf{t}' + \mathbf{b} = \mathbf{0},$$

$$\mathbf{T}' + \mathbf{B} - \mathbf{x}' \wedge \mathbf{t} = \mathbf{0}. \tag{2.6}$$

Substituting these relations into the expression (2.5) we obtain the stress power formula

---

[1] The inner product for skew-symmetric tensors has been defined as

$$\mathbf{B}\cdot\mathbf{W} = \frac{1}{2}\mathrm{tr}\,(\mathbf{B}\mathbf{W}^{\mathrm{T}}).$$

$$\mathcal{W} = \int_0^L [\mathbf{t} \cdot (\mathbf{v}' - \mathbf{W}\mathbf{x}') + \mathbf{T} \cdot \mathbf{W}' + \pi a + \varpi a'] \, ds$$
$$= \int_0^L [\mathbf{s} \cdot \dot{\mathbf{e}} + \mathbf{S} \cdot \dot{\mathbf{E}} + \pi a + \varpi a'] \, ds \tag{2.7}$$

where $\mathbf{s} := \mathbf{R}^T\mathbf{t}$ e $\mathbf{S} := \mathbf{R}^T\mathbf{T}\mathbf{R}$ and

$$\pi = \eta + \varpi'. \tag{2.8}$$

It is worth noting that this expression is a special case of (3.17) in [7].

Assuming that the material is elastic and homogeneous, it can be proved that the most general constitutive relations are of the form

$$\begin{aligned} \mathbf{s} &= \hat{\mathbf{s}}(\mathbf{e}, \mathbf{E}, \alpha, \beta), \\ \mathbf{S} &= \hat{\mathbf{S}}(\mathbf{e}, \mathbf{E}, \alpha, \beta), \\ \pi &= \hat{\pi}(\mathbf{e}, \mathbf{E}, \alpha, \beta), \\ \varpi &= \hat{\varpi}(\mathbf{e}, \mathbf{E}, \alpha, \beta). \end{aligned} \tag{2.9}$$

By choosing an orthonormal basis $\{\mathbf{D}_1, \mathbf{D}_2, \mathbf{D}_3\}$ of $\mathcal{V}$, such that $\mathbf{D}_3 = \mathbf{X}'$, we can introduce the components of $\mathbf{s}, \mathbf{S}, \mathbf{E}$, which will be used in the next section, as follows

$$\begin{aligned} \mathbf{s} &= Q_1\mathbf{D}_1 + Q_2\mathbf{D}_2 + N\mathbf{D}_3, \\ \mathbf{S} &= M_1 \mathbf{D}_2 \wedge \mathbf{D}_3 + M_2 \mathbf{D}_3 \wedge \mathbf{D}_1 + T\mathbf{D}_1 \wedge \mathbf{D}_2, \\ \mathbf{E} &= \mu_1\mathbf{D}_2 \wedge \mathbf{D}_3 + \mu_2\mathbf{D}_3 \wedge \mathbf{D}_1 + \tau\mathbf{D}_1 \wedge \mathbf{D}_2. \end{aligned} \tag{2.10}$$

It also turns out to be useful to define the vector $\mathbf{u} := \mathbf{x} - \mathbf{X}$ and its components v_i with respect to the basis

$$\mathbf{d}_i := \mathbf{R}\,\mathbf{D}_i, \quad i = 1, 2, 3. \tag{2.11}$$

3. Constrained model and comparison with Vlasov's model

We now want to show that, by assuming a suitable characterization of the material, Vlasov's model can be obtained as a special case of the continuum model described in the previous section.

Let us first note that, if we want to cast into our model the property that, as shown by experimental observations, the warping is an effect of the other strain measures, it seems reasonable to assume an internal constraint of the form

$$\alpha = k(\mathbf{e}, \mathbf{E}), \quad k(\mathbf{o}, \mathbf{O}) = 0. \tag{3.1}$$

For any objective scalar function k, this constraint turns out to be objective because any change of frame leaves both $\mathbf{e}$ and $\mathbf{E}$ unchanged.

By the axiom of determinism for constrained materials, which states that the stress is determined by the motion only to within an arbitrary additive part which does no work in any motion compatible with the constraint, the indetermined part of the stress, which will be denoted by a subscript c, corresponding to the above constraint has to satisfy the following relations

$$\begin{aligned} \mathbf{s}_c &= -\pi_c k_{\mathbf{e}}, \\ \mathbf{S}_c &= -\pi_c k_{\mathbf{E}}, \\ \varpi_c &= 0, \end{aligned} \tag{3.2}$$

where k_e and $k_{\mathbf{E}}$ stand for the derivatives of k with respect to its arguments.

Let us assume now a special form of the above constraint, which could be seen as reflecting more closely the observation that warping is related to torsion, by replacing (3.1) with

$$\alpha = \check{k}(\tau), \quad \check{k}(0) = 0. \tag{3.3}$$

In general we do not expect this constraint to model the observed property more accurately than the previous constraint. We only want to lay down the assumptions which lead our model to match Vlasov's model. In the same spirit we shall introduce all the assumptions below.

Besides the internal constraint (3.3) let us assume, as in Vlasov's model, the material to be shear-undeformable. This constraint is defined in our model by the relation

$$\mathbf{e} = \varepsilon \mathbf{X}'. \tag{3.4}$$

The indetermined part of the stress, corresponding to the constraints (3.4) and (3.3), must satisfy the following scalar relations

$$N_c = 0,\ M_{1c} = 0,\ \ M_{2c} = 0,\ \ T_c = -\ \pi_c(d\check{k}/d\tau),\ \varpi_c = 0, \tag{3.5}$$

while Q_{1c}, Q_{2c}, as well as π_c, can take any value.

In order to compare the two models let us consider an elastic beam, with a straight reference shape, subject to terminal loads which consist of an axial force and a couple at each end defined by $\mathbf{t}(L) = -\mathbf{t}(0) = \lambda f\, \mathbf{D}_3$, $\mathbf{T}(L) = -\mathbf{T}(0) = \lambda F\, \mathbf{D}_2 \wedge \mathbf{D}_3$, where f and F are constants and λ is a parameter.

Let us assume that, for $\lambda = \lambda_0 \neq 0$, a solution exists which is characterized by

$$\begin{aligned}
&\mu_2 = 0,\ \tau = 0,\\
&Q_1 = 0,\ M_2 = 0,\ T = 0,\\
&\pi_a = 0,\ \varpi_a = 0,
\end{aligned} \tag{3.6}$$

where the subscript a denotes the determined part of the stress. Linearization of the scalar form of equations (2.6) and (2.8) at the above solution yields, denoting by a tilde differentation with respect to λ, the following equations

$$\begin{aligned}
&\tilde{Q}_1' - N\,\tilde{\mu}_2 + Q_2\tilde{\tau} = 0,\\
&\tilde{Q}_2' + \mu_1\tilde{N} + N\,\tilde{\mu}_1 = 0,\\
&\tilde{N}' - \mu_1\,\tilde{Q}_2 - Q_2\,\tilde{\mu}_1 = 0,\\
&\tilde{M}_1' + (1+\varepsilon)\,\tilde{Q}_2 + Q_2\,\tilde{\varepsilon} = 0,\\
&\tilde{M}_2' + \mu_1\,\tilde{T} - M_1\,\tilde{\tau} - (1+\varepsilon)\,\tilde{Q}_1 = 0,\\
&\tilde{T}' + M_1\,\tilde{\mu}_2 - \mu_1\,\tilde{M}_2 = 0,\\
&\tilde{\pi}_c = -\ \tilde{\pi}_a\ + \tilde{\varpi}_a',
\end{aligned} \tag{3.7}$$

where, as a consequence of $(3.5)_3$,

$$\tilde{T} = \tilde{T}_a - \tilde{\pi}_c\,\frac{d\check{k}}{d\tau}\ . \tag{3.8}$$

By linearizing equation (2.4) and (3.4) we obtain as well

$$\tilde{\varepsilon} = \tilde{v}_3' - \mu_1 \tilde{v}_2,$$

$$\tilde{\mu}_1 = \frac{1}{1+\varepsilon}(\tilde{v}_2'' + \mu_1' \tilde{v}_3 + \mu_1 \tilde{v}_3') - \frac{\varepsilon'}{(1+\varepsilon)^2}(\tilde{v}_2' + \mu_1 \tilde{v}_3),$$

$$\tilde{\mu}_2 = \frac{\varepsilon'}{(1+\varepsilon)^2}\tilde{v}_1' - \frac{1}{1+\varepsilon}v_1'' - \mu_1 \tilde{\theta}, \tag{3.9}$$

$$\tilde{\tau} = -\tilde{\theta}' + \frac{\mu_1}{1+\varepsilon}\tilde{v}_1',$$

where $\tilde{\theta}$ denotes the component of $\tilde{\mathbf{R}}\mathbf{R}^T$ with respect to the first tensor of the basis $\{\mathbf{d}_2 \wedge \mathbf{d}_1, \mathbf{d}_3 \wedge \mathbf{d}_2, \mathbf{d}_1 \wedge \mathbf{d}_3\}$.

Let us now assume the constitutive functions to be analytic and their series expansion near the reference shape to be

$$N = \hat{N}(\varepsilon, \tau) = A\varepsilon + o(\tau),$$

$$M_1 = \hat{M}_1(\mu_1, \tau) = B_1\mu_1 + o(\tau),$$

$$M_2 = \hat{M}_2(\mu_2, \tau) = B_2\mu_2 + o(\tau),$$

$$T_a = \hat{T}(\varepsilon, \mu_1, \mu_2, \tau) = C\tau + (C_\varepsilon\varepsilon + C_{\mu_1}\mu_1 + C_{\mu_2}\mu_2)\tau + o(\tau), \tag{3.10}$$

$$\varpi_a = \hat{\varpi}(\beta) = D\beta + o(\beta),$$

$$Q_{1a} = 0, Q_{2a} = 0, \ \pi_a = 0,$$

where $A, B_1, B_2, C, C_\varepsilon, C_{\mu_1}, C_{\mu_2}, D$ are constants. Note that such specifications are consistent with the solution (3.6). Due to (3.10) and (3.8) equations (3.7) become

$$\tilde{Q}_1' - N\tilde{\mu}_2 + Q_2\tilde{\tau} = 0,$$

$$\tilde{Q}_2' + \mu_1\tilde{N} + N\tilde{\mu}_1 = 0,$$

$$A\tilde{\varepsilon}' - \mu_1\tilde{Q}_2 - Q_2\tilde{\mu}_1 = 0,$$

$$B_1\tilde{\mu}_1' + (1+\varepsilon)\tilde{Q}_2 + Q_2\tilde{\varepsilon} = 0, \tag{3.11}$$

$$B_2\tilde{\mu}_2' + \mu_1 (C + C_\varepsilon \varepsilon + C_{\mu_1}\mu_1)\tilde{\tau} - \mu_1 D\left(\frac{d\check{k}}{d\tau}\right)^2 \tilde{\tau}'' - M_1\tilde{\tau} - (1+\varepsilon)\tilde{Q}_1 = 0,$$

$$(C + C_\varepsilon \varepsilon + C_{\mu_1}\mu_1)\tilde{\tau}' - D\left(\frac{d\check{k}}{d\tau}\right)^2 \tilde{\tau}''' + (M_1 - B_2\mu_1)\tilde{\mu}_2 = 0.$$

At first glance equations (3.11) appear to be quite different from Vlasov's classical buckling equations (see [3], Ch. V (1.10)). But we must consider that, while in deriving equations (3.11) the exact solution (3.6) has been adopted, in the other case such a solution has been replaced by a fictitious one in which stresses are linearized near the reference shape and strains are assumed to be zero. Although clearly inconsistent, these assumptions are very usual in 'technical' theories. Only by accepting these assumptions the last three of equations (3.11), after substitution of (3.9), become

$$B_1 \tilde{v}_2'''' - N \tilde{v}_1'' = 0,$$

$$B_2 \tilde{v}_1'''' - N \tilde{v}_1'' - M_1 \tilde{\theta}'' = 0, \tag{3.12}$$

$$C_o \tilde{\theta}'''' - (C + \frac{C_\varepsilon}{A} N + \frac{C_{\mu_1}}{B_1} M_1)\tilde{\theta}'' - M_1 \tilde{v}_1'' = 0,$$

where $C_o := D(d\check{k}/d\tau)^2$.

These equations are equal to Vlasov's equations in [3], even though the constants have different meanings. In particular note that the last equation contains the term $C_o\tilde{\theta}''''$ corresponding to the term in Vlasov's equations due to the bimoment. What is illuminating here is that that term originates from nothing but a constraint reaction. Note also that the expression for the coefficient of $\tilde{\theta}''$ in the last equation arises from the assumption that the torsional component T of the contact couple depends on ε, μ_1 and μ_2 (hence for an hyperelastic material N, M_1 and M_2 must depend on τ as well).

Finally it is interesting to note that the constitutive relations in [8], which have been derived from a two-dimensional continuum through a projection method, satisfy the restrictions (3.10).

References

[1] Wagner, H., Verdrehung und Knickung von offenen Profilen, *Festschrift 25-Jahre*, T.H. Danzig (1929), translated in: Torsion and buckling of open

sections, *NACA TM* No. 807 (1936).

[2] Kappus, R., Drillknicken zentrich gedrückter Stäbe mit offenem Profil im elastischen Bereich, *Luftfahrtforschung*, **14** (1937), 444-457, translated in: Twisting failure of centrally loaded open-section columns in the elastic range, *NACA TM* No. 851 (1938).

[3] Vlasov, V.Z., *Thin-walled elastic beams*, Israel program for scientific translations, Monson, Jerusalem (1961).

[4] Epstein, M. and Murray, D.W., Three-dimensional large deformation analysis of thin walled beams, *Int. J. Solids Structures*, **12** (1976), 867-876.

[5] Epstein, M., Thin-walled beams as directed curves, *Acta Mechanica*, **33** (1979), 229-242.

[6] Capriz, G., A contribution to the theory of rods, *Riv. Mat. Univ. Parma* **7** (4) (1981).

[7] Green, A.E. and Laws, N., A general theory of rods, *Proceedings of the Royal Society, Series A Mathematical and Physical Sciences*, No. 1432 **293**, London (1966).

[8] Mollmann, H., Theory of thin-walled elastic beams with finite displacements, in *Finite Rotations in Structural Mechanics* (ed. W. Pietraszkiewicz), Euromech Colloquium 197, Jablonna, Poland (1985), Springer-Verlag (1986).

A. Tatone
Dipartimento di Ingegneria delle Strutture
Facolta di Ingegneria
Università di L'Aquila
I-67040 Monteluco di Roio (AQ)
ITALY

N. Rizzi
Dipartimento di Ingegneria Strutturale e Geotecnico
Università di Roma "La Saprienza",
via Endossiana 18
I-00184 Roma
ITALY

A.A. LIOLIOS

Upper and lower solution bounds in incremental elastoplasticity by the hypercircle method*

Abstract

By using some functional analysis concepts, two-sided solution bounds are obtained for the piecewise linearized problem in incremental elastoplasticity.

1. Introduction

Methods of obtaining upper and lower solution bounds for linear, equality boundary-value problems of mathematical physics and especialy of elastostatics are already well known, see e.g. [1]-[4]. A generalization of such a method, namely the hypercircle method [1], has been also obtained for inequality problems in unilateral elastostatics [5], [9], [10]. As known, the governing conditions in the inequality problems are equalities as well as inequalities. These problems can be treated mathematically by the variational inequality approach [4], [6].

The aim of this paper is to extend further the hypercircle method to structural elastoplasticity. So, a procedure is presented which provides upper and lower bounds for solution quantities, global as well as local ones, of the piecewise linearized static problem in incremental elastoplasticity. This bounding procedure is entirely different from other ones providing upper deformation bounds only in elastoplasticity ([11], [14]).

2. Problem conditions

For the problem formulation we follow the matrix operator one used by Maier ([7], [8], [11]). Thus the equilibrium and compatibility conditions are

$$\mathbf{G}^T\dot{\boldsymbol{\sigma}} = \dot{\mathbf{f}}, \qquad \mathbf{G}\,\dot{\mathbf{u}} = \dot{\mathbf{e}}, \qquad (1),(2)$$

where $\mathbf{G}$ is the compatibility rectangular matrix, $\mathbf{G}^T$ its transpose (equilibrium matrix) and $\boldsymbol{\sigma}, \mathbf{f}, \mathbf{u}, \mathbf{e}$ are the stress, load, displacement and total strain vectors, respectively. Dots over symbols denote rates.

---

* Dedicated to Professor E. Kröner on the occasion of his 70th birthday.

Further, the piecewise linearized constitutive relations are

$$\dot{\mathbf{e}} = \dot{\boldsymbol{\varepsilon}} + \dot{\mathbf{p}} + \dot{\mathbf{d}}\,, \tag{3}$$

$$\dot{\boldsymbol{\sigma}} = \mathbf{E}\,\dot{\boldsymbol{\varepsilon}}\,, \qquad \dot{\mathbf{p}} = \mathbf{N}\,\dot{\boldsymbol{\lambda}} \tag{4),(5}$$

$$\dot{\boldsymbol{\varphi}} = \mathbf{N}^{\mathrm{T}}\dot{\boldsymbol{\sigma}} - \mathbf{H}\dot{\boldsymbol{\lambda}}\,, \qquad \boldsymbol{\varphi} = \mathbf{N}^{\mathrm{T}}\boldsymbol{\sigma} - \mathbf{H}\boldsymbol{\lambda} - \mathbf{r}\,, \tag{6a,b}$$

$$\dot{\boldsymbol{\lambda}} \geq O, \qquad \dot{\boldsymbol{\varphi}} \leq 0O, \qquad \dot{\boldsymbol{\varphi}}^{\mathrm{T}}\dot{\boldsymbol{\lambda}} = 0. \tag{7a, b, c}$$

Here $\boldsymbol{\varepsilon}$, $\mathbf{p}$ and $\mathbf{d}$ are the pure elastic, plastic and imposed (e.g. thermal) strain vectors, respectively; $\boldsymbol{\varphi}$, $\boldsymbol{\lambda}$ and $\mathbf{r}$ are the yield-functions, plastic multiplier and non-negative initial yield capacity vectors, respectively; $\mathbf{N}$ is the rectangular matrix of outward normal vectors to the convex polyhedron of elastic domain, defined by $\boldsymbol{\varphi} \leq O$, in the stress space; $\mathbf{E}$ is the symmetric and positive definite elasticity matrix; and $\mathbf{H}$ is the symmetric work-hardening matrix. For various aspects concerning $\mathbf{H}$ (hardening, perfect-plastic, softening etc. behaviour) see [8]. Here we note that the problem of conditions (1)-(7) has a solution set $(\dot{\boldsymbol{\varepsilon}}, \dot{\mathbf{u}}, \dot{\mathbf{p}}, \dot{\boldsymbol{\lambda}}, \dot{\boldsymbol{\sigma}}, \dot{\boldsymbol{\varphi}})$ for a given action set $(\dot{\mathbf{f}}, \dot{\mathbf{d}})$ if the following condition holds:

$$\text{matrix } \mathbf{A} = \mathbf{H} - \mathbf{N}^{\mathrm{T}}\mathbf{Z}\,\mathbf{N} \text{ is positive-(semi) definite,} \tag{8}$$

where $\mathbf{Z}$ is the negative-(semi) definite matrix of the influence of plastic strains $\mathbf{p}$ on stresses $\boldsymbol{\sigma}$ assuming linear elastic behaviour. Finally, if the dots over symbols are dropped and (6b) instead of (6a) is used, then (1)-(7) concern the problem of holonomic plasticity (deformation theory).

3. Function space formulation of the problem

Due to (1)-(7), each pair $(\dot{\mathbf{u}}, \dot{\boldsymbol{\lambda}})$ or $(\dot{\boldsymbol{\sigma}}, \dot{\boldsymbol{\varphi}})$ can fully describe an elastoplastic state. With such a state is now associated an elenent S of a real Hilbert space H. As scalar product of two states S_1 and S_2 we define the elastoplastic energy

$$S_1 \cdot S_2 = \dot{\boldsymbol{\varepsilon}}_1\,\mathbf{E}\,\dot{\boldsymbol{\varepsilon}}_2 + \dot{\boldsymbol{\lambda}}_1\,\mathbf{H}\,\dot{\boldsymbol{\lambda}}_2\,. \tag{9}$$

This symmetric bilinear form is positive definite when condition (8) holds, as may be shown. Thus the squared norm $\| S \|^2$ equals the double of the sum of the elastic and plastic strain energy. Further, using (1)-(7) and the principle of virtual work, we obtain the following formulae for the incremental and the holonomic case, respectively:

$$S_1 \cdot S_2 = \dot{\mathbf{f}}_1^{\mathrm{T}} \dot{\mathbf{u}}_2 - \dot{\sigma}_1^{\mathrm{T}} \dot{\mathrm{d}}_2 - \dot{\boldsymbol{\varphi}}_1^{\mathrm{T}} \dot{\boldsymbol{\lambda}}_2 , \tag{10a}$$

$$S_1 \cdot S_2 = \mathbf{f}_1^{\mathrm{T}} \mathbf{u}_2 - \boldsymbol{\sigma}_1^{\mathrm{T}} \mathbf{d}_2 - (\boldsymbol{\varphi}_1 + \mathbf{r})^{\mathrm{T}} \boldsymbol{\lambda}_2 . \tag{10b}$$

Because of (7a, b) the solution state S must be sought in a convex subset of H. Usually S can be obtained analytically in a few cases of academic interest only. Therefore two-sided solution bounds are useful for most practical elastoplastic problems solved numerically. To obtain such bounds according to the hypercircle method [1], first we split the problem by introducing two classes of admissible states.

An elastoplastic state S_s with $\dot{\boldsymbol{\sigma}}_s$ and $\dot{\boldsymbol{\varphi}}_s$ satisfying (1), (4), (6a) and (7b) is called statically admissible. Similarly, a state S_k with $\dot{\mathbf{u}}_k$ and $\dot{\boldsymbol{\lambda}}_k$ satisfying (2)-(5) and (7a) is called kinematically admissible. Obviously, the sets H_s of all S_s and H_k of all S_k constitute two convex subsets in the Hilbert space H.

4. Two-sided global and local solution bounds

Using the properties of the admissible states, we can prove ([5], [10]) that for any pair (S_s, S_k) and the solution state S the following bivariational inequality holds:

$$(S - S_s) \cdot (S - S_k) \leq a, \quad a = -\dot{\boldsymbol{\varphi}}_s^{\mathrm{T}} \dot{\boldsymbol{\lambda}}_k \geq 0 . \tag{11}$$

From this inequality we obtain the equivalent one

$$(S - C)^2 \leq \mathrm{R}^2, \quad C = \frac{1}{2}(S_s + S_k), \ \mathrm{R}^2 = a + \frac{1}{4} \| S_s - S_k \|^2, \tag{12}$$

which geometrically interpreted means that in the Hilbert space H the solution state S lies on or inside the hypersphere (C, R). So we derive the following two-sided bounds for the generalized energy $W = S^2/2$:

$$(\| C \| - \mathrm{R})^2 \leq S^2 = 2\, W \, (\| C \| + \mathrm{R})^2 . \tag{13}$$

Other global two-sided solution bounds are obtained by using the procedure of [10]. So, introducing the functionals of potential and complementary energy, respectively,

$$\Pi_{\mathrm{p}}(S) = \frac{1}{2} \dot{\boldsymbol{\varepsilon}}^{\mathrm{T}} \mathbf{E} \, \dot{\boldsymbol{\varepsilon}} + \frac{1}{2} \dot{\boldsymbol{\lambda}}^{\mathrm{T}} \mathbf{H} \dot{\boldsymbol{\lambda}} - \dot{\mathbf{f}}^{\mathrm{T}} \dot{\mathbf{u}} , \tag{14}$$

$$\Pi_c(S) = \frac{1}{2}\,\dot{\boldsymbol{\sigma}}^T\,\mathbf{E}^{-1}\,\dot{\boldsymbol{\sigma}} + \frac{1}{2}\,\dot{\boldsymbol{\lambda}}^T\,\mathbf{H}\,\boldsymbol{\lambda} + \dot{\boldsymbol{\sigma}}^T\,\mathbf{d}, \tag{15}$$

we have the minimum principles inequalities and global bounds

$$-\,\Pi_c(S_s) \leq -\,\Pi_c(S) = \Pi_p(S) \leq \Pi_p(S_k). \tag{16}$$

As regards, now, local (pointwise) bounds for a kinematic, $F_k(\mathbf{x})$, or a static, $F_s(\mathbf{x})$, field quantity at the point $\mathbf{x}$, we introduce an appropriate 'safe' Green state $\hat{S}$, see e.g. [11]-[14]. So, excluding the singularity point $\mathbf{x}$, $\hat{S}$ must have a limited intensity P such that (1)-(7) are satisfied in finite (not incremental) form with $\mathbf{f} = \mathbf{0}$ and $\mathbf{d} = \mathbf{0}$. Further, to avoid infinite norms, $\hat{S}$ is conceived as the sum of a singular part $\overset{\infty}{S}$, corresponding to the singularity in the infinite continuum, and of a regular part $\tilde{S}$, satisfying the boundary etc. conditions of $\hat{S}$ ([2]-[4]):

$$\hat{S} = \overset{\infty}{S} + \tilde{S}\,. \tag{17}$$

In a way similar to that one for S, we introduce also for $\hat{S}$ statically, $\hat{S}_s$, and kinematically, $\hat{S}_k$, admissible states. Applying now the procedure of [5], first from (10) we obtain

$$\hat{S}_s.S = P{\cdot}F_k(\mathbf{x}) - \hat{\boldsymbol{\sigma}}_s^T\,\dot{\mathbf{d}} - (\hat{\boldsymbol{\varphi}}_s + \mathbf{r})^T\,\dot{\boldsymbol{\lambda}}\,, \tag{18}$$

$$S.\hat{S}_k = P{\cdot}F_s(\mathbf{x}) - \dot{\mathbf{f}}^T\,\hat{u}_k - \phi^T\,\hat{\lambda}_k. \tag{19}$$

Further we choose for $\hat{S}_s$ and $\hat{S}_k$ to satisfy, respectively,

$$\hat{\boldsymbol{\varphi}}_s = -\,\mathbf{r}, \qquad \hat{\boldsymbol{\lambda}}_k = \mathbf{0} \tag{20, (21)}$$

and we define

$$F(\mathbf{x}) = F_k(\mathbf{x}) - F_s(\mathbf{x}), \quad F_a(\mathbf{x}) = \dot{f}^T\,\hat{u}_k + \hat{\sigma}_s^T\,\dot{\mathbf{d}}\,. \tag{22), (23}$$

So, subtracting (19) from (18) and using Schwarz inequality, we obtain

$$|\,F(\mathbf{x}) - F_a(\mathbf{x}) - \frac{1}{2\,P}(S_s + S_k){\cdot}(\hat{S}_s - \hat{S}_k)\,| \leq \frac{\mathrm{R}}{P}.\,\|\,\hat{S}_s - \hat{S}_k\,\|\,. \tag{24}$$

5. Concluding remarks

Formulae (13), (16) and (24) provide global and local, respectively, two-sided solution bounds in terms of the admissible states. These states are constructed numerically by the finite element method on the basis of the minimum principles for the functionals (14), (15), see [8]. In formula (24), any allowable value for P satisfying (6b) can be used, but the best one is obtained by optimization under inequality constraints. Similar improvement of the bounds is further obtained by minimizing the bounds gaps R and $\| \hat{S}_s - \hat{S}_k \|$. For this purpose, convex combinations of the admissible states are used ([4], [6]). Thus, the numerical realization of the herein presented bounding procedure requires the use of available computer codes of the finite element method and nonlinear (quadratic) mathematical programming.

References

[1] Synge, J.L., *The Hypercircle in Mathematical Physics*, Cambridge University Press, 1957.

[2] Rieder, G., Eingrenzungen in der Elastizitäts- und Potentialtheorie, *ZAMM* **52**, T340–T347 (1972).

[3] Stumpf, H., *Eingrenzungsverfahren in der Elastomechanik*, Westdeutscher Verlag, Forschungsber. NRW No. 2116, Köln, 1970.

[4] Velte, W., *Direkte Methoden der Variationsrechnung*, B.G. Teubner Verlag, Stuttgart, 1976.

[5] Liolios, A.A., Upper and lower bounds in nonlinear and unilateral elastostatics, *ZAMM* **62**, T138–T139 (1982)

[6] Panagiotopoulos, P.D., *Inequality Problems in Mechanics and Applications. Convex and Nonconvex Energy Functions*, Birkhäuser, Basel, 1985.

[7] Capurso, M. and Maier, G., Incremental elastoplastic analysis and quadratic programming, Meccanica **5**, 107–116 (1970).

[8] Cohn, M.Z. and Maier, G. (eds.) Proc. NATO Adv. Study Inst., Univ. Waterloo, *Engineering Plasticity by Mathematical Programming*, Pergamon Press, New York, 1979.

[9] Collins, W.D., An extension of the method of the hypercircle to linear operator problems with unilateral constraints, *Proc. Roy. Soc. Edinb.* **85A**, 173–193 (1980).

[10] Liolios, A.A., A direct formulation of dual extremum principles in unilateral elastostatics by a generalization of the hypercircle method, *ZAMM* **65**, T348–T350 (1985).

[11] Maier, G., Upper bounds on deformations of elastic-workhardening

structures in the presence of dynamic and second-order geometric effects, *J. Struct. Mech.* **2**, 265-280 (1973).

[12] Ponter, A.R.S., An upper bound on the small displacements of elastic, perfectly plastic structures, *J. Appl. Mech.* **39**, 959-963 (1972).

[13] Capurso, M. Some upper bound principles to plastic strains in dynamic shakedown of elastoplastic structures, *J. Struct. Mech.* **7**, 1-20 (1979).

[14] Martin, J.B. Extended displacement bound theorems for work hardening continua subjected to impulsive loading, *Int. J. Solids and Structures* **2**, 9-26 (1966).

A. A. Liolios
Democritus University of Thrace
Department of Civil Engineering
Institute of Structural Mechanics
GR-67100 Xanthi
GREECE

A. KACZYŃSKI AND S.J. MATYSIAK

Thermal stresses in a periodic two-layered composite weakened by a row of interface cracks

1. Introduction

Interface cracks in view of their importance in numerous applications have received considerable attention in the literature devoted to fracture mechanics. Within the framework of linear thermoelasticity the interface crack problems have been investigated in [1]–[9].

In this paper we deal with the study of stress singularities caused by a finite or infinite system of cracks lying on an interface in a microperiodic two-layered thermoelastic space. The basis of our considerations is the homogenized model of microperiodic thermoelastic composites given in [10]–[12]. The equations of the homogenized model are expressed in terms of unknown macrodisplacements and macrotemperature and certain extra unknowns called kinematical and thermal microlocal parameters. Thus, it is possible to evaluate not only mean but also local values of deformation and temperature gradients in every material component of the composite.

We confine attention to the two-dimensional stationary problems of interface cracks in thermally insulated and free from tractions in microperiodic two-layered thermoelastic space, and suppose that the constant heat flux vector normal to the layering is applied at infinity.

2. Formulation of the problem

We consider a periodic laminated thermoelastic body in which each lamina is composed of two homogeneous isotropic thermoelastic layers. By (x, y, z) we define the Cartesian coordinate system such that the axis y is normal to the layering. Let l_1, l_2 be the thicknesses of the layers and δ be the thickness of each basic unit (lamina) of the composite, thus $\delta = l_1 + l_2$. By $\lambda_j, \mu_j, k_j, \beta_j/(\lambda_j + \frac{2}{3}\mu_j)$, $j = 1, 2$ we denote the Lamé constants, the coefficients of thermal conductivity and the coefficient of volume expansion of the subsequent layers, respectively.

According to the results presented in [13–15] in two-dimensional problems the displacement vector $\vec{u}$ and the temperature θ are assumed as follows:

$$\vec{u}(x, y) \simeq (u(x, y),\ v(x, y), 0), \tag{2.1}$$

$$\theta(x, y) \simeq \vartheta(x, y)$$

and the stresses $\sigma^{(j)}$ and the heat flux vector $\vec{q}^{(j)}$ in layers of the jth kind (with material constants $\lambda_j, \mu_j, k_j, \beta_j$) are given in the form

$$\begin{aligned}
\sigma_{yy}^{(j)} &\simeq (\lambda_j + 2\mu_j)(v_{,y} + h_j q) + \lambda_j u_{,x} - \beta_j \vartheta, \\
\sigma_{xy}^{(j)} &\simeq \mu_j(v_{,x} + u_{,y} + h_j p), \\
\sigma_{xx}^{(j)} &\simeq (\lambda_j + 2\mu_j)u_{,x} + \lambda_j(v_{,y} + h_j q) - \beta_j \vartheta, \\
\sigma_{zz}^{(j)} &\simeq \lambda_j(u_{,x} + v_{,y} + h_j q) - \beta_j \vartheta, \\
\vec{q}^{(j)} &\simeq -k_j(\vartheta_{,x}, \vartheta_{,y} + h_j \gamma, 0),
\end{aligned} \tag{2.2}$$

where

$$h_1 = 1, \quad h_2 = -\eta(1-\eta)^{-1}, \quad \eta = l_1/\delta, \tag{2.3}$$

and $(u, v, 0)$ and ϑ are unknown functions interpreted as macrodisplacement vector and macrotemperature, respectively. The unknowns p, q and γ stand for microlocal kinematical and thermal parameters and are related with the microperiodic structure of the body.

The equations of the homogenized model with microlocal parameters take the form ([13]-[15]):

$$\begin{aligned}
&(\tilde{\lambda} + \tilde{\mu})(u_{,xy} + v_{,yy}) + \tilde{\mu}(v_{,xx} + v_{,yy}) + [\mu] p_{,x} + ([\lambda] + 2[\mu]) q_{,y} - \tilde{\beta}\vartheta_{,y} = 0, \\
&(\tilde{\lambda} + \tilde{\mu})(u_{,xx} + v_{,xy}) + \tilde{\mu}(u_{,xx} + u_{,yy}) + [\mu] p_{,y} + [\lambda] q_{,x} - \tilde{\beta}\vartheta_{,x} = 0, \\
&\tilde{k}(\vartheta_{,xx} + \vartheta_{,yy}) + [k]\gamma_{,y} = 0, \\
&(\hat{\lambda} + 2\hat{\mu}) q = -[\lambda](u_{,x} + v_{,y}) - 2[\mu] v_{,y} + [\beta]\vartheta, \\
&\hat{\mu} p = -[\mu](u_{,y} + v_{,x}), \\
&\hat{k}\gamma = -[k]\vartheta_{,y}
\end{aligned} \tag{2.4}$$

where

$$(\tilde{\lambda}, \tilde{\mu}, \tilde{k}, \tilde{\beta}) = \eta(\lambda_1, \mu_1, k_1, \beta_1) + (1-\eta)(\lambda_2, \mu_2, k_2, \beta_2),$$

$$([\lambda], [\mu], [k], [\beta]) = \eta(\lambda_1 - \lambda_2, \mu_1 - \mu_2, k_1 - k_2, \beta_1 - \beta_2), \tag{2.5}$$

$$(\hat{\lambda}, \hat{\mu}, \hat{k}) = \eta(\lambda_1, \mu_1, k_1) + \eta^2(1-\eta)^{-1}(\lambda_2, \mu_2, k_2).$$

Within the framework of the homogenized model with microlocal parameters we consider the two-dimensional problem of laminated space weakened by interface cracks occupying the region

$$L = \bigcup_{k=1}^{n} L_k, \ L_k = \langle a_k, b_k \rangle \times \{0\} \quad = 1,\ldots,n, \tag{2.6}$$

see Fig. 1 (problem C_n). The boundary conditions are assumed to be

$$q_y^{(1)}(x, 0^+) = q_y^{(2)}(x, 0^-) = 0,$$

$$\sigma_{yy}^{(1)}(x, 0^+) = \sigma_{yy}^{(2)}(x, 0^-) = 0, \tag{2.7}$$

$$\sigma_{xy}^{(1)}(x, 0^+) = \sigma_{xy}^{(2)}(x, 0^-) = 0, \text{ for } x \in L,$$

and the conditions at the infinity

$$\lim_{r \to \infty} q^{(j)}(x, y) = (0, -q_0, 0), \quad r = (x^2 + y^2)^{0.5},$$

$$\lim_{r \to \infty} (\sigma_{yy}^{(j)}(x, y), \sigma_{xy}^{(j)}(x, y)) = (0, 0). \tag{2.8}$$

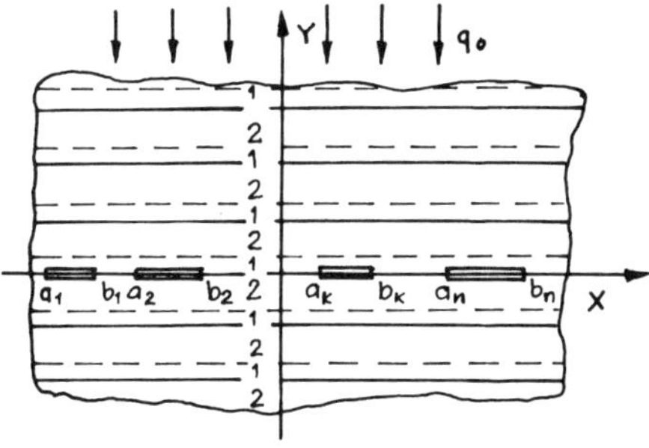

Figure 1. The scheme of middle cross-section of the body

3. Solution of the crack problems

The solution of the problem C_n can be obtained by using the complex potentials method and is presented in the paper [13]. We confine attention here to a discussion of the results from the standpoint of the linear fracture mechanics. As might be expected (see [14], [15]) the obtained solutions lead to typical non-oscillating singularities with the power -0.5 of certain thermal stresses, unlike the classical solutions of interface crack problems (see, for example, [1]). If we define the SIF (stress intensity factor) for the crack $L_j = \langle a_j, b_j \rangle \times \{0\}$ as follows:

$$(k_1(a_j), k_2(a_j)) = \lim_{x \to a_j^-} \sqrt{2\pi |x - a_j|}\,(\sigma_{yy}(x,0), \sigma_{xy}(x,0)),$$

$$(k_1(b_j), k_2(b_j)) = \lim_{x \to b_j^+} \sqrt{2\pi |x - b_j|}\,(\sigma_{yy}(x,0), \sigma_{xy}(x,0)), \tag{3.1}$$

then for the problem C_n we have ([13])

$$\forall j,\ k_1(a_j) = k_1(b_j) = 0,$$

$$k_2(a_j) = \frac{(-1)^{n+j+1}\sqrt{2\pi}\ \mathrm{i}\ P_{n-1}(a_j)}{t_*\sqrt{b_j - a_j}\ \prod\limits_{\substack{m=1\\ m\neq j}}^{n} \sqrt{|a_j - a_m||a_j - b_m|}} \tag{3.2}$$

$$k_2(b_j) = \frac{(-1)^{n-j}\sqrt{2\pi}\ \mathrm{i}\ P_{n-1}(b_j)}{t_*\sqrt{b_j - a_j}\ \prod\limits_{\substack{m=1\\ m\neq j}}^{n} \sqrt{|a_j - a_m||a_j - b_m|}}$$

where

$$\mathrm{i}^2 = -1,\ t_* = (A_1/A_2)^{0.25},$$

$$A_1 = \tilde{\lambda} + 2\tilde{\mu} - \frac{([\lambda]+2[\mu])^2}{\hat{\lambda} + 2\hat{\mu}}, \tag{3.3}$$

$$A_2 = \tilde{\lambda} + 2\tilde{\mu} - \frac{[\lambda]^2}{\hat{\lambda}+2\hat{\mu}},$$

and

$$P_{n-1}(x) = c_0 + c_1 + \cdots + c_{n-1}\, x^{n-1}\,, \tag{3.4}$$

where the coefficients $c_0, c_1, \ldots, c_{n-1}$ are determined by solving n linear equations given in [13].

Example 1. Problem C_1

In this problem involving $L = L_1 = \langle -a, a \rangle \times \{0\}$ we have

$$k_1(\pm a) = 0,\; k_2(\pm a) = \pm \frac{\sqrt{\pi}\, q_0\, a^{3/2}\, (A_2/A_1)^{0.5}\, \delta_t}{2\, k_0\, K} \tag{3.5}$$

and δ_t, K are the constants connected with material constants of the subsequent layers and defined in [13].

Example 2. Problem C_∞

Problem C_∞ treats of an infinite array L_∞ of equal cracks of length $2a$ speed at constant intervals l along the x-axis $(l > 2a)$. Hence, we set $a_k = k\, l - a$, $b_k = k\, l + a$, $k \in \{0, \pm 1, \pm 2, \ldots\}$. The SIF are given by

$$k_1(\pm a) = 0,\; k_2(\pm u) = \mp \frac{q_0\, \delta_t\, l^{3/2}\, \ln\left(\cos\frac{\pi a}{l}\right)}{k_0\, K\, t_*\, \left(\tan\left(\frac{\pi a}{l}\right)\right)^{0.5}}. \tag{3.6}$$

Remark

Assuming that the body under consideration is homogeneous $(\lambda_1 = \lambda_2, \mu_1 = \mu_2, k_1 = k_2, \beta_1 = \beta_2)$ the stress intensity factors given by (3.2), (3.5) and (3.6) agree with the corresponding SIF in the classical thermoelasticity [8].

References

[1] E.J. Brown and F. Erdogan, *Int. J. Engng. Sci.* **6** 517–529, 1968.

[2] I.V. Gajvas, *Thermal Stress. Struct.* **9**, 88–99, 1970 (in Russian).

[3] A.M. Sachenko *el al*. *Constr. and Polymers in Low Temperatures*, **47**, 1976.

[4] G.L. Bachman and L.A. Prusov, *Izv. AN SSSR, Mech. Solids* **6**, 146–148, 1976.

[5] C. Atkinson, *Int. J. Fract.* **13**, 807-820, 1977.
[6] N.I. Ioakimidis and I.A. Theocaris, *Rev. Roum. Sci. Techn.* **23**, 563-575, 1978.
[7] A.F. Mak *et al*, *Trans. ASME, J. Appl. Mech.* **47**, 347-350, 1980.
[8] G.S. Kit and M.G. Krivtzun, *Thermoelastic Plane Problems for Bodies with Cracks*, Naukova Dumka, Kiev, 1983 (in Russian).
[9] H.E. Williams, *J. Therm. Stress.* **8**, 183-203, 1985.
[10] Cz. Woźniak, *Int. J. Engng. Sci.* **25**, 483-499, 1987.
[11] Cz. Woźniak, *Bull. Pol. Ac.: Techn.* **35**, 143-153, 1987.
[12] S.J. Matysiak and Cz. Woźniak, *J. Techn. Phys.* **29**, 85-97, 1988.
[13] A. Kaczyński and S.J. Matysiak, *Int. J. Engn. Sci.* **27**, 131-147, 1989.
[14] A. Kaczyński and S.J. Matysiak, *Int. J. Fract.* **37**, 31-45, 1988.
[15] S.J. Matysiak, *Acta Mech.* **78**, 95-108, 1989.

A. Kaczyński
Institute of Mathematics
Warsaw Univerisiy of Technology,
Plac Jedności Robotniczej 1
00-661 Warsaw
POLAND

S.J. Matysiak
Institute of Hydrogeology and Engineering Geology,
University of Warsaw,
Al Zwirki i Wigury 93,
02-089 Warsaw
POLAND

A. KAVEH

Crossing number, genus and thickness of a graph for the flexibility analysis of structures

1. Introduction

The stress resultant distribution due to a given loading on a skeletal structure S, obtained by the flexibility method, is given by

$$r = [B_0 - B_1(B_1^t F_m B_1)^{-1} B_1^t F_m B_0] p, \qquad (1)$$

where F_m contains the flexibility of the individual members, and p is the load vector. Matrix B_0 can easily be found using a shortest route tree of S. Matrix B_1 can be formed from a maximal set of independent self-equilibrating stress systems (SESs), known as a *statical basis*. The dimension of this basis is the same as the degree of statical indeterminacy (DSI) of the structure, denoted by $\alpha(S)$. A statical basis can be formed on the elements of a cycle basis of the graph model S of the structure. On each cycle 3 or 6 SESs can be formed, depending on whether S is a planar or a space structure, respectively.

For an efficient flexibility analysis, the flexibility matrix of the structure, $G = B_1^t F_m B_1$, should be sparse, banded and well-conditioned. Here only sparseness of G is studied.

In this paper three topological invariants, namely crossing number, genus and thickness of a graph are applied to the flexibility analysis of skeletal structures. Different embeddings are employed for selecting cycle basis corresponding to highly sparse flexibility matrices.

For topological and graph theoretical definitions, the reader may refer to Spanier [1] and Harary [2], respectively.

2. Degree of statical indeterminacy and crossing number

The concept of statical indeterminacy is central to a clear understanding of the mechanics of a skeletal structure, when analysed by means of the flexibility method. In the following, two theorems are presented for calculating the DSI of space structures using the crossing numbers of their planar drawings.

Definitions. A *drawing* S^p of a graph S is a mapping of S into a surface. The nodes of S go into distinct nodes of S^p. A member and incidence nodes map into a homeomorphic image of the closed interval [0,1] with the relevant nodes as end points and the interior, a member, containing no nodes. A *good drawing* is one in which no two members incident with a common node have a common point, and no two members have more than one point in common. A common point of two members is a *crossing*. An *optimal* drawing in a given surface is one which exhibits the least possible crossings. The number of crossing points of S after drawing on a plane or sphere, S^P, is denoted by $\nu(S^P)$. For cases when the drawing is optimal, $\nu(S^P)$ becomes the crossing number.

Theorem A: *Let S be a space rigid-jointed frame. Then*

$$\alpha(S) = 6b_1(S) = 6[R_i(S^P) - \nu(S^P)], \tag{2}$$

where $R_i(S^P)$ is the number of internal regions of S^P; i.e. $R_i(S^P) = R(S^P) - 1$.

Theorem B: *Let S be a ball-jointed space truss. Then*

$$\alpha(S) = \nu(S^P) - M_c(S^P), \tag{3}$$

where $M_c(S^P)$ is the number of members required for full triangulation of S^P.

Simple proofs of the above theorems are given by the author [3]. Obviously it is advantageous to have an optimal drawing of S, since the number of countings becomes minimum in calculating $\alpha(S)$. For the case where $\nu(S^P) = 0$, the corresponding mesh basis of S^P forms a suitable cycle basis of S. However, surprisingly little is known about the crossing number of S, in general (White and Beineke [4]). Further research is required to develop a systematic approach for constructing an optimal drawing or at least a drawing with subminimal crossings.

3. Embeddings for selection of cycle bases

The quality of a mesh basis consisting of the cycles bounding the finite regions of a planar graph embedded in $\mathbb{R}^2$, suggest the extension of this approach to general nonplanar graphs, by embedding S on polyhedrons and manifolds with certain properties. These properties can be measured by using the homology groups $H_p(K, R)$. The underlying complexes, which measure the extent to which K has non-bounding p-cycles.

3.1. A two-dimensional polyhedron embedding.

Let S be the mathematical model of a structure which is a simple graph (1-complex).

The underlying polyhedron or geometric realization $|S|$ of S is also denoted by S. An *embedding* $f: S \to P$ is a homeomorphism of S into the polyhedron P. An embedding f is called a *regular 2-cell embedding* if the components of $[P - f(S)]$ are all regular 2-cells. Let $f(S)$ be dissected into a 1-complex isomorphic to the dissection of S. Then $f(S)$ and the components $[P - f(S)]$ form a dissection of P into a 2-dimensional cell-complex K.

3.2. Admissible embeddings

The cycles bounding the 2-cells of K are known as *regional cycles*. An admissible embedding f of S is one for which the regional cycles form a set of $b_1(S)$ independent cycles from $Z_1(K, R)$. It can be proved that the necessary and sufficient condition for $f: S \to P$ to be admissible is that the corresponding K be acyclic. A regular complex K is called *acyclic* if all homology groups of K are trivial. Equivalently, any cycle of $Z_1(K, R)$ bounds in K; i.e. $Z_1(K, R) = B_1(K, R)$, Cooke and Finney [5].

It is easy to show that a contractible complex is acyclic, hence a contractible embedding is admissible. If K is collapsible, then it is obviously contractible. Thus a collapsible embedding is also admissible.

3.3. Modified manifold embedding

Every graph is embeddable on some orientable surface. S can be drawn on the sphere and enough handles are added to eliminate crossings. This number is called the *genus* of the surface.

The genus $\gamma(S)$ of S is the minimum genus of the orientable surfaces on which S is embeddable. An embedding is *minimal* if the genus of surface is the same as $\gamma(S)$. Edmond's permutation technique provides a method for a 2-cell embedding of S in an orientable 2-manifold M. Youngs [6] has developed an algorithm for minimal embedding, by considering all the possible sets of vertex permutations of S, which is a lengthy and non-pratical approach. The author [7] developed a method in which the vertex permutations were determined during the process of embedding of S. However, this algorithm does not always lead to a minimal embedding.

An intuitive embedding may also be employed, which can only be useful for simple structural models, Henderson and Maunder [8]. After embedding S on a manifold M, $2\gamma(M)$ appropriate fillings and one perforation of order 2 is required to provide the admissibility condition.

In such an embedding, the quality of the selected cycle basis (bounding cycles) depends on the genus of M on which S is embedded. It is ideal to have a minimal embedding; however, there is no efficient systematic approach for this purpose.

AnS example Fig. 1 shows an embedding of S on a sphere with one handle; i.e. a torus. Modifications are made by two proper fillings (15, 19) and one perforation (A) of order 2.

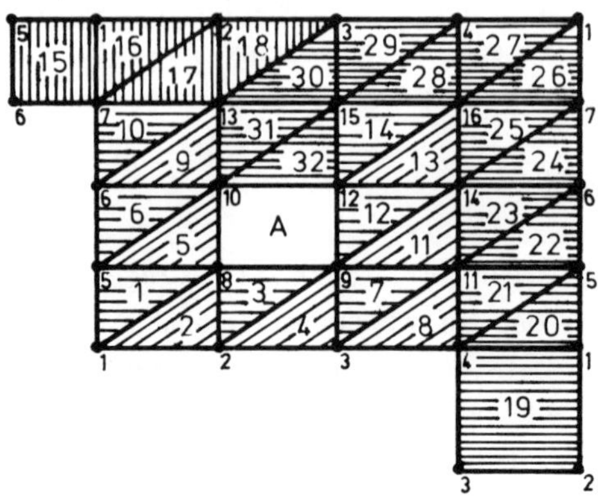

Figure 1. S embedded on a manifold

30 cycles of length 3 and 2 cycles of length 4 are selected as a minimal cycle basis of S.

4. Embedding S on a union of discs

S can be considered as the union of some planar connected non-separable subgraphs of S. The process of embedding is as follows:

Step 1: Identify a planar subgraph S_1 and embed it on a disc d_1 whose dissection K_1 is isomorphic to S_1.

Step 2: The second subgraph S_2 is identified such that the corresponding K_2 has a 2-cell with a free 1-face and $|K_1| \cap |K_2|$ is a connected subspace of the frontier of d_1.

Step 3: The process of Step 2 is continued and at the ith step K_i is joined to $K^{i-1} = \bigcup_{j=1}^{i-1} K_i$ with K_i having a free 1-face and $|K_i| \cap |K^{i-1}|$ is a connected subspace of the frontier of d_i. Obviously K^i collapses to K^{i-1}.

The process is terminated when all 1-cells of S are embedded in $K = \bigcup_{j=1}^{q} K_i$ which is collapsible.

It is ideal to embed S in a complex K with the minimum number of discs. This number for all possible collapsible embeddings is an invariant of S, known as the *thickness* of S.

Again only partial results are available on the thickness of a graph, White and Beineke [4]. Any systematic approach for embedding on a minimum number of discs should be advantageous for reducing the overlaps of the cycles.

5. An expansion process for cycle selection

A major difficulty in the formation of a collapsible complex has been its automation. Thus a reverse approach, namely an expansion process, has been adopted by the author [9]. This method starts from a specified node and generates connected independent cycles of the least lengths. The process terminates when $b_1(S)$ cycles are selected. This approach is a greedy type of algorithm which embeds S on a collapsible set of discs.

The above algorithm is order dependent like many other combinatorial algorithms and starting node influences the total length of the selected cycles. Further modifications have been made by allowing the selection of disjoint cycles (disjoint discs), Kaveh [10]. This method is successfully implemented and efficient generation of cycle bases have become feasible [11]. In the above approach, the total length of the cycles of a basis is reduced to decrease the overlaps of the cycles. However, such a basis, in general, does not lead to a cycle adjacency matrix D of maximal sparsity. A direct method for maximizing the sparsity of D is recently developed by the author [12]. For this purpose weights are assigned to the member of S indicating the length and the number of minimal cycles that can be generated on each member of S. Restrictions are then imposed on the selection process to generate cycles of least overlaps, often leading to optimal cycle bases. In this method measures can be assigned to the members of a structure representing their flexibilities or stiffnesses, and minimal measure cycles can then be selected to improve the conditioning of G.

Once an optimal or suboptimal cycle basis of S is generated, matrix B_1 corresponding to a highly sparse flexibility matrix G can easily be obtained. The bandwidth of G can also be reduced by ordering the selected cycles using the algorithm of the author [13].

6. Concluding remarks

In this paper, it is shown how three topological invariants, namely crossing number, genus and thickness of a graph are involved in structural mechanics. Any advances regarding the efficient and systematic computation of these invariants in the field of graph theory will naturally improve the efficiency of structural analysis by the flexibility method.

The advantage of the author's expansion approach lies in its simplicity for programming and capability of including material and metrical properties, in order to improve the conditioning as well as the sparsity of the overall flexibility matrices of structures.

Acknowledgement

The author is grateful to Prof. Dr F. Ziegler of TU-Wien for his kind and continuous encouragement in the course of this research.

References

[1] Spanier, E.H., *Algebraic Topology*, McGraw-Hill, 1966.

[2] Harary, F., *Graph Theory*, Addison Wesley, 1969.

[3] Kaveh, A., Topology and skeletal structures, *ZAMM*, **68** (1988), 347-353.

[4] White, A.T. and Beineke, L.W., Topological graph theory, in *Selected Topics in Graph Theory* (Ed. Beineke and Wilson), 1978.

[5] Cooke, G.E. and Finney, R.L., *Homology of Cell Complexes*, Princeton University Press, 1967.

[6] Youngs, J.W.T., Minimal embeddings and the genus of a graph, *J. Math. Mech.*, **12** (1963), 303-315.

[7] Kaveh, A. On 2-cell embedding of a graph, *Proc. 6th Nat. Math. Conf.,* Iran (1975), 106-113.

[8] Henderson, J.C. de C. and Maunder, E.A.W., A problem in applied topology, *J. Inst. Math. Applics*, **5** (1969), 254-269.

[9] Kaveh, A., *Application of Topology and Matroid Theory to the Flexibility Analysis of Structures*, Ph.D. Thesis, London University (1974).

[10] Kaveh, A., Improved cycle bases for the flexibility analysis of structures, *Comput. Meth. Appl. Mech. Engng.*, **9** (1976), 267-272.

[11] Kaveh, A., An efficient computer program for generating cycle bases for the flexibility analysis of structures, *Commun. Appl. Numer. Meths.*, **2** (1986), 339-344.

[12] Kaveh, A., Suboptimal cycle bases of graphs for the flexibility analysis of skeletal structures, *Comput. Meths. Appl. Mech. Engng.*, **71** (1988), 259-271.

[13] Kaveh, A., Ordering for bandwidth reduction, *Comput. Struct.*, **24** (1986), 413-420.

A. Kaveh
Iran University of Science and Technology
Narmak,
Tehran - 16
IRAN

P.D. PANAGIOTOPOULOS

Fractals in mechanics

The present paper deals with several new applications of the theory of fractals in mechanics. Until now most of the existing applications concern the calculation of fractal dimensions with respect to dynamical systems. In the present paper several other applications are presented which, from the standpoint of mechanics, seem to be of greater interest since they permit the study of difficult or yet unsolved mechanical problems from an entirely new point of view.

A set $A \subset R^n$ is called a fractal set if its Hausdorff dimension $\dim A$ is fractional or greater than n. For the definition of $\dim A$ we refer to [1], [2]. Let $\{X, d\}$ be a complete metric space with the metric d. We denote by $\mathcal{H}(X)$ the space of the compact subsets of X. If $d(A, B)$ is the distance between the sets $A \subset X$ and $B \subset X$ defined by the formula

$$d(A,B) = \max_{x \in A} \min_{y \in B} d(x,y), \tag{1}$$

then $\mathcal{H}(X)$ endowed with the metric $h(A,B) = \sup\{d(A,B); d(B,A)\}$ is a complete metric space which is called the space of fractals. An iterated function system (IFS) on X consists of N contractive mapings $w_n : X \to X$ with contractivity factors $0 \le s_n < 1$, $n = 1,...,N$. The following proposition holds [1].

Proposition. *Let* $\{X; w_n, n = 1,...,N\}$ *be an IFS. We define* $w_n : \mathcal{H}(X) \to \mathcal{H}(X)$ *by setting* $w_n(B) = \{w_n(x); x \in B\}$ $\forall B \in \mathcal{H}(X)$ *and let*

$$W(B) = w_1(B) \cup w_2(B) \cup ... \cup w_n(B) \quad \forall B \in \mathcal{H}(X). \tag{2}$$

Then W is a contraction mapping on $\mathcal{H}(X)$ with contractivity factor $s = \max\{s_1,...,s_N\}$. The unique fixed 'point' of W is the set $A \in (X)$ such that $A = W(A)$ and is given by the relation

$$A = \lim_{j \to \infty} w^j(B) \quad \forall B \in \mathcal{H}(X), \tag{3}$$

where $w^1(x) = w(x)$, $w^2(x) = w(w^1(x)),...,w^j(x) = w(w^{j-1}(x))$ are the forward iterates of w.

The set A is called the 'attractor' of the IFS $\{X; w_n\}$. First we present here a generalization of the above result which seems to be important to mechanical problems expressed in terms of variational and hemivariational inequalities, i.e. in nonsmooth mechanics, [3]-[5]. Since the aforementioned variational expressions are equivalent to multivalued differential or integral equations and when discretized to multivalued algebraic equations we are led to introduce multivalued iterated function systems (MIFS). We call $\{X; w_n, n = 1,...,N\}$ an MIFS, if $w_n(x)$ is nonempty closed set for every $x \in X$ and w_n is contractive in the multivalued sense, i.e.

$$h(w_n(x), w_n(y)) \le s_n d(x, y) \quad \forall x, y \in X \ 0 \le s_n < 1. \tag{4}$$

We denote by W the mapping defined in (2) which again is contractive with factor $s = \max\{s_1,...,s_N\}$. The following proposition holds, and constitutes a new result.

Proposition. *The mapping* $W : \mathcal{H}(X) \to \mathcal{H}(X)$ *has a fixed 'point', the set* A, *such that* $A \subset W(A)$. *Then*

$$A = \lim_{j \to \infty} w^j(B) \quad \forall B \in \mathcal{H}(X), \tag{5}$$

where $w^1(B) = W(B)$, $w^2(B) \in W(w^1(B))$, *etc. are the forward iterates of* W.

The set A is called the attractor of the MIFS. In all problems in mechanics giving rise to multivalued algebraic equations of the form $x \in F(x)$, as e.g. in the LCPs in elastoplasticity, this proposition can be applied. It gives additional information concerning the geometry of the iterations of the numerical scheme. Moreover it can be applied in order to show the existence of an attractor of the dynamic inclusion $x'(t) \in f(t, x(t))$. This will be presented elsewhere.

The theory of fractal interpolation has important applications in mechanics. Suppose that on R^2 for instance we have a set of data $\{x_i, F_i\}$ $i = 1,...,N$. We seek a fractal interpolation function $f : [x_0, x_N] \to R$, i.e. a fractal function f such that $f(x_i) = F_i$, $i = 1,...,N$. Following [1] we consider the IFS $\{R^2; w_n, n = 1,...,N\}$ defined by the 'shear transformation'

$$w_n \begin{pmatrix} x \\ y \end{pmatrix} = \begin{pmatrix} a_n & 0 \\ c_n & d_n \end{pmatrix} \begin{pmatrix} x \\ y \end{pmatrix} + \begin{pmatrix} e_n \\ f_n \end{pmatrix}, \tag{6}$$

and constrained by the data $(x_0 < x_1 < ... < x_N)$

$$w_n \begin{pmatrix} x_0 \\ F_0 \end{pmatrix} = \begin{pmatrix} x_{n-1} \\ F_{n-1} \end{pmatrix}, \quad w_n \begin{pmatrix} x_N \\ F_N \end{pmatrix} = \begin{pmatrix} x_n \\ F_n \end{pmatrix}, \quad n = 1,...,N. \tag{7}$$

Let the factors d_n, called scaling factors, satisfy $0 \leq d_n < 1$.

Then we can always obtain the relation

$$a_n = (x_n - x_{n-1})/(x_N - x_0) \tag{8a}$$

$$e_n = (x_N x_{n-1} - x_0 x_n)/(x_N - x_0) \tag{8b}$$

$$c_n = (F_n - F_{n-1})/(x_N - x_0) - d_n (F_N - F_0)/(x_N - x_0) \tag{8c}$$

$$f_n = (x_N F_{n-1} - x_0 F_n)/(x_N - x_0) - d_n(x_N F_0 - x_0 F_N)/(x_N - x_0). \tag{8d}$$

The following proposition holds, [1].

Proposition. *Let G be the attractor of the IFs defined by* (6), (7), (8). *Then G is the graph of a continuous function $f : [x_0, x_N] \to R$ interpolating the data $\{x_i, F_i\}$, $i = 1,..,N$. If $\mathcal{F}$ is the set of all continuous functions $\tilde{f} : [x_0, x_N] \to R$, then the sequence of functions $\tilde{f}_{n+1} = (T\tilde{f}_n)(x)$, where operator $T : \mathcal{F} \to \mathcal{F}$ is defined by*

$$T\tilde{f}(a_n x + e_n) = c_n x + d_n \tilde{f}(x) + f_n \quad x \in [x_0, x_N] \quad n = 1,\ldots,N, \tag{9}$$

converges to the attractor G.

Furthermore it can be easily shown that if $x_0,\ldots,x_N$ are equally spaced, then

$$\dim G = 1 + [\log \left(\sum_{n=1}^{N} | d_n | \right)/\log N].$$

(a) Fractal interpolation functions in the BEM. Suppose that we have to solve with respect to q the integral equation

$$\int_{x_0}^{x_N} k(x)q(x)\,\mathrm{d}x = r(x). \tag{10}$$

We define on $[x_0, x_N]$ the equally spaced points $x_0, x_1, x_2,\ldots,x_N$ and let $q_0 = F_0$, $q_1 = F_1,\ldots,q_N = F_N$. We assume instead of a piecewise linear or quadratic interpolation function a fractal interpolation function and let it be defined by the IFS (6), (7), (8). Accordingly, we consider through the operator T, instead of (10), the equation

$$\sum_{n=1}^{N} \int_{x_{n-1}}^{x_n} k(x)(c_n x + d_n q(x) + f_n)\, \mathrm{d}(a_n x + e_n) = r(x), \tag{11}$$

where $0 \leq d_n < 1$ and a_n, e_n, c_n, f_n satisfy Eqs. (8). If F_0 and F_N are given, satisfying (11) at $x_0, x_1, \ldots, X_N$ implies $N - 1$ linear equations with the $n - 1$ unknowns $F_1, \ldots, F_{N-1}$. Appropriate choice of the factors d_n improves the condition number of the system.

(b) Analogously we may consider fractal interpolation functions in the FEM.

(c) Calculation of structures with very complicated boundaries or with interfaces of very complicated geometry. This is the case, e.g. of fissurated structures with known fissure geometry. Suppose e.g. that we have a sawtooth irregular boundary in R^2 for a linear elastic body. From the coordinates $\{x_i, F_i\}$ of the boundary we define an IFS, e.g. according to (6), (7), (8), and we get the operator T. First we solve a structure having as boundary the straight line f_0 connecting the points $\{x_0, F_0\}$ with $\{x_N, F_N\}$, then the structure having as boundary $f_1 = Tf_0$, then the structure having as boundary the graph $f_2 = Tf_1$ and so on. The composition of two fixed point theorems (for T and for the linear elasticity operator) shows that this procedure converges to the solutions of the initial problem. The same procedure can be applied to the case of interfaces introducing unilaterial contact and friction boundary conditions. Then for each iterate $f_0, f_1, f_2, \ldots$ of the interface we have to solve a nonlinear problem (inequality problem) resulting from the Kakutani fixed point theorem. Numerically we get convergent results but the theoretical proof of convergence is still open.

(d) Calculation with real stress-strain laws in composite structures. These laws have a sawtooth form. Therefore they can be considered as attractors of an appropriate IFS. Thus through the operator T we may consider simple forms of stress-strain laws for which the numerical calculations are easy and which 'approximate' the initial law.

(e) The iterative procedures used for the numerical treatment of geometrically nonlinear problems lead to sets of fractal dimensions as noticed in [6]. The quality of the convergence of the algorithm is closely connected with the fractal dimension of the point set it generates. Making it more precise in terms of a quantitative result is still an open problem.

References

[1] M. Barnsley. *Fractals Everywhere,* Academic Press, 1988.

[2] K.J. Falconer. *The Geometry of Fractal Sets*, Cambridge University Press, 1985.

[3] P.D. Panagiotopoulos. *Inequality Problems in Mechanics and Applications. Convex and Nonconvex Energy Functions*, Birkhäuser Verlag, Basel, Boston 1985 (Russian transl. MIR Publ. Moscow, 1989).

[4] J.J. Moreau, P.D. Panagiotopoulos and G. Strang. *Topics in Nonsmooth Mechanics,* Birkhäuser Verlag, Basel, Boston, 1988.

[5] J.J. Moreau, P.D. Panagiotopoulos (eds). *Nonsmooth Mechanics and Applications*, CISM Lectures and Courses Vol. 302, Springer Verlag, Vienna and New York, 1988.

[6] P.D. Panagiotopoulos. *On the Fractal Nature of Mechanical Theories*, GAMM 1989 (to appear in ZAMM 1990).

P.D. Panagiotopoulos
Institut für Technische Mechanik
RWTH Aachen
D-5100 Aachen
GERMANY
and
Department of Civil Engineering
Aristotle University
GR-54006 Thessaloniki
GREECE

I. TROCH

Industrial robots: a control theoretical problem and mechanical models

1. Introduction

One of the main difficulties in connection with actuator control of a robotic manipulator stems from the couplings between joint movements. A control signal applied to one single actuator will normally make all joints move. Consequently, whenever only one or two joints are to move in a prescribed manner all actuators have to be active in order to cancel out unwanted movements.

Among the concepts being suggested to overcome this difficulty, nonlinear decoupling is a rather attractive one. It is based on the idea of designing first a nonlinear control law which is to be computed from the actual joint positions and joint velocities such that after its application the robotic arm with m degrees of freedom behaves like m independent systems (links) where the control signal of actuator j makes only joint j move and no other joint. Then, the control for performing the desired movement is added.

At the time being, nonlinear decoupling is based only on the mathematical model of the arm, the so-called drive equations, whereas the actuator dynamics are considered only in a secondary way. In the following this decopling problem shall be investigated from a different point of view, where also actuator dynamics are taken into account when deriving the decoupling law. Due to the implicit nature of the resulting mathematical model special investigations are required. A decoupling control is derived, which does not require numerical differentiation like the aforementioned usual approach. Decoupling is based on complete state feedback, i.e. not only the state of the mechanical arm but also the states of the joint actuators have to be used.

2. Modelling a robotic system

The starting point for the derivation of a mathematical model for the dynamic behaviour of a robotic arm consists of the so-called drive equations, which can be derived e.g. by applying the Newton–Euler approach. Collecting drive forces and moments in a vector k_{Dr} and defining the vectors p and $\dot{p}$ of relative joint positions and relative joint velocities respectively, allows description of the movement, [3], by

$$k_{Dr} = M'(p)\ddot{p} + h'(p, \dot{p}) \tag{1}$$

Secondly, the dynamic behaviour of the actuators has to be modelled. For the sake of simplicity it is assumed that the arm is equipped with d.c. motors. The mathematical model of these motors will be of the same structure for all three actuators. However, the various motor constants may have different values for the various actuators. A rather simple mathematical model, [3], [4], can be used, based on the state variables 'armature voltages' U_{Aj}, 'armature currents' I_{Aj} and 'relative angular speeds' σ_j of the rotor shafts in the respective link where actuator j is mounted. Inputs are 'control voltages' u_{Cj} and 'modified moments k_{Lj} of the load' (see (5)):

$$\begin{aligned} & T_{STj}\,\dot{U}_{Aj} + K_{STj}\,u_{Cj} \\ & L_{Aj}\,\dot{I}_{Aj} + R_{Aj}\,I_{Aj} = U_{Aj} + K_j\,\sigma_j \qquad j = 1,\ldots,m. \\ & I_{Rj}\,\dot{\sigma}_j = K_j\,I_{Aj} - k_{Lj} \end{aligned} \tag{2}$$

Collecting U_{Aj}, I_{Aj}, σ_j in a state vector z_j and using appropriate matrix notation results in

$$\dot{z}_j = A_j\,z_j + b_j\,u_{Cj} + d_j\,k_{Lj} \qquad j = 1,\ldots,m. \tag{3}$$

Considering the vectors z_j as elements of a vector z yields finally

$$\dot{z} = Az + Bu + Dk_L, \tag{4}$$

where all three matrices A, B and D are block diagonal, indicating that couplings can stem only from the inputs u or k_L.

Thirdly, gearings have to be modelled. For the present study it is sufficient to model gearings as simple proportional elements with gear ratios as constants of proportionality (collected in a diagonal matrix N). This results in

$$\begin{aligned} & N\dot{p} = T'z \\ & k_L - k_{GF} - k_R = N^{-1}\,k_{Dr} = N^{-1}\,[M'(p)\ddot{p} + h'(p,\dot{p})\,], \end{aligned} \tag{5}$$

where k_{GF} denotes the moments of friction of the gearings referred to the rotor axes and k_R some torques to the rotor shafts caused by the absolute angular accelerations of the links, [2], [3]. Usually, each joint velocity is proportional to one actuator state (for d.c. motors to the relative angular speed of the respective rotor shaft). Therefore, in general T' is block diagonal with projection matrices as elements. The state of the robotic system at any time t is completely characterized by the vectors p and z, the dynamic behaviour is given by equations (4) and (5). Combining all relevant equations and introducing appropriate abbreviations results finally in the model

$$\dot{z} = Az + Bu + Dk_L = Az + Bu + D\,[M(p)\ddot{p} + h(p, \dot{p})\,]$$
$$\dot{p} = Tz. \tag{6}$$

This demonstrates that couplings are present whenever the vector k_L of the moments of load is not completely decoupled, i.e. for every robot being different from a Cartesian one. Consequently, decoupling controls are of great importance.

3. Decoupling control for robots

In the standard literature, [1], [3], [6], nonlinear decoupling or the equivalent technique of the inverse system are based only on the drive equations (1) whereas actuator dynamics are not considered at this stage of the investigations (however, they have to be taken into account when setting up the concrete control input). Now, k_L is considered as control for which a law

$$k_L = \hat{k}_L = M(p)w + h(p, \dot{p}) \tag{7}$$

is chosen with w as new control input. A chain of decoupled integrators for the influence of w on p results. Realization of this procedure requires the computation of the actual control input u in such a way that the computed moments of load are compensated leading to rather involved computations including numerical differentiation. Fig. 1 indicates the system structure resulting before this transfer to the input u is carried out.

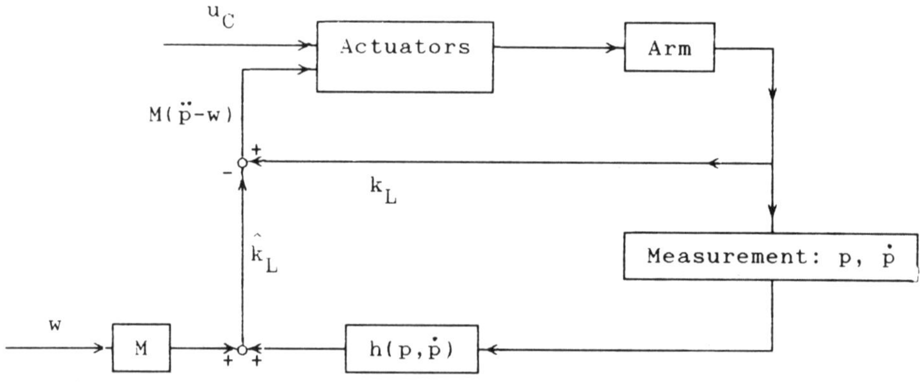

Figure 1. Decoupling based only on drive equations

Therefore, a direct approach to the decoupling problem will be considered in the following, starting with the model equations (6). Differentiating with respect to time yields

$$\ddot{p} = T\dot{z} = T\{ Az + Bu + D[M(p)\ddot{p} + h(p,\dot{p})]\}. \tag{8}$$

It can be easily recognized that system (6) is of implicit structure from which some additional problems result. Standard algorithms cannot be applied without modifications. Therefore, a more direct approach will be taken. Obviously, a decoupling control

$$u = (TB)^{-1}[w - TAz - TDM(p)\ddot{p} - TDh(p,\dot{p})] \tag{9}$$

can be designed whenever the matrix TB is regular. Applying it yields

$$\ddot{p} = w \tag{10}$$

i.e. decoupled chains of integrators. Applying control (9) to the relation (6) for the actuator dynamics indicates that internal couplings resulting from the moments of load are cancelled out by counteracting couplings contained in the control (9).

Control law (9) suffers from the drawback that it requires not only state feedback (i.e. feedback of p and $\dot{p}$) but also feedback of the accelerations $\ddot{p}$. This difficulty can be overcome by transforming (8) to

$$[I = TDM(p)]\ddot{p} = T\{ Az + Bu + Dh(p,\dot{p})\}. \tag{11}$$

This suggests application of the control law

$$u = (TB)^{-1}\{[I - TDM(p)]w - TAz - TD\,h(p,\dot{p})\}, \tag{12}$$

yielding

$$[I - TDM(p)](\ddot{p} - w) = 0 \tag{13}$$

and consequently, again (10) results, being the unique solution in case the coefficient matrix $[I - TDM(p)]$ is regular. Regularity of this matrix is the generic situation (with respect to the parameters of T, D, M and position p). This can be seen from the consideration that small changes in the parameters will make all eigenvalues of $TDM(p)$ different from 1.

Now, the second case has to be investigated, i.e. the case of TB being singular. For the sake of simplicity, it shall be asumed first that $TB = 0$ holds. Differentiating (8) with respect to time then yields with $\dot{z}$ from (6)

$$\dddot{p} = TDM(p)\dddot{p} + TABu + TA^2z + h_1(p,\dot{p},\ddot{p}) \tag{14}$$

with obvious abbreviation h_1. Evidently, (14) is of the same structure as (8) and (11). Consequently, in case TAB is regular, proceeding analogously to (9) or – of greater interest - to (12) is possible. A system results which again behaves like (13) and consequently like (10), i.e. a decoupled chain of integrators - however, now of third order. In case the matrix TAB is singular differentiation has to be repeated until a regular coefficient matrix TA^NB of u results for some natural number N (being less than the number of rows of A) whereas $TA^kB = 0$ for all $k < N$. Decoupling is not possible, in case such a number N does not exist. However, for properly chosen actuators, from this procedure a control of the type

$$u = (TA^NB)^{-1}\,[\,I - TDM(p)\,]\,w + g(p, \dot{p}, \ddot{p}, \ldots, p^{(N+1)})$$

$$= M_N(p)w + g(p, \dot{p}, \dddot{p}, \ldots, p^{(N+1)}) \tag{15}$$

is derived, which depends on relative joint positions and their derivatives up to the order $N+1$. The resulting input/output behaviour is given by

$$p^{(N+2)} = w. \tag{16}$$

Consequently, the control (15) depends in general on all state variables, whereas the weighting matrix of w depends only on relative joint positions. The system structure corresponds to the one indicated in Fig. 2. Concerning the requested measurement of states it should be noted that because of (8), derivatives of p up to order $N + 1$ can be replaced by actuator states. However, this requires inversion of the matrix $[I - TDM(p)]$. In case not all required measurements are possible, a dynamic observer can be used to reconstruct unmeasurable states.

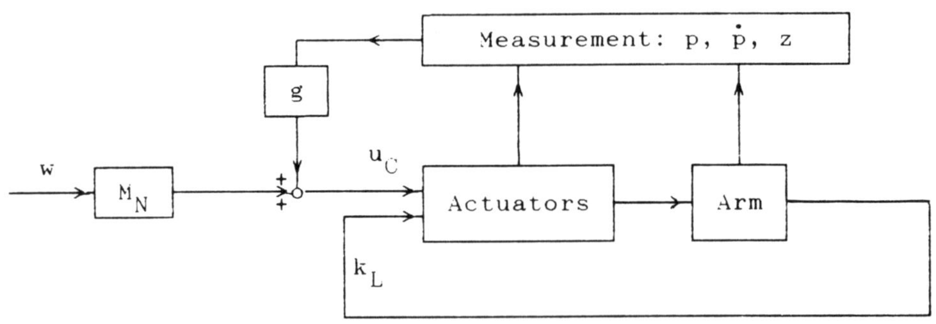

Figure 2. Decoupling taking actuator dynamics into account

Now, a remark is appropriate concerning the case where the matrix TB or one of the subsequent matrices TA^kB is singular but non-vanishing. The basic idea remains unchanged, but it is now applied to the individual actuator equations (3). This is possible because all matrices TA^kB are block diagonal. A decoupling control can be designed whenever the coefficient of u_{Cj} becomes different from zero for the first time. The procedure stops when either a value being different from zero is found or when the exponent of A_j equals the number n_j of rows of that sub-matrix A_j. The resulting control is of the same type as above. The order of the resulting subsystem (i.e. the number of integrators in the jth chain) is given by

$$N_j = \min \{ k \mid c_j^T A_j^k b_j \neq 0 \} \leq n_j. \tag{17}$$

However, it should be noted that all numbers N_j are equal is the generic case, because normally all actuators of a robot are of the same type (either all d.c. motors, or all electro-hydraulic etc.). This also justifies the above procedure concerning the matrices TB, TAB,..., because the discussed alternatives concern the generic case.

It is interesting to note that in the generic as well as in the non-generic case, the conditions for being able to find a decoupling control are equivalent to the conditions for complete output controllability of the system

$$\dot{z} = Az + Bu, \qquad y = Tz\,. \tag{18}$$

This indicates that the problem of whether or not a decoupling control can be derived concerns a structural property of the actuators being used for joint movement and is not only a structural property of the arm.

Structural properties of the arm are important for the answer to the question whether or not this decoupling control can be based only on states or whether also derivatives of state variables are required. The important property is that the arm can be described in a form such that the resulting moments of load are of the structure (5) with accelerations appearing linearly. Further details as well as an example are given in [7], [8].

4. Conclusions

The problem of designing a decoupling control for a robotic manipulator was investigated. The study is based on a mathematical model for the dynamic behaviour of the arm and the joint actuators contrary to earlier studies which were based only on the dynamic equations for the arm and neglected actuator dynamics.

Based on the assumption of linear actuator models (which is rather frequently true) it could be demonstrated that

(i) Decoupling is possible provided the joint actuators are completely output controllable (where the output equals relative joint velocity). Concerning the structural properties of the arm, the only condition of relevance for the decoupling problem is that joint accelerations appear linearly in the drive equations (and consequently in the relations for the moments of load).

(ii) Generically, the decoupling control is of the type (15) leading to a system of m decoupled chains of integrators.

(iii) Generically each of these chains has the same number of integrators.

For actuators described by affine models similar results may be obtained, [7]. This is important, because most actuators being nowadays in use can be modelled either by linear models or at least by affine models.

The approach to the decoupling problem for robots presented in this paper offers the possibility of avoiding numerical differentiation and to overcome by this at least parts of the troubles connected with nonlinear decoupling in robot control.

References

[1] M. Brady *et al.* (eds.), *Robot Motion, Planning and Control*, MIT Press, Cambridge, Mass., 1983.

[2] K. Desoyer, Kinematics and kinetics of robots - A short survey. In: L. Basanez, G. Ferrate and G.N. Saridis (Eds.), *Robot Control*, Pergamon Press, 1985, 419-424.

[3] K. Desoyer, P. Kopacek and I. Troch, I., *Industrieroboter und Handhabungsgeraete*. Oldenbourg, Muenchen, (1985).

[4] O. Foellinger, *Regelungstechnik*, Elitera, Berlin, 1978.

[5] J. Fossard and C. Gueguen, *Multivariable System Control*, North-Holland, Amsterdam (2nd Printing), 1979.

[6] E. Freund, Fast nonlinear control with arbitrary pole-placement for industrial robots and manipulators, *Int. J. of Robotics Res.* **1** (1982), 65-78.

[7] I. Troch, Decoupling control of robots revisited. In preparation.

[8] I. Troch, A new view on the decoupling problem for industrial robots (to appear in Robotica, 1990)).

[9] H. Unbehauen, *Regelungstechnik I-III*, Vieweg, Braunschweig, 1985.

[10] M. Vukobratovic and D. Stokic, *Control of Manipulation Robots. Theory and Application*, Springer, Berlin, 1982.

I. Troch
Institut für Technische Mathematik
Technical University of Vienna (Inst. 114)
Wiedner Hauptstraße 8-10
A-1040 Wien
AUSTRIA

C. WOŹNIAK

On the micromechanical modelling of solids: an application of Ω-calculus

Abstract

There is a variety of approximate (effective) theories for micro-nonhomogeneous media endowed with a periodic structure, [1]. The aim of the note is to show how effective theories for certain deterministic but in general non-periodic micro-nonhomogeneous material structures can be formulated. To this end the basic ideas of Laugwitz Ω-calculus (see [2] and the mathematical preliminaries below) are applied.

Notation. Throughout the note we tacitly assume that indices A, E and a run over $1,...,M$, $1,...,N$ and $1,...,m$, respectively. The summation convention with respect to the index a holds. Symbol n stands for an arbitrary positive integer and Ω is an infinite natural number (see below).

1. Mathematical preliminaries

Following [2] we outline the basic ideas of Laugwitz Ω-calculus, which represents a certain simple and direct approach to the nonstandard analysis, [3]. Let T be a specified mathematical theory, i.e. a collection of true statements formulated in a formal language L. The expressions of L are formed from an alphabet A. Adding to A a new number constant Ω we obtain a new theory $T\langle\Omega\rangle$ as a collection of statement for which we define the following notion of 'true':

Basic definition (BD). Let $S(n)$ for each $n = 1,2,...$ be a statement formulated in L. If $S(n)$ is true (belongs to T) for a.e. (almost every) n then $S(\Omega)$ is true (belongs to $T\langle\Omega\rangle$).

Let $\mathrm{id}(n) = n$ be the identity sequence and n_0 stand for an arbitrary but fixed natural number. Since $n > n_0$ for a.e. n, then from (BD) we obtain $\mathrm{id}(\Omega) = \Omega$ and $\Omega > n_0$ for any n_0. Hence Ω can be treated as an infinitely large natural number. Let $a(n)$ be a sequence of reals tending to zero and k be an arbitrary but fixed positive number. Because of $|a(n)| < k$ for a.e. n, then (BD) yields $|a(\Omega)| < k$ and hence $a(\Omega)$ is an infinitely small positive number. In a set of real numbers involved in $T\langle\Omega\rangle$ we introduce an indiscernibility relation $\simeq$, setting $\alpha \simeq \beta$ if

$\alpha - \beta$ is an infinitely small number. Let A_0 be an arbitrary entity involved in T and $A(n) = A_0$ be a constant sequence. Then from (BD) we obtain $A(\Omega) = A_0$; in this case an entity $A(\Omega)$ will be called standard.

2. Analysis

We consider a body $\mathcal{B}$ made of M hyperelastic materials that are specified in the undeformed state by (constant) mass densities ρ_A and strain energy functions $\varepsilon_A(F)$, F being a deformation gradient. Let $\mathcal{B}$ in the undeformed state occupy a region B and let B_A be a part (open set) of B occupied by the Ath material component. We assume that $\mathcal{B}$ is a micro-nonhomogeneous body, i.e. there exists a positive constant δ such that: (i) δ is sufficiently small relative to the smallest characteristic length dimension of B, (ii) every ball of a radius δ included in B has a non-empty intersection with each B_A.

In order to formulate an effective description of $\mathcal{B}$ (i.e. plausible from the engineering standpoint) we introduce some auxiliary concepts. Let $\kappa : B \to \mathbb{R}^3$ be a smooth invertible mapping which assigns to every $x \in B$ the coordinates $X \equiv (X^1, X^2, X^3) \in x(B)$. Let $\varepsilon^1, \varepsilon^2, \varepsilon^3$ be the increments of coordinates X^1, X^2, X^3, respectively, and define $a_i \equiv (\varepsilon^1 \delta_{i1}, \varepsilon^2 \delta_{i2}, \varepsilon^3 \delta_{i3})$, $i = 1, 2, 3$, and $\Delta \equiv 0.5[(-\varepsilon^1,\varepsilon^1) \times (-\varepsilon^2,\varepsilon^2) \times (-\varepsilon^3,\varepsilon^3)]$. Moreover, for each positive integer n, let us introduce in B a lattice

$$\mathcal{L}^{(n)} \equiv \{z \in B : \kappa(z) = (\Sigma\, a_i \nu_i)/n,\ \nu_i = 0, \pm 1, \pm 2, ...\}$$

and a family of cells

$$C^{(n)}(z) \equiv \{x \in B : \kappa(x) - \kappa(z) \in \Delta/n\},\quad z \in \mathcal{L}^{(n)}.$$

We shall assume that $\bar{B} = \bigcup \overline{C^{(1)}(z)}$, $z \in \mathcal{L}^{(1)}$. For every $X \in \kappa(B)$ let $\Delta = \bigcup \overline{\Delta_A(X)}$ be a decomposition of Δ into M disjoint polyhedrons $\Delta_A(X)$, such that the multi-functions $\kappa(B) \ni X \to \Delta_A(X)$, $A = 1,...,M$, are sufficiently regular and can be treated, from the numerical standpoint, as constant in every $Z + \Delta \subset \kappa(B)$. Finally define

$$C_A^{(n)}(z) \equiv \{x \in C^{(n)}(z) : \kappa(x) - \kappa(z) \in \Delta_A(X)/n,\ X = \kappa(z)\},\quad z \in \mathcal{L}^{(n)}.$$

Now assume that each set $B_A^{(1)} \equiv \bigcup C_A^{(1)}(z)$, $z \in \mathcal{L}^{(1)}$, represents a certain 'filtrated' (averaged and regularized) distribution of the Ath material component of the body $\mathcal{B}$. Since $\bar{B} = \bigcup \overline{B_A^{(1)}}$, then we pass from $\mathcal{B}$ to a new 'filtrated' body $\mathcal{B}^{(1)}$, which in the undeformed state also occupies a region B, but its Ath material component occupies a part $B_A^{(1)}$ of B. The choice of a mapping $\kappa(\cdot)$ and multifunctions $\Delta_A(\cdot)$, satisfying the conditions mentioned above, will be called a filtration procedure. Setting aside all particulars concerning a filtration procedure (they will be discussed elsewhere) we shall restrict ourselves to micro-nonhomogeneous bodies $\mathcal{B}$ that satisfy the following heuristic.

Filtration assumption. Under the proper choice of the filtration procedure the macro-response of a body $\mathcal{B}$ can be approximated by the macro-response of the 'filtrated' body $\mathcal{B}^{(1)}$.

By the macro-response we mean here an averaged (over every cell $C^{(1)}(z)$, $z \in \mathcal{L}^{(1)}$) distribution of displacements, strains and stresses in every material component of a body. Examples of the filtration procedure can be found in [4].

Let $\mathcal{B}^{(n)}$ be a body which in the undeformed state occupies the region B and its Ath material component (specified at the beginning of this section) occupies a part $B_A^{(n)} \equiv \bigcup C_A^{(n)}(z)$, $z \in \mathcal{L}^{(n)}$ of B. It can be seen that $\mathcal{B}^{(n)}$ is a body with a micro-nonhomogenity which is, roughly speaking, n times finer than that of a body $\mathcal{B}^{(1)}$. Now, using the approach proposed in [5], we shall postulate the following:

Micromodelling hypothesis. If the wavelength of excitations for a body $\mathcal{B}^{(1)}$ is sufficiently large related to the maximum length dimension of every cell $C^{(1)}(z)$, $z \in \mathcal{L}^{(1)}$, then a macro-response of this body can be approximated by a macro-response of each body $\mathcal{B}^{(n)}$, $n = 2,3,...$, under the same excitations.

Setting $\mathcal{B}^{(\Omega)} \equiv \mathcal{B}^{(n)}$ for $n = \Omega$, from the basic definition we obtain

Micromodelling lemma. *If the micromodelling hypothesis holds then a macro-response of a body* $\mathcal{B}^{(1)}$ *can be approximated by a macro-response of a 'nonstandard' body* $\mathcal{B}^{(\Omega)}$.

Notice that a body $\mathcal{B}^{(\Omega)}$ is endowed with, roughly speaking, an infinitely fine micro-nonhomogeneous material structure.

Let us introduce a surjection $\varphi - \{1,...,N\} \rightarrow \{1,...,M\}$, $N \geq M$, and finite element decompositions

$$\bar{\Delta} = \bigcup_{E=1}^{N} \overline{\Delta^{E}(X)}, \quad X = \kappa(x), \quad x \in B, \tag{1}$$

of Δ into N disjoint elements (tetrahedrons) $\Delta^E(X)$, such that the condition $\overline{\Delta_A(X)}$, $E \in \varphi^{-1}(A)$ holds for every A. Let us also introduce for each $X \in \kappa(B)$ a sequence of m real-valued functions $h_a(X, \cdot)$, defined and continuous on $\bar{\Delta}$, equal to zero on

$\partial\Delta$ and linear in every $\Delta^E(X)$. It is assumed that multifunctions $\Delta^E(\cdot)$ and mappings $h_0(\cdot, w)$, $w \in \bar{\Delta}$, are sufficiently regular and from the numerical standpoint can be treated as constant in every $Z + \Delta \subset \kappa(B)$. Setting

$$\Lambda_a^E(X) \equiv \nabla_w h_a(X, w) \text{ for every } w \in \Delta^E(X), \tag{2}$$
$$\eta_E(X) \equiv \text{vol}\, \Delta^E(X)/\text{vol}\, \Delta,$$

we shall also assume that $\Lambda_a^1(X)\eta_1(X) + ... + \Lambda_a^N(X)\eta_N(X) = 0$. Now we introduce

Microlocal approximation assumption. Deformations of a body $\mathcal{B}^{(\Omega)}$ under arbitrary (standard) excitation can be expected in a class of deformation functions $\chi^{(\Omega)} : \kappa(B) \to \mathbb{R}^3$ given by

$$\chi^{(\Omega)}(X) = p(X) + \frac{1}{\Omega} h_a(Z, \Omega(X - Z))q^a(X),\ X \in Z + \Delta/\Omega,\ \ Z \in \kappa(\mathcal{L}^{(\Omega)}),$$

where $h_a(Z, \cdot), Z \in \kappa(B)$ are the known functions and $p : \kappa(B) \to \mathbb{R}^3$, $q^a : \kappa(B) \to \mathbb{R}^3$ are an arbitrary (standard) smooth invertible mapping and an arbitrary differentiable (standard) vector field, respectively.

Following [5] we refer to $p(\cdot)$ and $q^a(\cdot)$ as a macrodeformation and a microlocal parameter field, respectively. The postulated *a priori* functions $h_a(X, \cdot)$, $X \in \kappa(B)$ are called the shape functions and play a role similar to that of the shape functions in the finite element method.

Setting $q(X) \equiv (q^1(X),...,q^m(X))$, $I(X) \equiv \det \nabla\kappa^{-1}(X)$, $X \in \kappa(B)$, let us define

$$\tilde{\rho}(X) \equiv I(X) \sum_{E=1}^{N} \eta_E(X)\rho_{\varphi(E)},\ \tilde{\varepsilon}(X, \nabla p, q) \equiv I(X) \sum_{E=1}^{N} \eta_E(X)\rho_{\varphi(E)}\, \varepsilon_{\varphi(E)}\, (\nabla p + \Lambda_a^E(X) \otimes q^a), \tag{3}$$

and introduce the functionals

$$\tilde{I} \equiv \int_{\kappa(B)} \tilde{\varepsilon}(X, \nabla p(X), q(X))dX,\ \tilde{\kappa} \equiv \int_{\kappa(B)} \frac{1}{2} \tilde{\rho}(X)v^2(X)dX;\ \ dX \equiv dX^1dX^2dX^3$$

where $v(\cdot)$ is an arbitrary (standard) continuous vector field. Let us also introduce the internal and kinetic energy of the body $\mathcal{B}^{(\Omega)}$, represented by the integrals

$$I = \sum_{E=1}^{N} \int_{\kappa(B_A^{(\Omega)})} I(X)\rho_A\varepsilon_A(\nabla\chi^{(\Omega)}(X))dX,\ \kappa = \frac{1}{2} \sum_{A=1}^{M} \int_{\kappa(B_A^{(\Omega)})} I(X)\rho_A\, v^2(X)dX,$$

respectively. The last step on our line of modelling is given by

Micromodelling theorem. *For an arbitrary deformation function* $\chi^{(\Omega)}(\cdot)$ *satisfying the microlocal approximation assumption and for an arbitrary (standard) continuous vector field* $v(\cdot)$ *defined on* $\kappa(B)$, *the internal and kinetic energy of a body* $\mathcal{B}^{(\Omega)}$ *satisfy the indiscernibility conditions:* $I \simeq \tilde{I}$, $\mathcal{K} \simeq \tilde{\mathcal{K}}$.

Thus we have arrived at the conclusion that the macro-response of a micro-nonhomogeneous body $\mathcal{B}$ can be approximated by the response of a certain 'oriented' body $\tilde{\mathcal{B}}$ (with $3+3n$ degrees of freedom $p(\cdot)$ and $q(\cdot)$) with a mass density $\tilde{\rho}(\cdot)$ and a strain energy function $\tilde{\varepsilon}(\cdot)$ determined by the formulae (3). We also obtain

$$\chi^{(1)}(X) \approx p(X), X \in \kappa(B);$$

$$\nabla\chi^{(1)}(X) \approx \nabla p(X) + \Lambda_a^E(X) \otimes q^a(X), X \in Z + \Delta^E(Z), Z \in \mathcal{L}^{(1)}, \tag{4}$$

where an approximation $\approx$ has to be understood in the sense of the micromodelling hypothesis and $\chi^{(1)}(\cdot)$ is a deformation of a body $\mathcal{B}^{(1)}$.

3. Results

If the functions $I(\cdot)$, $\eta_E(\cdot)$, $\Lambda_a^E(\cdot)$ have wavelengths not much smaller than the smallest characteristic length dimension of $\kappa(B)$, then $\tilde{\mathcal{B}}$ can be taken as an effective (in general macro-nonhomogeneous) model of a micro-nonhomogeneous body $\mathcal{B}$. For every time instant $\tau \in [\tau_0, \tau_f]$ let $\mathcal{B}$ be subject to body forces $b(X, \tau)$, $X \in \kappa(B)$ and surface tractions $s(X, \tau)$, $X \subset \Gamma \subset \partial\kappa(B)$. Introducing the action functional $\tilde{K} - \tilde{I}$ we derive the equations of motion

$$\text{Div}\, \tilde{T}(X, \tau) + \tilde{\rho}(X)b(x, \tau) = \tilde{\rho}(X)\ddot{p}(X, \tau), \quad h(X, \tau) = 0; \; X \in \kappa(B), \tag{5}$$

the constitutive equations

$$\tilde{T}(X, \tau) = \frac{\partial\tilde{\varepsilon}(X, \nabla p(X, \tau), q(X, \tau))}{\partial\nabla p(X, \tau)}, \quad h(X, \tau) = -\frac{\partial\tilde{\varepsilon}(X, \nabla p(X, \tau), q(X, \tau))}{\partial q(X, \tau)}; \; X \in \kappa(B) \tag{6}$$

and the traction boundary conditions $\tilde{T}(X, \tau)\, n(X) = s(X, \tau)$, $X \subset \Gamma$, where $n(X)$ is the unit normal outward to Γ at X. Eqs. (5), (6) are the governing equations of $\tilde{\mathcal{B}}$ and together with the formulae (4) are a basis for solutions to diverse engineering problems. If $q(X, \tau)$ can be uniquely obtained from $(5)_2$, $(6)_2$, i.e. if $q = \tilde{q}(X, \nabla p)$, then setting $\varepsilon_{\text{eff}}(X, \nabla p) \equiv \tilde{\varepsilon}(X, \nabla p, \tilde{q}(X, \nabla p))$, we obtain

$$\text{Div}\ \tilde{T}(X,\tau) + \tilde{\rho}(X)\ b(X,\tau) = \tilde{\rho}(X)\ddot{p}(X,\tau),\ \tilde{T}(X,\tau) = \frac{\partial \varepsilon_{\text{eff}}(X, \nabla p(X,\tau))}{\partial \nabla p(X,\tau)};\ X \in \kappa(B). \quad (7)$$

The functions $\varepsilon_{\text{eff}}(X,\cdot)$, $X \in \kappa(B)$ describe the local (effective) macroproperties of a micro-nonhomogeneous body $\mathcal{B}$, provided that $\tilde{\mathcal{B}}$ is an effective model of $\mathcal{B}$.

The example of application of the proposed line of modelling to nonperiodic multilayered laminates can be found in [4].

It can be shown that if $\mathcal{B}$ has a periodic structure in B, then $\tilde{\mathcal{B}}$ is a homogeneous body and the proposed method of modelling reduces to the homogenization approach discussed in [5].

References

[1] G. Herrmann, R.K. Kaul and T.J. Delph, Review on approximate theories for composites, *Arch. Mech.* **28** (1976), 405.

[2] C. Schmieden and D. Laugwitz, Eine Erweiterung der Infinitesimalrechnung, *Math. Z.* **69** (1958), 1.

[3] A. Robinson, *Nonstandard Analysis*, North-Holland, Amsterdam 1966.

[4] E. Wierzbicki and C. Woźniak, Macro-modelling of nonperiodic multilayered elastic media, submitted.

[5] C. Woźniak, A nonstandard method of modelling of thermoelastic periodic composites, *Int. J.K. Engng. Sci.* **25** (1987), 483.

C. Woźniak
Miedzynarodowa 58 m.63
PL-03-922 Warszawa
POLAND